PLANT INVASIONS

GENERAL ASPECTS AND SPECIAL PROBLEMS

PLANT INVASIONS

GENERAL ASPECTS AND SPECIAL PROBLEMS

Edited by Petr Pyšek, Karel Prach,
Marcel Rejmánek and Max Wade

SPB Academic Publishing, Amsterdam

CIP-DATA KONINKLIJKE BIBLIOTHEEK, DEN HAAG

Plant

Plant invasions : general aspects and special problems /
ed. by Petr Pysek ... [et al.]. - Amsterdam : SPB
Academic Publishing. - Ill.
With index, ref.
ISBN 90-5103-097-5
Subject headings: plant invasions / plant ecology.

P.O. Box 11188, 1001 GD Amsterdam, The Netherlands

ISBN 90-5103-097-5

TABLE OF CONTENTS

PREFACE

In the last decade, biological invasions have been increasingly recognized as one field of contemporary ecology which integrates an extraordinary challenge to basic science with serious practical implications. Although the geographical emphasis in the field has been on those areas of the world which are most vulnerable to plant invasion (Australia, New Zealand, South Africa, Pacific Islands, North America), introduced plants are undoubtedly becoming a serious threat worldwide and Europe is no exception in this respect.

This book summarizes the results of the international workshop held in Kostelec nad Černými lesy, Czech Republic, September 16-19, 1993, and represents a sequel to the previous volume (de Waal L. *et al.*, Ecology and management of invasive riverside plants, John Wiley and Sons, 1994). The topics covered relate both to theoretical problems associated with the process of plant invasions and to practical management of those invaders that are most aggressive with an emphasis on temperate Europe. The papers in this volume also cover a wide range of habitas and geographical areas and indicate the global importance of the problems involved.

Generally, throughout the entire volume, the invaders are understood as plants alien to the area of interest. Our view of terminology associated with plant invasion studies is outlined in the introductory section where it is treated in more detail.

We would like to thank Miroslav Martiš of the Institute of Applied Ecology, Kostelec nad Černými lesy, Czech Republic, for support during the workshop and to the following collegues for their involvement in reviewing some of the papers: David J. Beerling (Sheffield), Pierre Binggeli (Ulster), Jan Lepš (České Budějovice), Leoš Klimeš (Třeboň), Zdeněk Skála (Průhonice), Petr Šmilauer (České Budějovice), Miroslav Šrůtek (Třeboň).

Petr Pyšek, Karel Prach, Marcel Rejmánek, Max Wade

LIST OF CONTRIBUTORS

Michael S. Adams
Botany Department, University of Wisconsin, Madison, Wisconsin, USA

John P. Bailey
Department of Botany, Leicester University, Leicester LE1 7RH, United Kingdom

David J. Beerling
Department of Animal and Plant Sciences, University of Sheffield, PO Box 601, Sheffield S10 2UQ, United Kingdom

John L. Bramley
Ecology Research Group, Canterbury Christ Church College, Canterbury, Kent CT1 1QU, United Kingdom

John H. Brock
School of Planning and Landscape Architecture, Arizona State University, Tempe, AZ 85287-2005, USA

Susana Calvelo
Centro Regional Universitario Bariloche, cc 1336, 8400 Bariloche, Argentina

Lois E. Child
International Centre of Landscape Ecology, Department of Geography, Loughborough University, Loughborough, Leicesteshire LE11 3TU, United Kingdom

Louise C. de Waal
School of Applied Sciences, Division of Environmental Science, University of Wolverhampton, Wolverhampton WW1 1SB, United Kingdom

Georges B.J. Dussart
Ecology Research Group, Canterbury Christ Church College, Canterbury, Kent CT1 1QU, United Kingdom

Keith R. Edwards
Institute for Environmental Studies, University of Wisconsin, Madison, Wisconsin, USA

M. Teresa Ferreira
Departamento de Engenharia Florestal, Instituto Superior de Agronomia, Tapada da Ajuda, 1399 Lisboa Codex, Portugal

Miriam Gobbi
Centro Regional Universitario Bariloche, cc 1336, 8400 Bariloche, Argentina

Rod H. Gritten
Snowdonia National Park, Penrhyndeudraeth, Gwynedd, LL48 6LS,
United Kingdom

Tomáš Herben
Institute of Botany, Academy of Sciences of the Czech Republic, CZ-252 43
Průhonice, Czech Republic

Ingo Kowarik
University of Hannover, Department of Landscape Architecture and
Environmental Studies, Plant Ecology, Herrenhäuser Straße 2, D 30419 Hannover,
Germany

Jan Květ
Plant Ecology Section, Institute of Botany, Academy of Sciences of the Czech
Republic, Dukelská 145, CZ-379 01 Třeboň, Czech Republic

I.S. Moreira
Departamento de Botânica e Engenharia Biológica, Instituto Superior de
Agronomia, Tapada da Ajuda, 1399 Lisboa Codex, Portugal

Karel Prach
Faculty of Biological Sciences, University of South Bohemia, Branišovská 31,
CZ-370 01 České Budějovice, Czech Republic

Javier Puntieri
Centro Regional Universitario Bariloche, cc 1336, 8400 Bariloche, Argentina

Petr Pyšek
Institute of Applied Ecology, University of Agriculture Prague, CZ-281 63
Kostelec nad Černými lesy, Czech Republic

Jo T. Reeve
Ecology Research Group, Canterbury Christ Church College, Canterbury, Kent
CT1 1QU, United Kingdom

Marcel Rejmánek
Section of Evolution and Ecology, University of California, Davis, CA 95616,
USA

David Spencer-Jones
Water Wise Consultancy, Plowmans, Park Road, Forest Row, Sussex RH18 5BX,
United Kingdom

Uwe Starfinger
Institut für Ökologie, Technische Universität Berlin, Schmidt-Ott Straße 1,
D 12165 Berlin, Germany

Herbert Sukopp
Institut für Ökologie, Technische Universität Berlin, Schmidt-Ott Straße 1,
D 12165 Berlin, Germany

Petr Šmilauer
Faculty of Biological Sciences, University of South Bohemia, Branišovská 31,
CZ-370 01 České Budějovice, Czech Republic

Ulla Vogt Andersen
Royal Veterinary and Agricultural University, Botanical Section, Rolighedsvej 23,
DK-1958 Frederiksberg C, Denmark

Max Wade
International Centre of Landscape Ecology, Department of Geography,
Loughborough University, Loughborough, Leicesteshire LE11 3TU,
United Kingdom

I

GENERAL ASPECTS OF PLANT INVASIONS

WHAT MAKES A SPECIES INVASIVE?

Marcel Rejmánek
Section of Evolution and Ecology, University of California, Davis, CA 95616, USA

Abstract

Attempts to predict which species will become invaders and those which will not, represent one of the main areas of interest in the study of biological invasions. At present, however, only very limited generalizations are available, based on plant physiology, genetics, demography, species behaviour in other countries, or behaviour of congeneric species. Here I report that invasiveness of pines (genus *Pinus*) and, very likely, other woody species of seed plants in disturbed landscapes, is predictable on the basis of a small number of attributes: small mean seed mass, short juvenile period, and short mean interval between large seed crops. Moreover, vertebrate dispersal is responsible for success of many woody invaders in disturbed as well as 'undisturbed' habitats. As for herbaceous species, their primary (native) latitudinal range seems to be the best predictor of invasiveness, at least for species introduced from Eurasia to North America.

Introduction

Weeds, invaders, and colonizers are three closely related but not identical concepts (Williamson 1993). Although their definition and relationship are often only vaguely described (Baker and Stebbins 1965; Moore 1975; Di Castri 1990), they reflect three different viewpoints: anthropocentric (weeds are plants growing where they are not desired), ecological (colonizers appear early in successional series), and biogeographical (invaders are spreading into areas where they are not native).

Some plant species are weeds because they are poisonous but they are not colonizers or invaders (*Datisca glomerata*, *Aconitum* spp.). Some historically very important agronomic weeds are poor colonizers so are now endangered species in some European countries (*e.g., Agrostemma githago* and *Centaurea cyanus*; Sevensson and Wigren 1986). Colonizers are, in general, successful invaders, but only in disturbed environments (Bazzaz 1986; Rejmánek 1989). Many invaders certainly become very serious weeds but others can hardly be classified as weeds in the usual sense (*Lygodium japonicum* in floodplain forests of Louisiana). However, all invaders in protected areas are usually classified as weeds (sometimes 'ecological' or 'environmental weeds') because in national parks and similar areas non-native species often interfere with the major management goal, *i.e.*, protection of native biota.

Obviously, there is a considerable overlap between these three categories (weeds, invaders, colonizers), and many plant species can be cross-classified as belonging to all three (Fig. 1). It is this overlap which guarantees that generalizations made about one category can be helpful in understanding the performance of many (but not all) species belonging to the other two. In California, for example, about 75% of weeds are classified as invaders and over 85% of invaders can be classified as colonizers. In the British flora, however, only about 11% of weeds are classified as invaders (89% are presumably native) and only 47% of invaders are classified as colonizers (Williamson 1993).

Plant Invasions - General Aspects and Special Problems, pp. 3-13
edited by P. Pyšek, K. Prach, M. Rejmánek and M. Wade

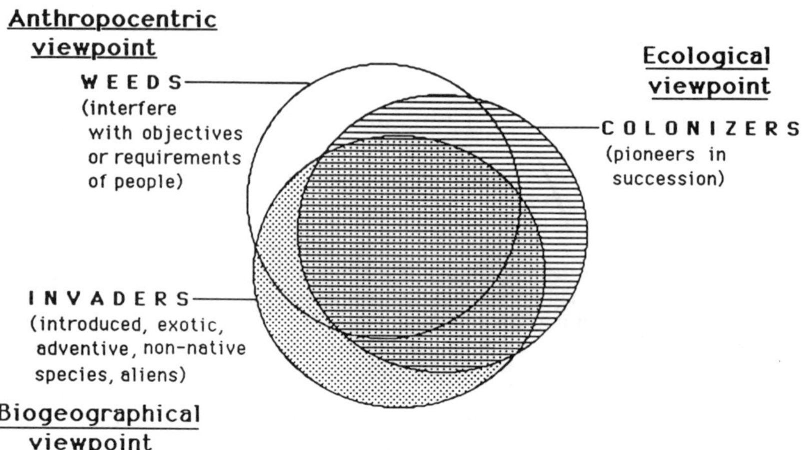

Fig. 1. Weeds, colonizers, and invaders are overlapping but not identical concepts reflecting three different vievpoints.

Historically, there was little concern about the negative effects of exotic species introductions. Recently, however, 'biological pollution' has become nearly as alarming as chemical pollution in many areas of the earth (McKnight 1993). The negative effects of exotic mammals were recognized rather early (*e.g.*, rodents, ungulates, and predators on oceanic islands). Also, a lot of attention has been paid to exotic agronomic and range weeds, especially in North America, since the beginning of this century (Parish 1920; Rejmánek and Randall 1994). However, it has been realized only recently that plant invaders may have far reaching ecosystem-level effects. *Ammophila arenaria* (European beach grass) was introduced for stabilization of coastal dunes in California and Oregon and is drastically changing the geomorphology of the dunes, creating an environment hostile for several native rare species (Aptekar *et al.* 1993). The actinorrhizal nitrogen-fixer *Myrica faya* alters primary successional ecosystems in Hawaii Volcanoes National Park by quadrupling inputs of nitrogen (Vitousek 1990). *Tamarix* species have drawn down the water table in many riparian and wetland sites in the U.S. Southwest and Northern Teritory, Australia (Brock 1994; Howe and Knopf 1991). In many parts of the world, the introduction of productive, fuel-producing but fire-tolerant grasses (*e.g.*, *Andropogon virginicus*, *Schizachyrium condensatum*) has led to an increase in the frequency of fires and the elimination of fire-sensitive natives (D'Antonio and Vitousek 1992). Exotic pines (*Pinus radiata*, *P. patula*, *P. halepensis*, *etc.*) and wattles (*Acacia cyclops*, *A. longifolia*, *A. saligna*) are changing species-rich native vegetation (fynbos) to homogeneous forests and scrubs depauperate of species in southern Africa (Richardson *et al.* 1989; Stirton 1987).

We need predictive theories which can help us set priorities for the control of introduced invasive weeds and allow us to predict the risk of future invasions. Unfortunately, pressing questions like "what attributes make some species more invasive?" or "what makes some ecosystems more invasible than others?" still do not have satisfactory answers (Baker and Stebbins 1965; Kornberg and Williamson 1987; Drake *et al.* 1989; Groves and Di Castri 1991; Barrett 1992; Perrins *et al.* 1992a,b; Scott and Panetta 1993; Wiliamson and Fitter 1994). Useful generalizations are hard to develop because most of the data come from contingent qualitative observations. Moreover, data on failed invasion attempts are usually not available. At present, only very limited generalizations are available, based on plant physiology, genetics, demography, species behaviour in other countries, or behaviour of congeneric species. Not surprisingly, the lack of progress in this area has fostered pessimism regarding the prospect of predicting which organisms are likely to become successful invaders (Crawley 1987; Roy 1990).

In this paper I will first show that invasiveness of pines (genus *Pinus*) and, very likely, other woody species of seed plants in disturbed landscapes, is predictable on the basis of a small number of simple biological characters. Then I will report that analyses of statistical relationships between primary (native) and secondary (adventive) distribution ranges of species introduced from Eurasia to North America can provide some understanding and generate interesting hypotheses concerning the nature of successful herbaceous invaders.

Invasive pines have predictable characters

Characters responsible for remarkable differences in invasiveness of some pine species were analyzed by Rejmánek and Richardson (in preparation). There are several reasons why the genus *Pinus* represents a unique opportunity for this type of study. Pines form a clearly defined genus in the Northern Hemisphere with about 100 species. Pines are economically important, or at least promising, and many species have been introduced to almost all countries with climates reasonably similar to areas of their native distribution. Life history characters of many species have been studied in detail. Reliable records about individual introductions in terms of failures, survival, growth, regeneration, and spread are extensive. There are species of pines spreading spontaneously in many countries (especially in the Southern Hemisphere) and species frequently planted but never reported as spreading. Finally, pine reproduction biology is relatively simple, and underlying trends which can be masked by intricacies of pollination and seed dispersal in angiosperms can be more easily determined.

Our analyses of characters responsible for species invasiveness were based on data available for 24 well known and frequently cultivated pine species (Table 1). I classified 12 of them as invasive *a priori* (reported as spontaneously spreading on at least two continents) and 12 as non-invasive (planted on at least three continents but never reported as spreading). Ten life-history characters were included in the analysis initially: mean height, maximum height, minimum juvenile period, mean longevity, mean seed mass, seed-wing loading index, average percentage of germination, mean interval between large seed crops, degree of serotiny, and fire tolerance index.

A simple discriminant analysis (Manly 1986; Huberty 1994) was performed using these characters as predictors of membership in the two groups. Only three characters contributed significantly to the discriminant function and consistently maximized the difference between the two groups: √(mean seed mass), √(minimum juvenile period), and mean interval between large seed crops (Table 1). The stability of the classification was further checked by 500 cross-validation runs where only 6 species (50%) from each group were chosen randomly each time. Results indicate an unusual robustness of the classification (Rejmánek and Richardson, in preparation).

Two variables incorporated into the discriminant function are rather straightfor-

Table 1. Discriminant analysis of invasiveness in frequently cultivated species of 'hard' (subgenus *Pinus*) and 'soft' (subgenus *Strobus*) pines. Based on empirical evidence, 12 species were *a priori* classified as invasive and 12 as non-invasive. A fuction (Z) discriminating most successfuly between these two groups combines mean seed mass in mg (M), mean interval between large seed crops in years (C), and minimum juvenile period in years (J): $Z = 19.77 - 0.51\sqrt{M} - 3.14\sqrt{J} - 1.21C$; F=23.38, $p<0.001$, relative contributions of $\sqrt{M}$, $\sqrt{J}$, and C are 42.7%, 40.7% and 17.1%, respectively.

		Pinus			*Strobus*	
		Species	Z score		Species	Z score
Invasive species						
	Shore	*contorta* ssp. *contorta*	11.41			
	Monterey	*radiata*	9.27			
	Jack	*banksiana*	8.85			
	Aleppo	*halepensis*	8.21			
	Bishop	*muricata*	8.14			
	Maritime	*pinaster*	7.46			
	Mexican weeping	*patula*	7.30			
	Scots	*sylvestris*	7.12			
	Slash	*elliotii*	4.33			
	Austrian	*nigra*	1.33	Eastern white	*strobus*	3.46
	Ponderosa	*ponderosa*	0.29			
Non-invasive species						
	Caribbean	*caribea*	-0.47	Limber	*flexilis*	-2.77
	Red	*resinosa*	-1.78	Pinyon	*edulis*	-8.94
	Coulter	*coulteri*	-4.06	Swiss stone	*cembra*	-10.31
	Apache	*engelmannii*	-4.45	Mexican pinyon	*cembroides*	-10.49
	Longleaf	*palustris*	-6.36	Sugar	*lambertiana*	-12.35
	Torrey	*torreyana*	-7.62			
	Digger	*sabiniana*	-7.97			

ward: short juvenile period and short interval between large seed crops mean early and constant reproduction. Small mean seed mass seems to be associated with several potentially important phenomena: larger number of seeds produced, better dispersal, high initial germinability, and shorter chilling period needed to overcome dormancy. The three selected variables point to an underlying r-K selection continuum (early-late successional roles) along which invasive-non-invasive pine species are situated. The fact that 'r-strategists' are the best invaders is not surprising because the overwhelming majority of biological invasions take place in human- and/or naturally-disturbed habitats. Our modern landscape is mainly disturbed landscape.

In this context, it is important to mention that Wakamiya *et al.* (1993) reported significant positive correlation between nuclear DNA content and both seed mass and length of juvenile period for 18 North American *Pinus* species. The majority of the DNA in the genomes of most vascular plant species consists of repetitive base-sequences that are not transcribed to proteins. Copying these sequences may slow meiosis, mitosis, and seedling growth (Grime *et al.* 1988). Low nuclear DNA content seems to be a result of selection for short minimum generation time in time-limited environments (Bennett 1987; Goin *et al.* 1968).

Invasive pines are clearly concentrated in the subgenus *Pinus* (*Diploxylon*) and non-invasive species in the subgenus *Strobus* (*Haploxylon*). Membership in a subgenus therefore can be used as a first indication of possible invasiveness. The five most invasive pines, as they are known from literature and our own experience (*P. radiata, contorta, halepensis, patula, pinaster*), have the highest discriminant scores and in the course of all cross-validation runs were always correctly classified as invasive (Rejmánek and Richardson, in preparation).

I next applied the discriminant function (Table 1) to 34 different and, in general, less often cultivated species in the same genus. Potentially invasive species were again concentrated in the subgenus *Pinus*. The fact that at least some of the species with positive discriminant scores are potentially invasive is indicated by a number of published records of natural regeneration in countries of introduction. In fact, some of these species (*P. kesiya, thunbergiana, taeda*) could already be classified as invasive. Natural regeneration and spread of two 'non-invasive' species with large seeds (*P. koraiensis, pinea*) are facilitated by their dispersal by native or introduced squirrels. Not surprisingly, the general trend revealed by the discriminant function can be modified by species- and/or habitat-specific factors.

There are reasons to believe that the discriminant function derived in this study may be applicable to other groups of woody seed plants. Among gymnosperms, all species in the frequently cultivated genus *Araucaria* are correctly classified as non-invasive. On the other hand, *Picea sitchensis* (Sitka spruce) and *Larix decidua* (European larch) are correctly predicted to be invasive. An application of the discriminant function to 40 of the most invasive woody angiosperm species from 40 different genera resulted in correct classification of 38 species (Rejmánek and Richardson, in preparation). Only *Melia azedarach* and *Maesopsis eminii* were incorrectly classified as non-invasive. Efficient bird, bat, and primate dispersal seems to be responsible for this discrepancy. Using the 'pine discriminant function', many frequently cultivated but non-invasive angiosperm species (*Acer saccharum*, *Aesculus* spp., *Aleurites molucana*, *Camellia japonica*, *Carya* spp., *Corylus* spp., *Fagus* spp., *Juglans* spp., *Magnolia* spp., *Quercus* spp., *Thevetia peruviana*, *Swietenia macrophylla*) are correctly classified as non-invasive. However, some non-invasive species of *Populus* (*P. tremula, tremuloides*) are classified as invasive. The short seed viability and high seedling mortality brought about by the slow growth of seedling primary roots in these species prevent them from becoming invasive. In general, it appears that invasiveness of woody species with dry fruits and mean seed mass <2.0 mg (*Populus* spp., *Salix* spp., *Betula* spp., *Alnus* spp., *Melaleuca quinquenervia*, *Tamarix* spp.) is very often limited to wet habitats. Vertebrate dispersal is responsible for the success of many woody invaders in disturbed as well as 'undisturbed' habitats (Bass 1990; Gade 1976; Hayashida 1989; Richardson *et al.* 1990; Sallabanks 1993; Timmis and Williams 1987; Vitousek and Walker 1989). Over 60% of exotic woody species invading primary tropical forests have fleshy fruits and are dispersed by native or introduced vertebrates (Rejmánek 1994). This fact is taken into account in Table 2.

Ants may assist in invasions of plants that produce seeds with elaiosomes (Smith 1989; Bossard 1991). Self-pollination or pollination niche separation can be important in initial stages of some invasions (Baker 1967; Parrish and Bazzaz 1978). Several structural and physiological characters like bark thickness (Richardson *et al.* 1990), symbiotic nitrogen fixation (Vitousek and Walker 1989), shade tolerance (Jones and McLeod 1989), or stem photosynthesis (Bossard and Rejmánek 1992) may contribute to the success of some invaders in extreme environments. Nevertheless, keeping possible exceptions in mind, the discriminant function derived from simple demographic parameters of invasive and non-invasive pines, together with tentative general rules summarized in Table 2, represents the first general screening tool for detection of invasive woody seed plants.

Table 2. Tentative general rules for detection of invasive woody seed plants based on values of the discriminant function Z (see Table 1), seed mass values, and presence or absence of opportunities for vertebrate dispersal.

		Opportunities for vertebrate dispersal	
		Absent	Present
Z>0	Dry fruits and seed mass >2 mg	Likely invasive	Very likely invasive
	Dry fruits and seed mass <2 mg	Likely invasive in wet habitats	
	Fleshy fruits	Unlikely invasive	
Z<0		Non-invasive	Possibly invasive

Analyses of geographic ranges

Initially, comparison of native and adventive distributions of plants seems to be a simple matter. However, quantification and rigorous analyses of geographic ranges are beset by many problems (Gaston 1991; Ricklefs and Latham 1992). Moreover, until publication of the 2nd edition of Hultén's Atlas (Hultén and Fries 1986), there had not been any satisfactory attempt to present native and adventive distributions in a consistent way for a large number of species. Only this publication allows us to analyze primary and secondary distributional patterns of many originally European species growing north of the tropic of Cancer.

I chose two of the largest families of angiosperms - Gramineae and Compositae - for this study, using primary distributions in Eurasia and northern Africa as they were published in the Atlas. Then I checked longitudinal limits of secondary distributions in North America using about 20 recently published local floras, making corrections for about 40% of the species. Unfortunately, I had to exclude some species (*e.g., Digitaria sanquinalis*) because of uncertain southern limits either in Eurasia or in North America. Finally, for all remaining species I made comparisons of their primary longitudinal ranges in Eurasia with secondary ones in eastern and western North America.

First, mean values of primary (native) latitudinal ranges of species naturalized in North America are significantly larger (about 10°) than those of species which have never been reported as naturalized in North America (Table 3). Remarkably, the two families provided almost identical results. Second, all regressions of secondary latitudinal ranges (in eastern and western North America) on primary latitudinal ranges (in Eurasia and northern Africa) are positive and highly significant (Figs. 2 and 3).

There is, however, a large amount of variance in the the secondary latitudinal ranges which is not explained by the primary latitudinal ranges (R^2 varies between 0.21 and 0.47). Diversity of life forms certainly contributes to the unexplained vari-

Table 3. Mean primary latitudinal ranges (°) of Gramineae and Compositae species in Eurasia and northern Africa calculated separately for species which are and are not naturalized in North America. Means in rows are significantly different (two tailed t-test, $p<0.001$).

	Mean primary latitudinal range	
	Species naturalized in North America	Species not naturalized in North America
Gramineae	36.7 (SD=7.3, n=61)	25.6 (SD= 8.0, n=42)
Compositae	34.6 (SD=7.2, n=64)	25.3 (SD=10.4, n=63)

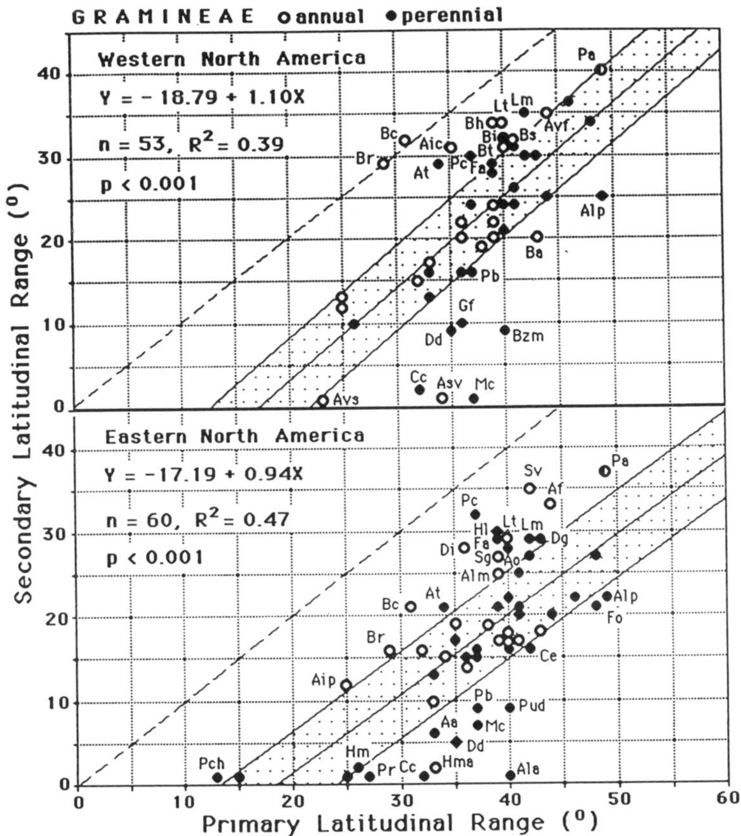

Fig. 2. Relationship between primary (Eurasia/Africa) and secondary (eastern and western North America) latitudinal ranges of grasses introduced to North America. Dashed lines represent identical latitudinal ranges in compared regions. Species *above* ±5° intervals along regression lines: Aic = *Aira caryophyllea*, Aip = *A. praecox*, Alm = *Alopecurus myosuroides*, Ao = *Anthoxanthum odoratum*, At = *Agrostis tenuis*, Avf = *Avena fatua*, Bc = *Bromus commutatus*, Bh = *B. hordeaceus* (*B. mollis*), Bi = *B. inermis*, Br = *B. ramosus*, Bs = *B. secalinus*, Bt = *B. tectorum*, Dg = *Dactylis glomerata*, Di = *Digitaria ischaemum*, Fa = *Festuca arundinacea*, Hl = *Holcus lanatus*, Lm = *Lolium multiflorum*, Lt = *L. temulentum*, Pa = *Poa annua*, Pc = *P. compressa*, Pch = *P. chaixii*, Sg = *Setaria glauca*, Sv = *S. viridis*. Species *below* ±5° intervals along regression lines: Aa = *Ammophila arenaria*, Ala = *Alopecurus arundinaceus*, Alp = *A. pratensis*, Asv = *Apera spica-venti*, Avs = *Avena strigosa*, Ba = *Bromus arvensis*, Bzm = *Briza media*, Cc = *Corynephorus canescens*, Ce = *Calamagrostis epigeios*, Dd = *Danthonia decumbens*, Fo = *Festuca ovina*, Gf = *Glyceria fluitans*, Hm = *Holcus mollis*, Hma = *Hordeum marinum*, Mc = *Molinia caerulea*, Pb = *Poa bulbosa*, Pr = *P. remota*, Pud = *Puccinellia distans*.

ance. Slopes of regression lines are greater for annuals and biennials than for perennials but, not significantly. Residence time (time since the introduction) and number of introductions could confound latitudinal agreement for some species.

Some inherent biological differences can play a role as well. Compare *Poa bulbosa*, *Alopecurus pratensis*, *Corynephorus canescens*, and *Molinia caerulea* whose secondary latitudinal ranges are below ±5° intervals along regression lines with *Poa compressa*, *Lolium multiflorum*, *Agrostis tenuis*, and *Festuca arundinacea* with secondary latitudinal ranges above ±5° intervals along regression lines for both eastern and western North America. Also, compare *Filago arvensis*, *Eupatorium cannabinum*, *Lactuca muralis*, and *Tussilago farfara* below with *Crepis tectorum*, *Hypochoeris*

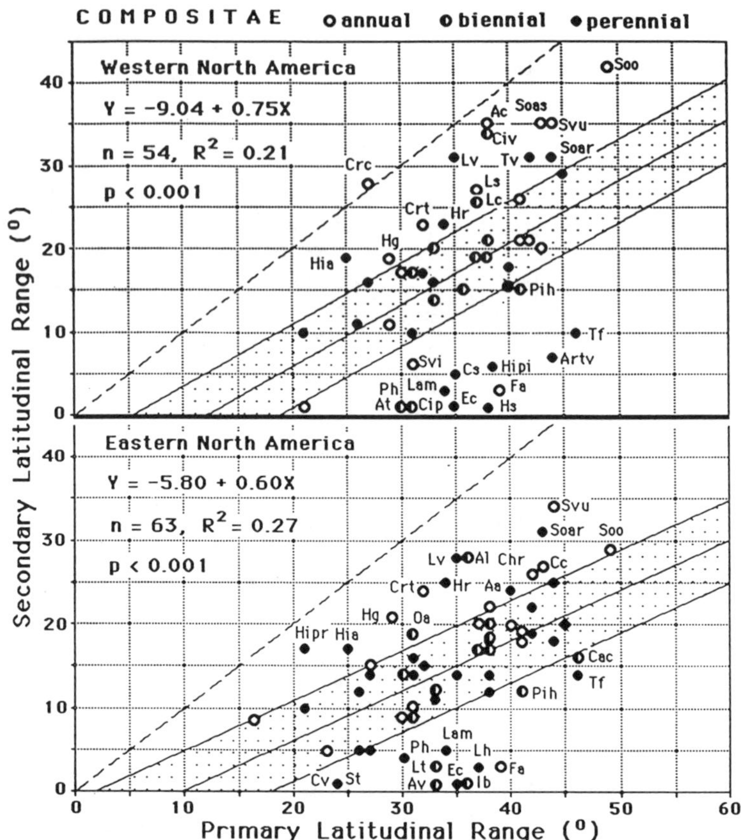

Fig. 3. Relationship between primary (Eurasia/Africa) and secondary (eastern and western North America) latitudinal ranges of Compositae species introduced to North America. Dashed lines represent identical latitudinal ranges in compared regions. Species *above* ±5° intervals along regression lines: Aa = *Artemisia absinthium*, Ac = *Anthemis cotula*, Al = *Arctium lappa*, Cc = *Centaurea cyanus*, Chr = *Chamomilla recutita*, Civ = *Cirsium vulgare*, Crc = *Crepis capillaris*, Crt = *C. tectorum*, Hg = *Hypochoeris glabra*, Hr = *H. radicata*, Hia = *Hieracium aurantiacum*, Hipr = *H. pratense*, Lc = *Lapsana communis*, Ls = *Lactuca serriola*, Lv = *Leucanthemum vulgare*, Oa = *Onopordon acanthium*, Soas = *Sonchus asper*, Soar = *S. arvensis*, Soo = *S. oleraceus*, Svu = *Senecio vulgaris*, Tv = *Tanacetum vulgare*. Species *below* ±5° intervals along regression lines: Artv = *Artemisia vulgaris* ssp. *vulgaris*, At = *Arctium tomentosum*, Av = *A. vulgare*, Cac = *Carduus crispus*, Cip = *Cirsium palustre*, Cs = *Centaurea scabiosa*, Cv = *Carlina vulgaris*, Ec = *Eupatorium cannabinum*, Fa = *Filago arvensis*, Hipi = *Hieracium pilosella*, Hs = *H.* group *Silvaciformia*, Ib = *Inula britanica*, Lam = *Lactuca muralis*, Lh = *Leontodon hispidus*, Lt = *L. taraxacoides*, Pih = *Picris hieracioides*, Ph = *Petasites hybridus*, Svi = *Senecio viscosus*, St = *Serratula tinctoria*, Tf = *Tussilago farfara*.

radicata, *Sonchus arvensis*, and *Hieracium aurantiacum* above ±5° intervals. From this analysis it follows that several potentially serious invaders still have a limited distribution in North America (*Apera spica-venti*, *Bromus arvensis*, *Calamagrostis epigeios*, *Carduus crispus*, *Tussilago farfara*).

There are two explanations why there should be a positive correlation between primary and secondary geographic ranges. Forcella and Wood (1984) and Forcella *et al.* (1984) concluded that the positive relation between area of native distribution and invading capacity arose from the fact that the propagules of widespread species have a higher probability of transport to other countries or continents. On the other hand, Noble (1989) and Roy *et al.* (1991) were of the opinion that, with the considerable increase in intercontinental exchange since the beginning of this century, invasion by a species depends more on the interaction between its biological properties and those of the recipient region than on the probability of reaching that region. Roy *et al.* (1991) suggested that the same biological traits that enable some species to spread across their native continents (and across different climatic zones) also make them able to invade new continents. Only rigorous introduction experiments and/or analyses of well documented invasions over long periods of time (decades) may help to determine which of these two explanations is more likely to be correct. Very likely, however, we will discover that they are not mutually exclusive (Scott and Panetta 1993).

Analyses of primary and secondary geographic ranges can generate stimulating hypotheses, but these should be used with caution for identifying potential invaders. Returning to pines, *Pinus radiata* and *P. patula* are among the most invasive weeds of the Southern Hemisphere (Richardson *et al.* 1994); their native geographic ranges, however, are very narrow (Critchfield and Little 1966). In general, it is true that invasive pine species (Table 1) have greater mean longitudinal range (16.3°, SD=9.6, n=12) than non-invasive pine species (10.4°, SD=5.2, n=12). However, the difference is not significant (p>0.05). As we have already shown, a few essentially demographic attributes seem to be responsible for the striking differences between invasive and non-invasive pine species.

Summary

1. Invasiveness of species in the genus *Pinus* is negatively related to their mean seed mass, minimum juvenile period, and mean interval between large seed crops. A robust discriminant function, combining these three variables can be used not only for detection of invasive pines but also for preliminary screening of invasive woody species in other groups of seed plants.
2. Low nuclear DNA content (genome size) seems to be a result of selection for short minimum generation time and, therefore, may be associated with plant invasiveness in disturbed landscapes.
3. Vertebrate dispersal is responsible for the success of many woody invaders in disturbed as well as 'undisturbed' habitats. Tentative general rules for detection of invasive woody seed plants are summarized in Table 2.
4. The primary (native) latitudinal range and range of climates in source areas seem to be the best predictors of invasiveness for herbaceous species known so far.

Acknowledgments

I thank Tamara Kan, Michael Barbour, Maureen Stanton, Frederic Hrusa, and Kristina Van Katwyk for their help with the data analyses and critical comments on earlier drafts of this paper.

References

Aptekar, R., Rejmánek, M. and Elmore, C. 1993. The Ecology and Control of *Ammophila arenaria* in State Parks of Northern California. 1992 Technical Progress Report to the Department of Parks and Recreation, Northern Region, Santa Rosa, California.

Baker, H.G. 1967. The evolution of weedy taxa in the *Eupatorium microstemon* species aggregate. Taxon 16: 293-300.

Baker, H.G. and Stebbins, G.L. (eds.) 1965. The Genetics of Colonizing Species. Academic Press, New York.

Barrett, S.C.H. 1992. Genetics of weed invasions. In: S.K. Jain and L.W. Botsford (eds.), Applied Population Biology, pp. 91-119. Kluwer Academic Publ., Dordrecht.

Bass, D.A. 1990. Dispersal of an introduced shrub (*Crataegus monogyna*) by the brush-tailed possum (*Trichosurus vulpecula*). Austral. J. Ecol. 15: 227-229.

Bazzaz, F.A. 1986. Life history of colonizing plants: some demographic, genetic, and physiological features. In: H.A. Mooney and J.A. Drake (eds.), Ecology of Biological Invasions of North America and Hawaii, pp. 96-110. Springer-Verlag, New York.

Bennett, M.D. 1987. Variation in genomic form in plants and its ecological implications. New Phytol. 106 (Suppl.): 177-200.

Bossard, C.C. 1991. The role of habitat disturbance, seed predation and ant dispersal on establishment of the exotic shrub *Cytisus scoparius* in California. Am. Midl. Nat. 126: 1-13.

Bossard, C.C. and Rejmánek, M. 1992. Why have green stems? Funct. Ecol. 6: 197-205.

Brock, J.H. 1994. *Tamarix* spp. (salt cedar), an invasive exotic woody plant in arid and semi-arid riparian habitats of the western USA. In: L.C. de Waal, L.E. Child, P.M. Wade and J.H. Brock (eds.), Ecology and Management of Invasive Riverside Plants, pp. 27-44. John Wiley and Sons, Chichester.

Crawley, M.J. 1987. What makes a community invasible? In: A.J. Gray, M.J. Crawley and P.J. Edwards (eds.), Colonization, Succession, and Stability, pp. 429-453. Blackwell Scientific Publ., Oxford.

Critchfield, W.B. and Little, E.L. 1966. Geographic Distribution of the Pines of the World. Misc. Publ. 991. U.S. Department of Agriculture, Forest Service, Washington, D.C.

D'Antonio, C.M. and Vitousek, P.M. 1992. Biological invasions by exotic grasses, the grass/fire cycle, and global change. Ann. Rev. Ecol. Syst. 23: 63-87.

Di Castri, A.J. 1990. On invading species and invaded ecosystems: the interplay of historical chance and biological necessity. In: A.J. di Castri, A.J. Hansen and M. Debussche (eds.), Biological Invasions in Europe and the Mediterranean Basin, pp. 3-16. Kluwer Academic Publ., Dordrecht.

Drake, J.A., Mooney, H.A., Di Castri, F., Groves, R.H., Kruger, F.J., Rejmánek, M. and Williamson, M. (eds.) 1989. Biological Invasions. A Global Perspective. John Wiley and Sons, Chichester.

Forcella, F. and Wood, J.T. 1984. Colonization potentials of alien weeds are related to their 'native' distributions: implications for plant quarantine. J. Austral. Inst. Agricult. Sci. 50: 35-40.

Forcella, F., Wood J.T. and Dillon, S.P. 1984. Characteristics distinguishing invasive weeds within *Echium* (Bugloss). Weed Res. 26: 351-364.

Gade, D.W. 1976. Naturalization of plant aliens: the volunteer orange in Paraguay. J. Biogeogr. 3: 269-279.

Gaston, K.J. 1991. How large is a species' geographic range? Oikos 61: 434-438.

Goin, O.B., Goin, C.J. and Bachmann, K. 1968. DNA and amphibian life history. Copeia 3: 31-47.

Grime, J.P., Hodgson, J.G. and Hunt, R. 1988. Comparative Plant Ecology. Unwin Hyman, London.

Groves, R.H. and Di Castri, F. (eds.) 1991. Biogeography of Mediterranean Invasions. Cambridge University Preass, Cambridge.

Hayashida, M. 1989. Seed dispersal by red squirrels and subsequent establishment of Korean pine. For. Ecol. Managem. 28: 115-129.

Howe, W.H. and Knopf, F.L. 1991. On the imminent decline of Rio Grande cottonwoods in central New Mexico. Southw. Natur. 36: 218-224.

Huberty, C.J. 1994. Applied Discriminant Analysis. John Wiley and Sons, New York.

Hultén, E. and Fries, M. 1986. Atlas of North European Vascular Plants North of the Tropic of Cancer. Vols. 1, 2, 3. Koeltz Scientific Books, Königstein.

Jones, R.H. and McLeod, K.W. 1990. Growth and photosynthetic responses to a range of light environments in Chinese tallowtree and Carolina ash seedlings. Forest Sci. 36: 851-862.

Kornberg, F.R.S. and Williamson, M.H. (eds.) 1987. Quantitative Aspects of the Ecology of Biological Invasions. The Royal Society, London.

Manly, B.F.J. 1986. Multivariate Statistical Methods. A Primer. Chapman and Hall, London.

McKnight B.N. (ed.) 1993. Biological Pollution. The Control and Impact of Invasive Exotic Species. Indiana Academy of Science, Indianapolis.

Moore, R.M. 1975. An ecologist's concept of noxious weed: plant outlaw? J. Austral. Inst. Agricult. Sci. 41: 119-121.

Noble, I.R. 1989. Attributes of invaders and the invading process: terrestrial vascular plants. In: J.A. Drake *et al.* (eds.), Biological Invasions. A Global Perspective, pp. 301-313. John Wiley and Sons, Chichester.
Parrish, J.A.D. and Bazzaz, F.A. 1978. Pollination niche separation in a winter annual community. Oecologia 35: 133-140.
Parish, S.B. 1920. The immigrant plants of southern California. Bull. South. Calif. Acad. Sci. 14: 3-30.
Perrins, J., Williamson, M. and Fitter, A. 1992a. A survey of differing views of weed classification: implications for regulation of introductions. Biol. Conserv. 59: 47-56.
Perrins, J., Williamson, M. and Fitter, A. 1992b. Do annual weeds have predictable characters? Acta Oecol. 13: 517-533.
Rejmánek, M. 1989. Invasibility of plant communities. In: J.A. Drake *et al.* (eds.), Biological Invasions. A Global Perspective, pp. 369-388. John Wiley and Sons, Chichester.
Rejmánek, M. 1994. Species richness and resistance to invasions. In: G.H. Orians, R. Dirzo and J.H. Cushman (eds.), Diversity and Processes in Tropical Forest Ecosystems. Springer-Verlag, New York (in press).
Rejmánek, M. and Randall, J.M. 1994. Invasive alien plants in California: 1993 summary and comparison with other areas in North America. Madroño 41: 161-177.
Richardson, D.M., Cowling, R.M. and Le Maitre, D.C. 1990. Assessing the risk of invasive success in *Pinus* and *Banksia* in South African mountain fynbos. J. Veget. Sci. 1: 629-642.
Richardson, D.M., Macdonald, I.A.W. and Forsyth, G.G. 1989. Reductions in plant species richness under stands of alien trees and shrubs in the fynbos biome. South Afr. For. J. 149: 1-8.
Richardson, D.M., Williams, P.A. and Hobbs, R.J. 1994. Pine invasions in the Southern hemisphere: determinants of spread and invadibility. J. Biogeogr. (in press).
Ricklefs, R.E. and Latham, R.E. 1992. Intercontinental correlation of geographical ranges suggests stasis in ecological traits of relict genera of temperate perennial herbs. Am. Nat. 139: 1305-1321.
Roy, J. 1990. In search of the characteristics of plant invaders. In: A.J. di Castri, A.J. Hansen and M. Debussche (eds.), Biological Invasions in Europe and the Mediterranean Basin, pp. 335-352. Kluwer Academic Publ., Dordrecht.
Roy, J., Navas, M.L. and Sonié, L. 1991. Invasion by annual brome grasses: a case study challenging the homoclime approach to invasions. In: R.H. Groves and F. di Castri (eds.), Biogeography of Mediterranean Invasions, pp. 207-224. Cambridge University Press, Cambridge.
Sallabanks, R. 1993. Fruiting plant attractiveness to avian seed dispersers: native *vs.* invasive *Crataegus* in western Oregon. Madroño 40: 108-116.
Scott, J.K. and Panetta, F.F. 1993. Predicting the Australian weed status of southern African plants. J. Biogeogr. 20: 87-93.
Sevenson, R. and Wigren, M. 1986. A changing flora - a matter of human concern. Acta Univ. Uppsala, Symb. Bot. Uppsala. 27: 241-251.
Smith, J.M.B. 1989. An example of ant-assisted plant invasion. Austral. J. Ecol. 14: 247-250.
Stirton, C.H. (ed.) 1987. Plant Invaders: Beautiful, but Dangerous. The Department of Nature and Environmental Conservation of the Cape Provincial Administration, Cape Town.
Timmins, S.M. and Williams, P.A. 1987. Characteristics of problem weeds in New Zealand's protected natural areas. In: D.A. Saunders *et al.* (eds.), Nature Conservation: The Role of Remnants of Native Vegetation, pp. 241-247. Surrey Beatty & Sons in association with CSIRO and CALM.
Vitousek, P.M. 1990. Biological invasions and ecosystem processes: towards an integration of population biology and ecosystem studies. Oikos 57: 7-13.
Vitousek, P.M. and Walker, L.R. 1989. Biological invasion by *Myrica faya* in Hawaii: plant demography, nitrogen fixation, ecosystem effects. Ecol. Monogr. 59: 247-265.
Wakamiya, I., Newton, R.J., Johnston, J.S. and Price, H.J. 1993. Genome size and environmental factors in the genus Pinus. Am. J. Bot. 80: 1235-1241.
Williamson, M. 1993. Invaders, weeds and the risk from genetically manipulated organisms. Experientia 49: 219-224.
Williamson, M. and Fitter, A. 1994. The varying success of invaders. In: J. Carey, P. Moyle, M. Rejmánek and G. Vermeij (eds.), Invasion Biology. Princeton University Press (in press).

TIME LAGS IN BIOLOGICAL INVASIONS WITH REGARD TO THE SUCCESS AND FAILURE OF ALIEN SPECIES

Ingo Kowarik
University of Hannover, Department of Landscape Architecture and Environmental Studies, Plant Ecology, Herrenhäuser Straße 2, D 30419 Hannover, Germany

Abstract

The historical reconstruction of invasion dynamics of woody species alien to Brandenburg, Germany revealed that 210 species began to spread within the period 1780 to 1990. Lag phases between the first release for cultivation and the beginning of spontaneous spread were reconstructed for 184 species. On average, there was a time lag of 147 years between the introduction to Brandenburg and the initiation of invasion (170 years for trees, 131 for shrubs). The first occurrence of a seedling in the area was defined as the start of a biological invasion without regard to the subsequent success or failure of this invasion. Species' success and failure were evaluated in terms of naturalization and frequency of the 3150 woody species introduced to Germany. It was established that successful invaders are not necessarily quicker in starting invasions than less successful species and that less than 10% of the introduced species begin to invade, 2% become established, and 1% may successfully invade the natural vegetation. When applying this 10:2:1 rule to assess the invasion risk, it has to be considered that the ratio between introductions and starting invasions is not constant, but has increased through time: in 1780 only about 3% of the introduced species had invaded, compared to 7.4% in 1990. Consequently, the number of invasions breaking out will still increase, even if no additional species are introduced. The analysis of timing of biological invasions cannot be explained sufficiently by deterministic factors referring to species' life history. As discussed for some species, the successful invasion can be due to an interaction of directional climatic changes (rising temperatures since 1850) and of stochastically induced shifts in the availability and accessibility of safe sites.

Introduction

There is abundant information on successful invasions by plant species but a fundamental lack of data on unsuccessful or failed introductions (Mooney and Drake 1989). This is particularily true of studies addressing the whole invasion process, including the history of first introduction and subsequent spread. There are some well documented case-studies showing the increased success of invading species over a longer time period (*e.g.*, Wein 1939/40; Jäger 1977, 1986; Trepl 1984; Mack 1986; Guillerm *et al.* 1990; Kornas 1990; Pyšek 1991), but these studies are biased due to the known success of the species studied: they had already spread conspicuously or were suspected of causing trouble in land management or nature conservation.

Lag phases in population growth are recognized preceding the successful invasion of an alien species (Hengeveld 1989), but in comparison to animal invasions, little data exist on selected plant species (*e.g.*, Jäger 1988; Pyšek and Prach 1993). Generalizations on the length of lag phases are necessary for the understanding of the history of population growth (Hengeveld 1987), though even this is not yet possible because most studies are confined to those species which have already spread conspicuously after having overcome their lag phase. The present paper considers the time lags of both successful (in terms of naturalization and frequency) and unsuccessful species.

Plant Invasions - General Aspects and Special Problems, pp. 15-38
edited by P. Pyšek, K. Prach, M. Rejmánek and M. Wade

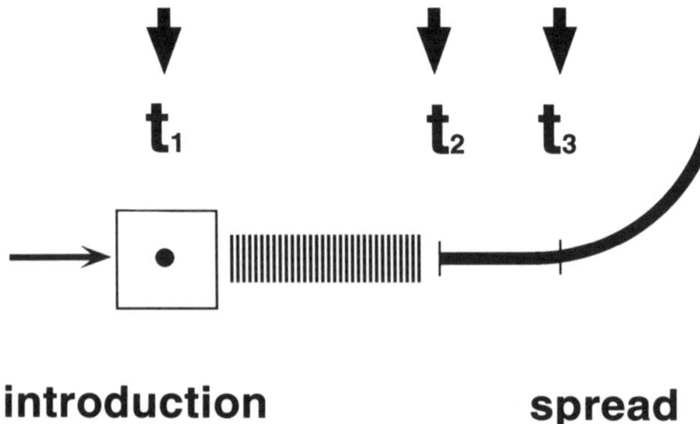

Fig. 1. The process of biological invasions including two kinds of lag phases: *(a)* the period between the first introduction to an area and the first spread (t2-t1), *(b)* the period preceding the switch to a significantly higher rate of population growth (t3-t2).

Two kinds of lag phases may be distinguished within the process of biological invasion (Fig. 1). After their first spread, most species need time to establish populations before a shift in the population growth rate opens the window to a more expansive phase of invasion (t3 - t2 in Fig. 1). Those species however, which had been brought in as ornamentals or crops had already surpassed a previous lag phase: the time between their first release for cultivation and the beginning of spontaneous spread (t2 - t1 in Fig. 1). The present paper focuses mainly on this lag phase as the first barrier between the initial introduction as a predisposition, and the first spontaneous spread as a starting point of a potentially successful invasion.

Aiming to provide a more general insight into the early history of success and failure of intentionally introduced species, the following questions are addressed: *(1)* How many of the deliberately introduced species were able to start an invasion by escaping from cultivation? *(2)* How long was the time lag between the first introduction and the beginning of spread? *(3)* Is there a relationship between the failure or success during the invasion process and the ability of a species to spread rapidly after introduction? In the discussion, the known implications of differences in species' life history for a successful invasion are only briefly considered. The focus lies on the role of both directional environmental changes and of stochastic events in order to elucidate the role of chance and timing in biological invasions. The historical reconstruction of invasion dynamics of all woody species which have been reported as escaped from cultivation in Brandenburg, Germany, were used as a suitable data set for this attempt.

Methods

This paper is based on historical reconstructions of introduction and spreading dynamics of woody species which are not native to Brandenburg, Germany. It covers about the last four hundred years. For several reasons, trees, shrubs, and woody climbers are well suited for such a study: *(a)* with only few exceptions, alien woody species have been brought in and planted deliberately, and the history of these introductions is well documented compared to unintentionally introduced species; *(b)* selecting the species studied by their life form admits a high taxonomic variety in the data set and prevents focusing on taxa which are known to be better pre-adapted than others to successful invasions; *(c)* the species are selected mainly because of aesthetic or economically useful attributes, that is the first introduction of woody species to a new area is not a result of successful invasion. In contrast, even the first appearance of alien herbs introduced unintentionally with exotic seeds or fruits, for example, can be assumed as a first step to successful invasion.

In this paper, the term 'introduction' refers to the first cultivation of an alien woody species in Brandenburg. 'Aliens' (= non-native, non-indigenous) are species occurring in an area in which they have not evolved since the last Ice Age *and* whose introduction or immigration was supported deliberately or involuntarily by human activities.* This includes species from other continents as well as species native to neighbouring regions. *Picea abies,* for example, is only indigenous to southern Brandenburg but escaped from cultivation in other parts of this country. The terms 'invasion' and 'spread of alien species' are used synonymously. They refer to the whole process of range extension of alien species, including its very beginning. Consequently, the first occurrence of a seedling is defined as the start of a biological invasion without regard to the subsequent success or failure of this invasion. However, because of the tradition in natural history, most data on the beginning of invasions actually refer to individuals already well established.

Published and unpublished horticultural and floristic data were used to reconstruct both the time of local introductions to Brandenburg and the first spontaneous emergence of species within this area. Beginning with the '*hortus lusatiae*' by J. Franke, the written sources date back to 1594. For those species whose time of introduction to Brandenburg could not be reconstructed, Goeze's (1916) data covering central Europe were used (including some corrections by Wein 1930, 1931). The first record of the spontaneous, *e.g.*, non-cultivated, appearance of a species was defined as the beginning of an invasion. The flora of Willdenow (1787) was used as the first source with safe distinctions between cultivated and spontaneously spreading species.

Some limitations of floristic data for historical reconstructions and ecological analysis, emphasized by Pyšek and Prach (1993), are also valid for this study. The quality of information may depend on the changing interest of botanists in selected species, and there may be a delay between introduction and spread, respectively, and the information thereon. Nevertheless, it can be assumed that the results are far from anecdotal. The gardening tradition in Brandenburg, including Berlin, is extensive and well documented in the horticultural and forest literature since the 16th century. The intensity of floristic research is high (Scholz 1987; Sukopp 1987) and has resulted in temporally and spatially overlying data.

---

*This definition of alien species is in the tradition of Thellung (1912, 1918/19, see also Trepl 1990). It also follows the definition of Roy (1990) but enlarges it by two additions: first, by the reference to the last Ice Age. This is necessary in order to exclude species as natives which had formerly evolved in the area but became extinct during the colder periods. (Some foresters, for example, call taxa such as *Ginkgo biloba* or *Pseudotsuga* native to Europe.) Re-introduced they should be treated as aliens because if they occurred in the area before or during the last Ice Age, it was not under the present conditions as the climate was different from today (Webb 1985). One of the important consequences is the lacking phase of coevolution with native species in the period after the last glaciation. Second, by the connection to the role of humans enabling the introduction or immigration of alien species. This reference is necessary because most native species did not evolve *in situ* either, but arrived during the period since the last Ice Age, as already stated by Egler (1961). In contrast to alien species, however, the immigration of natives has not necessarily been supported by humans (see Pyšek 1995 for a detailed discussion).

Commonly, the success of invasion is expressed by the rate of population growth over time or the gain of area occupied (Elton 1958; Hengeveld 1989). Appropriate to the data character, in this study the invasion success has been evaluated by the central European approach to naturalization. This concept had been promulgated by Thellung (1912, pp. 622ff.) who referred mainly to the classic work of De Candolle (1855). An alien species is naturalized in the sense of Thellung when it "demonstrates all the characteristics of a wild indigenous plant, *i.e.*, growth and reproduction with natural means of reproduction (seeds, bulbs, tubers *etc.*) without the direct assistance of humans, more or less frequent occurrence at suitable sites, and maintenance for a number of years (including years with unusual climatic phenomena)" (translation in Sukopp and Trepl 1987).

In this tradition, those alien species which took at least the first step toward spreading were grouped as: *(a)* species which began to spread without having become established ('ephemerophytes'; terminology following Schroeder 1969), *(b)* species which are only established in man-made plant communities and which would disappear with the cessation of human activities ('epekophytes'), *(c)* species which are established in the natural vegetation and which would continue to exist even if human influence ceased ('agriophytes'). A fourth group includes the species which became extinct after an unsuccessful invasion. As a second parameter of success, the species were ranked according to their frequency. Because of the substantial information available, both frequency and naturalization were evaluated for species spreading in the area of Berlin, which has been studied extensively during the last decades (for an overview see Sukopp 1990). The complete data set, including times of introduction and of first spread for all species, as well as their ranking according to naturalization and frequency, has been published in Kowarik (1992).

Results

Introductions and subsequent invasions

The exact number of woody species introduced to central Europe is unknown. The data set compiled by Goeze (1916) however, indicates the dimension, listing 2645 species introduced to the nemoral Europe. The temporal stratification of species with different origins reflects the history of voyages and discoveries (Fig. 2). At first, species from other parts of Europe and from the Mediterranean were cultivated, among them some which had already been dispersed by the Romans (*e.g.*, *Juglans regia, Castanea sativa*).

At the end of the 18th century, introductions from North America increased evidently. They had been enhanced by both a paradigmatic shift in garden style and the beginnings of sustainable forestry. The modern landscape garden contained more room for trees and shrubs compared to the traditional baroque garden (Kiermeier 1988). Additionally, American species were tested with great expectations as forest trees (*e.g.*, Von Burgsdorf 1787, 1806). In the second half of the last century, the opening of eastern Asia resulted in an exponential growth of introduced species, mainly from China and Japan (Bretschneider 1898). Some of these found limited use in forestry, but have been planted extensively to the present day in parks and gardens (*e.g.*, *Rhododendron* spp., *Cotoneaster* spp.).

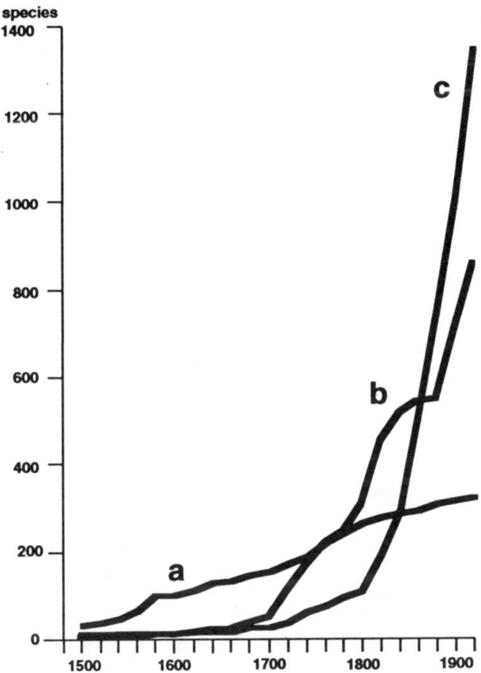

Fig. 2. Introductions of woody species to nemoral Europe during the period between 1500 and 1916. The species are grouped according to their origin in *(a)* other parts of Europe (including the Mediterranean, $n = 309$), *(b)* America ($n = 857$), and *(c)* central and east Asia ($n = 1351$) (cumulative curves; calculated after Goeze 1916; additional 128 species introduced from west Asia and of cultural or unknown origin are not shown).

The historical analysis of floristic data from between 1787 and 1990 revealed that in Brandenburg 210 woody species invaded beyond their original native area. Most of these species were shrubs (132, 62.9%), 70 were trees (33.3%) and 8 were woody climbers (3.8%). 65 species originated in other parts of nemoral and meridional Europe, 63 came from North America, 43 from central and east Asia, the other 39 came from west Asia or were of cultural or unknown origin.

Relating the 210 spreading species to the 2645 introduced species listed by Goeze (1916) or to the 3150 alien woody species currently cultivated in German parks (Kowarik 1992, from the data set of Bartels *et al.* 1981) showed that about 7-8% of these introduced species started spreading in Brandenburg. Assuming that fewer species were cultivated in Brandenburg than had been brought into central Europe, the probability of beginning invasions must be higher.

For 184 of that 210 species the dates of the beginning of invasion were identified (Fig. 3). (Among the 26 species which are not considered in Fig. 3, 10 are native to some parts of Brandenburg, but are invading other parts as aliens, 4 were being already cultivated before 1594, 8 *Rubus*-species were introduced incidentally as weeds instead of having escaped from cultivation, and 4 were doubtful cases.) The number of new species starting to spread has increased to the present day, but the gains are far from regular. Two periods can be distinguished during which higher numbers of newly invading species were reported: the second half of both the 19th and the 20th centuries.

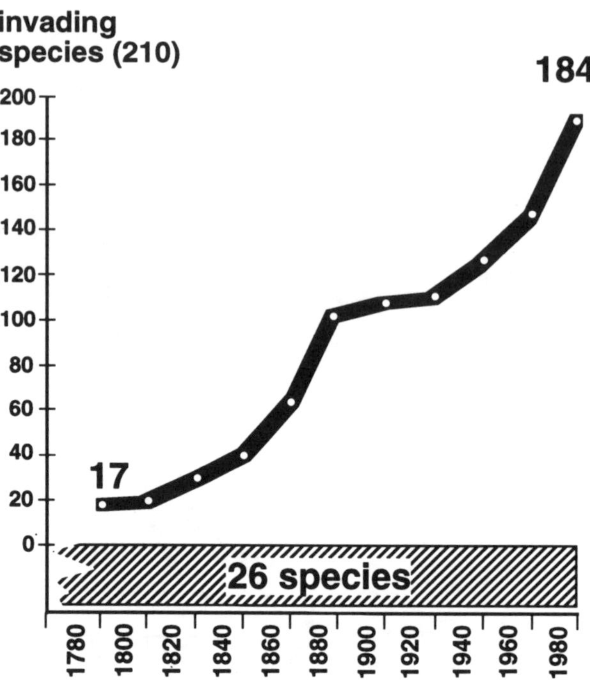

Fig. 3. Reconstruction of the beginning of invasion of woody species introduced to Brandenburg (cumulative curve for 184 of 210 species; 26 species were not differentiated temporally; see text for explanations).

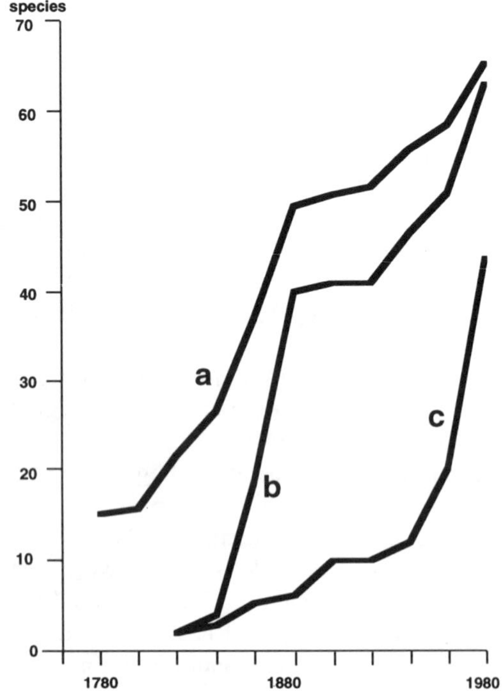

Fig. 4. Reconstruction of the beginning of invasion of woody species (1787-1990, cumulative curves) differentiated according to their origin in *(a)* other parts of Europe (including the Mediterranean), *(b)* America, and *(c)* central and eastern Asia. (Species native to more than one or two areas were counted more than once; species from west Asia and of cultural or unknown origin are not shown.)

Lag phases preceding invasions

In Fig. 4, three groups of invading species, originating in central or east Asia, North America, or in other parts of Europe (including the Mediterranean), were differentiated according to the time of the beginning of invasion. Comparing this figure with the temporal stratification of the total introductions from these areas (Fig. 2) shows a coincidental temporal pattern. Most of the European species spread earlier than American species, which were introduced later, followed finally by those from east Asia which had been introduced mainly in the 19th century. Obviously, the history of beginning invasions echoes the history of introductions.

For 184 species, time lags were calculated between the first introduction of a species to Brandenburg and the first record announcing its becoming invasive. These time lags have a broad range (Fig. 5a). Only 6% of the 184 species began to spread within 50 years after their first cultivation, 25% lagged up to 100 years, 51% up to 200 years, 14% up to 300 years and 4% invaded only after more than three centuries following their first introduction to Brandenburg. On average, there was a delay in invasion of 147 years after the first release of a species. In Fig. 5b, the species are differentiated by their life forms as trees and shrubs (including woody climbers). The pattern is similar, but shrubs began to spread more quickly ($\bar{x} = 131$ years) than trees ($\bar{x} = 170$ years). Through time, the ratio between introductions and beginning invasions was not a constant (Table 1). Calculations for those species which originated in other parts of Europe, North America, or east Asia show that at the end of the 18th century 3.0% of the introduced species began to be invasive. 100 years later this percentage has almost doubled to 5.3%, and, at present, it has increased to 7.4%. Supposing that the number of species introduced in the second half of the 20th century is negligible (Jäger 1988), Goeze's data for 1916 are also taken for the calculations for 1940 and 1990. Differentiating the data according to species' origin reveals obvious differences (Table 1). In the whole period between 1780 and 1990, 21% of

Table 1. Varying ratio *(c)* through time between cumulative number of woody species introductions to central Europe *(a)* and cumulative number of invasions breaking out in Brandenburg *(b)*. c = b/a × 100. The numbers of introductions have been calculated using the data of Goeze (1916). Supposing that the number of species introduced in the second half of the 20th century is negligible, Goeze's data for 1916 are also taken for the calculations for 1940 and 1990. The data are differentiated according to species' origin. Arranged by I: other parts of Europe (including the Mediterranean), II: North America, III: east Asia. Species from west Asia and those of cultural origin were not considered.

	1580	1680	1780	1820	1860	1900	1940	1990
I. Europe								
a. introductions	93	133	225	267	277	296	309	309
b. invasions	?	?	15	22	37	51	56	65
c. ratio b/a	?	?	6.7	8.2	13.4	17.2	18.1	21.0
II. North America								
a. introductions	7	51	268	449	562	713	857	857
b. invasions	?	?	0	2	18	41	47	93
c. ratio b/a	?	?	0	0.4	3.2	5.8	5.5	10.9
III. East Asia								
a. introductions	11	15	79	193	494	1004	1351	1351
b. invasions	?	?	0	2	6	10	12	43
c. ratio b/a	?	?	0	1.0	1.2	1.0	0.9	3.2
I+II+III								
a. introductions	111	199	572	909	1333	2013	2517	2517
b. invasions	?	?	17	29	65	107	122	185
c. ratio b/a	?	?	3.0	3.2	4.9	5.3	4.8	7.4

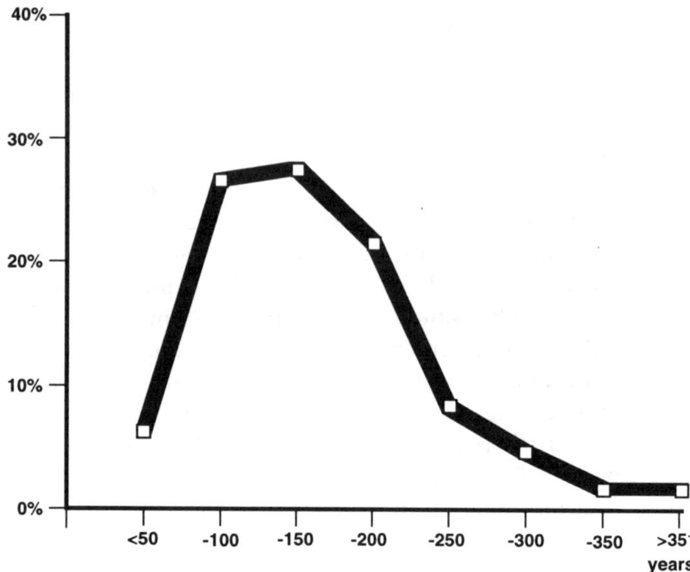

Fig. 5a.

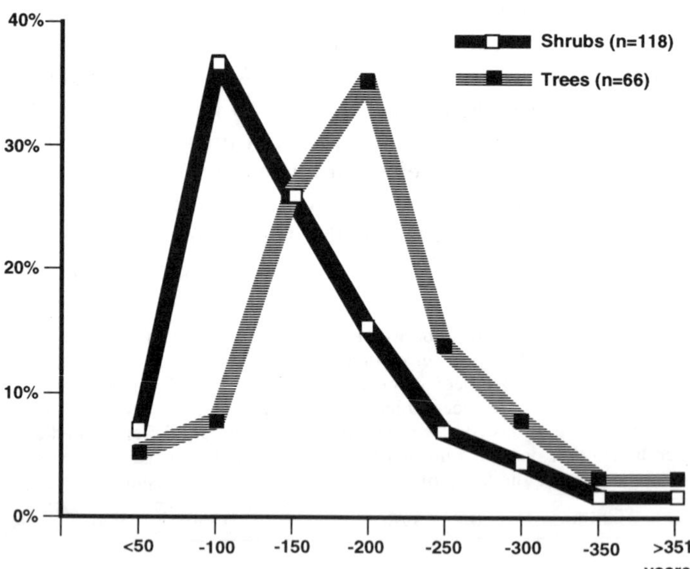

Fig. 5b.

Fig. 5. Time lags between the introduction and beginning of invasion of woody species in Brandenburg. *(a)* Calculation for 184 woody species grouped in 50-year-classes, *(b)* differentiation between trees (n = 66) and shrubs (including woody climbers, n = 118).

the European species began to spread, but only 10.9% of the American and 3.2% of the species from east Asia. In each of these three groups, the ratio between introductions and beginning invasions changes directionally through time, but with a delay in American and Asian species which were generally introduced later. (The slight decrease in the ratio of species from east Asia at the beginning of the 20th century is due to the extensive introductions at this time as shown in Fig. 2.) Considering the variation in the length of time lags (Fig. 5a, b), it can be suggested that the trend in more recently introduced American and, especially, east Asian species will still be upwards. In consequence, the number of invasions breaking out will increase, even if no additional species were introduced.

Success and failure

Most of the introduced woody species have remained unsuccessful in starting invasions: only 210 woody species spread outside their original range. For 182 of these species spreading in Berlin, the success or failure after the first escape from cultivation could be evaluated (Fig. 6). 29 or 16% of the species which began to invade failed (group I): 2 trees, 25 shrubs, and 2 woody climbers have become extinct. In the late 19th century, the performance of these species has been quite different: according to Bolle (1887) most of them had always been rare (*e.g.*, *Myrica cerifera* and *Menispermum canadense*), sometimes however, performing "quite like an indigenous plant" (*Spiraea tomentosa*). Others, however, are known to have been abundant (*Rhus radicans*) or even apparently permanently established (*e.g.*, *Robinia hispida*).

For 97 species (53%), success or failure has not yet been decided (group II): 41 trees, 55 shrubs, and 1 climber still occur as ephemerophytes in the wild without having become established. About a third of these initial invasions have led to a permanent establishment of 25 trees, 29 shrubs, and 2 climbers. Of these, 34 species (19%) have become naturalized as epekophytes in man-made vegetation (group III), and 22 species (12%) are considered to persist as agriophytes in natural vegetation even if human impact on these stands ceases (group IV).

Relating these data to the number of alien species currently cultivated in Germany (3150 species) or to the 2645 introductions to central Europe listed by Goeze (1916), shows that up till now only about 7% of these introduced species have begun to spread. About 2% have established successfully, and less than 1% have become permanent members of the natural vegetation.

Average time-lags were calculated for each of the groups I-IV, the largest being

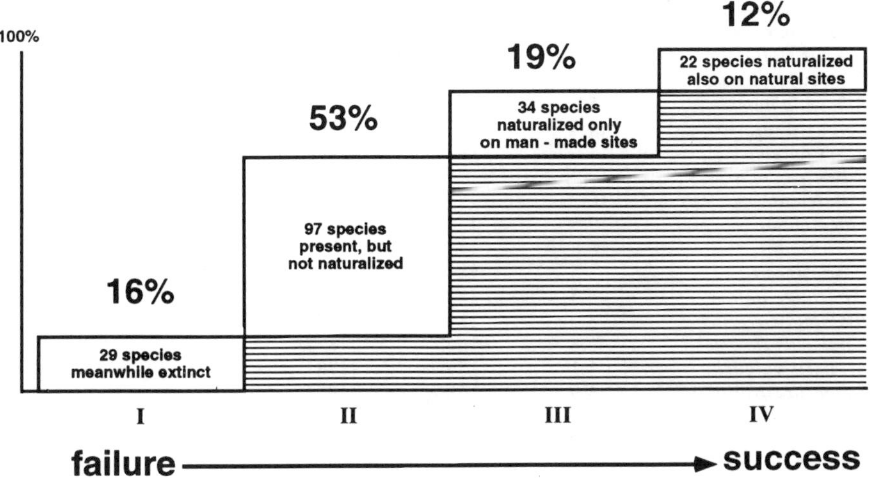

Fig. 6. Success and failure of 182 woody species which had been reported as starting invasion in Berlin in the period between 1787 and 1990 (grouped according to their degree of naturalization: I: extinct species, II: ephemerophytes, III: epekophytes, IV: agriophytes).

for the ephemerophytes (group II), which lagged 159 years on average. No distinct differences were found however, between the different groups. The results for groups I, III, and IV only range between 133 and 137 years, indicating that there is no relationship between the success in naturalization and the lag phase between introduction

Table 2. The most frequent trees, shrubs and woody climbers spreading in Brandenburg, Germany outside their original range. Data on introduction and beginning invasion refer to the area of Brandenburg. The species are ranked according to decreasing frequency in Berlin (see Kowarik 1992: Table 15, 17) The species are native to: 1: other parts of nemoral Europe, 2: meridional Europe, 3: western Asia, 4: central/northern Asia, 5: east Asia, 6: North America. x: Species of cultural origin.

	Relative frequency	First cultivation	First spread	Time-lag	Origin
Trees					
Robinia pseudoacacia	66.9	1672	1824	152	6
Acer negundo	61.0	1736	1919	183	6
Prunus serotina	58.5	1796	1825	29	6
Aesculus hippocastanum	42.6	1663	1787	124	2
Quercus rubra	37.2	1773	1887	114	6
Ailanthus altissima	34.9	1780	1902	122	5
Populus x *canadensis*	17.9	1787	1952	165	x
Prunus domestica	14.6	1594	1787	193	3
Pyrus communis	14.1	1594	1787	193	1-3
Juglans regia	14.1	<1594	1968	>374	2-3
Prunus mahaleb	13.8	1785	1839	54	1-3
Laburnum anagyroides	13.8	1663	1861	198	1
Sorbus intermedia	7.4	1796	1908	112	1
Celtis occidentalis	6.2	1785	1957	172	6
Ulmus pumila	5.1	1796	1985	189	4-5
Hippophaë rhamnoides	4.9	1663	1883	220	1-5
Elaeagnus angustifolia	4.1	1736	1883	147	3
Quercus cerris	2.8	1796	1957	161	2
Larix decidua	2.6	1594	1887	293	1
Shrubs					
Mahonia aquifolium	41.4	1822	1860	38	6
Syringa vulgaris	31.6	1663	1787	124	1
Symphoricarpos albus	30.7	1822	1887	65	6
Ligustrum vulgare	28.5	1594	1787	193	1-3
Philadelphus coronarius	21.9	1656	1839	183	2-3
Lycium barbarum	18.0	1769	1839	70	5
Cornus stolonifera	17.5	1785	1861	76	6
Prunus persica	16.1	<1594	1965	>371	5
Ribes alpinum	15.6	1736	1827	91	1-2
Lonicera tatarica	14.4	1770	1864	94	4
Ribes aureum	13.9	1822	1883	61	6
Ribes grossularia	12.9	1594	1787	193	1
Colutea arborescens	10.2	1594	1859	265	2
Vinca minor	9.5	1582	1787	205	1-2
Cornus alba	8.8	1773	1857	84	4
Rosa rugosa	8.5	1841	1960	119	5
Viburnum lantana	8.5	1736	1864	128	1-3
Buddleja davidii	8.0	1896	1952	56	5
Sambucus racemosa	6.8	1663	1857	194	1
Berberis vulgaris	6.1	1594	1787	193	1-3
Sorbaria sorbifolia	4.9	1796	1904	108	5
Rubus laciniatus	4.6	1808	1885	77	x
Caragana arborescens	3.9	1769	1964	195	4
Woody climbers					
*Clematis vitalba**	-	1663	1883	220	1-2
Parthenocissus inserta	-	1663	1884	221	6

*1787 as time of first spread (Kowarik 1992) has to be considered as doubtful.

and initiation of invasion.

The same is true for success in terms of frequency. The ranking of the most successful trees, shrubs, and climbers according to their frequency in Berlin (Table 2) does not correlate with increasing time lags. The three most frequent tree species, all native to North America, took the first step toward invasion between 29 years (*Prunus serotina*) and 183 years (*Acer negundo*) after their first cultivation in Brandenburg. High differences are also found in shrub species, and the frequent woody climbers were only reported as spreading about 220 years after their first cultivation in Brandenburg.

Discussion

The risk of invasions: or testing the '10:10 rule'

The risk of biological invasions has sometimes been estimated by relating the number of more or less successful invasions to the number of total introductions (Kowarik and Sukopp 1986; Williamson and Brown 1986; Weeda 1987; Di Castri 1989). Referring to the British flora, Williamson promulgated a '10:10 rule' as a rule of thumb, which means that 10% of the introduced species become established, and 10% of established species become pests (Williamson 1993). This approach is somewhat ambiguous because it combines two different dimensions of success. Ecologically, it refers to a species' capability of establishing self-sustaining populations. The denoting of species as 'pests' however, results from an anthropocentrical evaluation of unwanted effects caused by an invasive species. But was is unwanted? Perrins *et al.* (1992) demonstrated the vagueness of categorizing species as weeds or pests by revealing large variation in the evaluations made by agriculturists or by conservation-

Table 3. Probability of initiated, succeeded, and failed invasions of alien woody species. The numbers of invading, established, and extinct woody species in Brandenburg/Berlin are related to the total introductions to Germany. Additionally, data on vascular plants which are considered as alien to central Europe are included for comparison.

	Brandenburg (woody species)		Berlin (woody species)		Central Europe (vascular plants)	
	No	%	No	%	No	%
Introduced	3150	100.0[1]	?		12000	100.0[5]
Producing feral individuals	2214	70.3[1]	?			
Able to start invasions	210	>6.7[2]	182	100.0[3]	?	
Extinct after invasion	>32	>1.0[4]	29	15.9[3]	?	
Present, without established population	>114	>3.6[4]	97	53.3[3]	?	
Self-sustaining population, established	>64	>2.0[4]	56	30.8[3]	417	3.5[6]
Established in natural vegetation	>32	<1.0[4]	22	12.1[3]	228	1.9[7]

[1]Kowarik (1992) calculating the data compiled by Bartels *et al.* (1991) for Germany
[2]referring to the area of Brandenburg
[3]referring to the area of Berlin
[4]estimated by considering the results from Berlin
[5]after Sukopp (1976) referring to central Europe
[6]after Jäger (1988) referring to the area covered by Rothmaler (1976)
[7]after Lohmeyer and Sukopp (1992) referring to central Europe

ists. Considering that some species such as *Colchicum autumnale,* which are currently listed in 'Red Data Books', were persecuted as pests only some decades ago (R. Kirsch-Stracke, personal communication, with reference to Northrhine-Westphalia, Germany), it could be advantageous to differentiate species' invasion success, at least for ecological considerations, exclusively in an ecological perspective. For this purpose, a species' ability to establish self-sustaining populations is considered in this paper. The results are summarized in Table 3.

The numbers in Table 3 are inconclusive, because the success of invading woody species is related to different geographical levels. But even if "roughly 10% means between 5 and 20%" (Williamson 1993: 220), there is little support from this study for the '10:10 rule'. About 7% of the woody species introduced to Germany started their invasion in Brandenburg during the last 200 years, but 16% of the invasions failed through extinction, and this rate is usually unknown or overlooked. Supposing that fewer species had been introduced to Brandenburg than to Germany, the invasion rate is probably higher than 7% and may fit the 10% category. However, most of the invading species are far from being established. Only about one third were able to establish self-sustaining populations, and among these 12% became permanent members of the natural vegetation. Relating these results to the total introductions, a '10:2:1 rule' could be suggested. (It also holds when only species which have become feral are considered as introduced, as done surprisingly by Williamson in 1993.) The '10:2:1 rule' means that roughly 10% (probably less) of the introduced woody species begin to spread, about 2% become established (*i.e.*, significantly less than 5%), and half of these species become members of the natural vegetation. It could be argued that there might exist a dimensional shift in the invasion success of woody species compared to other life forms. Data available for alien vascular plants in central Europe, however, confirm an approximate rate of establishment which is obviously lower than 5% of the total introductions (Table 3, third column), and this fits the dimensions of the '10:2:1' rule. In other areas the relationship between introduced, invading and established species may be different.

Rules derived from the ratio between introduced and spreading species might be suitable for quantifying the actual invasion success, but they would only be adequate to predict the future risk of invasions if the ratio between introductions and invasions was constant. The analyses of the invasion dynamics through time however, shows that the contrary is true (see Table 1). Considering, additionally, the large variation in time lags between the first introduction and the initiation of invasion, the following generalisation can be proposed: the longer an introduced species is present in an area, the higher the probability of its starting an invasion. In consequence, the number of invasions breaking out is expected to increase, even if no additional species were introduced.

These results may stimulate the debate on the question whether or not the number of new invasions breaking out will decrease in this century and in the future (see discussion in Trepl and Sukopp 1993). Although the number of deliberate introductions of new species is negligible in this century, compared to the 18th and 19th centuries (Jäger 1988), the number of new invasions has obviously increased during this century (Fig. 3). In consequence, the ratio between introduced and invading species will change in the future as it has done in the past.

As Table 2 shows, successful species are not necessarily quick in starting invasions. They may become very successful despite lagging longer than less successful invaders. Thus, the performance of a species actually failing in spreading or in the establishment of self-sustaining populations could hide its future success.

Why is the lag phase so long?

Even if through time the probability of invasions has increased, time itself cannot be considered a cause of such biological change (Johnstone 1986). Three kinds of factors which may influence the timing of invasion break outs are tackled: first, some intrinsic factors refering to species' life history traits, second, the role of climatic changes, and, third, the availability of safe sites. The causes to be discussed need not be mutually exclusive, and there are others usually analysed as attributes of successful invaders or of ecosystems susceptible to invasions (Orians 1986; Crawley 1987; Rejmánek 1989; Trepl 1990). In recent years, the 'rule of less ruleness' in explaning the invasion success of different species is accepted more often (Williamson and Brown 1986; Crawley 1989; Roy 1990). Consequently, no attempt is made to find new predictable factors, but attention is paid to the limits in predicting the success or failure of invasions, which is often done by referring to deterministic attributes of species or environments.

The role of intrinsic factors

The minimal lag phase between the first release of a species and the start of an invasion is defined by the time needed to produce the first propagules. The more quickly a species can switch to the reproductive phase of its life cycle, the higher the probability of entering into invasion. Annuals and monocarpic perennials are not necessarily better invaders than perennials (Bazzaz 1986; Noble 1989; Roy 1990; Pyšek and Prach 1993), but they are supposed to be quicker in starting invasions. The lengthened lags of tree as compared to shrub species, which generally breed more quickly, confirm this hypothesis also for woody species (Fig. 5b). Rejmánek (1995) explains the varying invasion success of several *Pinus* species partly by differences in the maturation period. The initial and subsequent growth of a population is affected by other life history traits such as the quantity of seed production, specificity of germination requirements, and the mortality of seedlings. The failure or long lag phases of species which were only rarely planted can partly be explained by the concept of minimum viable populations (Richter-Dyn and Goel 1982). When seedlings occur only at a few sites and have become locally extinct, the chance is small that this first attempt at invasion will be observed.

It is doubtful that the large differences in time lags in those species which have been cultivated more frequently can be explained by this approach. Among the most frequent tree species listed in Table 1, are both pioneer and late successional species. The regeneration of *Prunus serotina*, for example, is enhanced by natural or man-made disturbance (Skeen 1976; Starfinger 1991), and this species is able to set seed after only 7 years (Starfinger 1990). In contrast, *Quercus rubra,* as a later successional species, only begins to fruit when 25 years old, and does not produce acorns abundantly until it is 50 years old (Fowells 1965). Thus, the invasion of *Prunus serotina*, which began almost four times earlier than that of *Quercus rubra* can be explained, at least partly, by the species' different life history traits. This explanation does not fit other frequent invaders, which behave like pioneer species in their areas of original distribution as well as in Brandenburg: *Robinia pseudoacacia* (Boring and Swank 1984; Kowarik 1990a), *Acer negundo* (Sachse 1991, 1992), and *Ailanthus altissima* (Pan and Bassuk 1986; Kowarik and Böcker 1984). All of them are able to set seed within one to two decades, all are prolific seed producers, and all have been planted frequently since long before invasion started. However, *Ailanthus* lagged about 120 years, *Robinia* 150 years and *Acer negundo* 180 years (Table 2). Although the later invasion success of these species was promoted by an increasing availability of safe sites (see discussion below), a lack of these would probably not have excluded an earlier start of invasion. All three species spread in sites which were also available in former times (*e.g.*, parks and gardens, road embankments).

Many but not all successful invaders may be characterized by a high level of genetic plasticity (*e.g.*, Baker 1965; Barrett and Richardson 1986; Bazzaz 1986). Unfortunately, there is little information on the role of genetic adaptation in opening invasion windows for species which lag for long periods after their first introduction. In Australia, changes in flowering time have been found in divergent strains of *Trifolium subterraneum* (Cocks and Phillips 1979), and a switch to earlier flowering is also supposed to be a key factor in the dramatic invasion success of *Senecio inaequidens* during the recent decades in Europe (Werner *et al.* 1991). Another example is *Cynodon dactylon,* which is damaged by frost in the colder parts of Germany. In Berlin, one population with an increased winter hardiness was found and crossed successfully with an American population (Burton and Monson 1978).

As for the woody species considered in this study however, there is no information on genetic adaptation which may explain large time lags. In *Ailanthus altissima,* there is a high genetic variation in both American and Chinese populations (Feret and Bryant 1974). A comparison of significantly different growth and seed characteristics of several seed sources revealed a lack of correlation to climatic and edaphic features indicating that *Ailanthus* has not adapted to macro-environments in America (Feret 1985). In consequence, long lag phases which have been reconstructed for *Ailanthus* and other early successional species may not be sufficiently explained by genetic adaptation or other intrinsic factors.

The role of climatic changes

Erkamo (1956) demonstrated that the rising temperatures in the first half of this century caused responses in the performance of single plant species as well as at the community level. In the last decade it has became evident in the international discussion that ecological theories on changes at the population or community level are insufficient when referring to climate as a constant. Climatic parameters change in a directional way through time, and these changes may affect plant communities, even if only over ecologically short periods such as decades or a few centuries are considered as the time scale (Davis 1986).

In consequence, Hengeveld (1987) stressed the necessity of non-equilibrium models leaving room for the history of populations expressed by the size of time lags. Davis (1986) illustrated big differences in both the magnitude and the timing of species' responses to the same climatic trend. She referred mainly to changes in abundance, demography, and geographical distribution of native species. Different response patterns to climatic changes however, may also be a key factor in determining the variation in lag phases preceding the start of biological invasions.

In this paper, the history of introduction and the initiation of invasion of alien woody species has been reconstructed with reference to the last 400 years. During this period several directional climatic changes have to be considered. The 'Little Ice Age' began around A.D. 1250, ending in the 17th and 18th century. After about 1850 however, the trend changed (Davis 1986). The aggregate change in temperature appears to be between 0.5°C and 0.7°C over the last century, but the rate of increase has been substantially higher in the last decade than at any previous time (Woodwell 1990). Davis (1986) stressed that community changes within forests occur less rapidly than the advances of the tree line, which are reported mainly from northern countries. Even if there is less evidence for changing frequencies in already existing populations of forest trees, the temperature increases since the 1850s may be decisive for those species whose regeneration by seed had previously been prevented by a less favourable climate. This suggestion is supported by Erkamo (1956) who analysed the capability of woody species to regenerate by seeds. He found that from the beginning to the middle of this century, a number of species extended their 'regeneration front' several latitudes northwards. This is true both for species such as *Quercus robur,* which is native to the south of Finland and for alien species such as the North American *Amelanchier spicata.*

For Berlin, as with other big cities, the effects of the climatic rewarming since the 1850s have been exacerbated by the rise in temperature of the urban climate. In the case of Berlin this involved a shift from a city with about 170,000 inhabitants in 1800 to a metropolitan area with 3.7 million inhabitants in 1910. Calculations of the increased warming effect of the urban climate are: 0.2°C for 1798-1804, 0.7°C for 1831-1837, and 1.4°C for 1886-1898 (Schlaak 1982, citing calculations by R. Scherhag). For the period 1961-1980, there was a difference in the annual mean air-temperature of more than 2°C between the centre and the surroundings of Berlin. This warming is correlated with a significant reduction of frost days (<64 days in the centre, >102 days in the surroundings, Von Stülpnagel 1987; Von Stülpnagel *et al.* 1990).

In consequence, the invasion of alien species with higher temperature requirements can be expected to be promoted, especially in big cities where the general trend to warmer temperatures is intensified by the effects of the urban climate. This hypothesis is supported by Fig. 3, showing an obvious increase in woody species starting to invade since the middle of the last century, coinciding with the changes in temperature: In 1756-1847, the winters were colder by -0.7°C compared to 1848-1907, and there was a significantly higher frequency in extremely cold winters in the period before 1846 (Hellmann 1910, 1917).

Similar trends such as in Fig. 3, showing an exponential gain of new species in the second half of the 19th century, have been observed for annual and perennial herbs in Berlin (Scholz 1960) and for all established alien plant species in Germany (Jäger 1977). They had been explained mainly by a huge increase in incidental introductions and subsequent dispersal, promoted both by the new traffic systems and the increasing commercial exchange of goods, which coincided with a diversification of sites available in the urban environment (Sukopp 1976, Jäger 1977, Kowarik 1990b). Many of the alien invaders are native to warmer areas (Erkamo 1956, Scholz 1960), and they are considered to profit from a more favourable local climate, even on a small spatial scale, as Sarisaalo-Taubert (1963) found for small Finnish towns. Pyšek *et al.* (1995) described requirements for temperature increasing for successful invaders of seminatural habitats through those invading man-made sites to unsuccessful aliens.

In contrast to weeds whose introduction and spread has been supported by a complex of diverse factors connected with urbanization and industrialization, the effects of climatic changes on the invasion of woody species which had been planted long before can be more clearly distinguished. Many woody species had already been cultivated during the 'Little Ice Age'. Thus, it is conceivable that for species with a long lag phase the changes in climate may have removed previous barriers to invasion. A cooler climate may prevent the invasion of a cultivated species in two ways: first, low temperatures may prevent seeds from germinating or seedlings from establishing as the performance of *Pinus longaeva* shows in California: for 200 years this species ceased to establish new seedlings, until the climatic conditions became more favourable (LaMarche 1973). This example illustrates how effectively a switch from a cooler to a warmer climate may affect the establishment of seedlings. The vegetation period may be either too short or too cool to allow the complete ripening of fruits.

In central Europe, the invasion success of *Robinia pseudoacacia* is higher in areas subjected to a subcontinental or submediterranean climate as compared with areas under oceanic influence (Kohler 1963; Kohler and Sukopp 1964). The same is true for *Ailanthus altissima:* in the Mediterranean or Pannonian regions, *Ailanthus* has spread abundantly through a broad range of sites (Kowarik 1983, Gutte *et al.* 1987). In central Europe however, it is virtually confined to warmer regions or to urban-industrial sites with a more favorable local climate (Kowarik and Böcker 1984). Considering that the continental climate is characterized by a longer growing period with an increased heat sum, it can be assumed that the spreading of *Robinia* has been promoted by the change to a more favorable climate in the 19th century. The invasion was noticed in 1824, but only in the second part of the last century did this species become frequent (Bolle 1887). The initial spread of *Ailanthus* was only noticed at the beginning of this century, although this species had been cultivated frequently since the end of the 18th century. Actually, it is more frequent in the warmed up zones of Berlin than on the urban fringe, and in Brandenburg it has been reported mainly in cities (Kowarik and Böcker 1984, Kowarik 1992, see also Table 3 in Kowarik 1995).

Apart from these species, *Prunus laurocerasus* and *Buddleja davidii* exemplify the possible limitations on invasions by low winter temperatures. *Buddleja* has been invasive since 1952, but it is actually less frequent in Berlin than in regions with milder winters (Kreh 1953; Burton 1983; Kunick 1990; Schmitz 1991). *Prunus laurocerasus* has been cultivated since 1663, but the first seedlings were not observed in Berlin until 1982. In regions with milder winter temperatures this species obviously succeeded better: it is common in London (Burton 1983), in cities in the northwest of Germany (personal observation), and has become a permanent member of natural forests in the Insubrian region (Gianoni *et al.* 1988). In Zurich, it has only begun to invade forests during recent years (Landolt 1991). Other species which are considered as thermophilous (indicator values for temperature >7 in Ellenberg *et al.* 1991; time lags in parenthesis) are *Laburnum anagyroides* (198 years), *Quercus cerris* (161 years) and the submediterranean *Colutea arborescens* (Kowarik 1985), which is native to the warmest sites in southwest Germany (Sebold *et al.* 1990). In

Brandenburg, 265 years elapsed before the spread of *Colutea* began in 1859. There is even larger time lag in *Vitis vinifera* and in *Juglans regia,* both of which have been cultivated in Brandenburg since 1200 A.D. (A. Brande, personal communication), but were first reported as invading in 1860 and 1968, respectively. The warmer climate may also have encouraged the spread of the frequently occuring *Prunus persica* and of the rarer *P. armeniaca,* both of which spread 300-400 years after their first cultivation (but see discussion below).

The example of *Syringa vulgaris* shows that the mode of spreading also has to be considered. This species was reported as an invasive early on (Willdenow 1787), but it usually spreads by clonal growth enlarging previous plantations. Even on rocky outcrops in the Rhine valley with a favourable local climate, fruits of *Syringa* do not ripen regularly (Lohmeyer and Sukopp 1992). In Brandenburg, *Syringa* spreads mainly by root suckers, but recently some established shrubs were discovered fruiting along abandoned railway areas in Berlin. It is assumed that the fruit ripening on these sites is promoted by the warmer urban climate. In this case, climatic changes would have enabled a species to enlarge its repertoire of spreading strategies.

The role of safe sites

There is a broad agreement that a biological invasion could not start without safe sites available for germination (Harper 1977; Johnstone 1986), and that invasion is faciliated by open communities subjected to natural or man-made disturbance (Trepl 1983; Fox and Fox 1986; Crawley 1987; Kowarik 1990b). There is some evidence that the lack of safe sites may have prevented some species from an earlier invasion. *Hippophaë rhamnoides* for example, occurs in central Europe originally on calcareous sites in dunes at the lake shore and on gravel banks of alpine rivers (Ellenberg 1988). Planted since 1663 in Brandenburg, its first spread was reported about 200 years later. It invades sand-pits where calcareous soils are exposed, and is spreading in urban areas (Kowarik 1992). From this spectrum of settled sites it can be concluded that the limited availability of open calcareous sites in the natural landscape of Brandenburg had prevented any earlier invasion by this species.

The demolition of larger parts of Berlin's centre during the Second World War resulted in a sudden shift in open space available for the colonization by plants. Responses to these changes on the community level have been studied in many cities which were bombed during the war. Many annual and perennial herbs conspicuously invaded the new sites (*e.g.*, Salisbury 1943; Fitter 1946; Scholz 1956; Sukopp 1971). There is considerable evidence that the invasion rates of several woody species were amplified by the availability of new sites. The population growth of *Buddleja davidii* exploded in bombed cities (not in Berlin, but in cities with less cold winters such as Stuttgart and London (Kreh 1952; Burton 1983). In Berlin, there was an obvious increase in the population growth in *Ailanthus, Robinia, Acer negundo,* and also in *Clematis vitalba* (Scholz 1956; Kohler and Sukopp 1964). *Clematis* is not native to Brandenburg, but it is very frequent in other regions of Germany, growing on rich basic soils (Ellenberg 1988). It has been cultivated since 1663, but was first reported as spreading, but only rarely, after 1880. One of the first sites settled was a construction area near a railway station (Bünger 1884), and probably, a lack of suitable sites had prevented *Clematis* from beginning its invasion earlier.

The switch of *Ailanthus, Robinia, Acer negundo,* and *Clematis vitalba* to an exponential population growth since the 1950s could have resulted for several reasons: firstly, the effects of climatic changes, already discussed for *Ailanthus* and *Robinia*; secondly, the concept of the minimum viable population could explain why species whose invasion started some decades earlier needed some time to stabilize a founder population. In the case of these species, however, the sudden increase in the number of safe sites which became available after the war could be the key factor, despite the fact that the invasion by these species started before.

The case of *Platanus hybrida* illustrates the role of changing site qualities. In Berlin, *Platanus* has long been commonly planted, but was reported to spread only after the 1950s. Because moisture is required for germination (Brennenstuhl 1990), invasions of *Platanus* are virtually confined to the slopes of canals or channelized rivers which are moistened occasionally by waves (Kowarik 1984). Formerly, the urban canals had been used much more intensively for transport purposes, and it can be assumed that safe sites for *Platanus* only became available when, after the Second World War, the frequency of use decreased coincidentally with a decrease in maintenance.

Other sites may have become suitable for invasion due to the general trend toward eutrophication brought about by atmospheric deposition of nitrogen (*e.g.*, Ellenberg 1985, Van Breemen and Van Dijk 1988). *Acer pseudoplatanus* and *A. platanoides,* both of which were formerly rare natives to Brandenburg, became more frequent even on sites formerly poor in nutrients (Sachse 1989). Correspondingly, *Sambucus racemosa* invaded oak-pine-forests on poor sandy soils in Berlin and elsewhere in Brandenburg, although this species had been confined to other parts of Germany as a common forest species on better soils (Ellenberg 1988).

Notwithstanding the existence of safe sites, invasion can only start if these sites became accessible (Heimans 1954). This statement is self-evident, though sometimes underestimated, despite being able to explain a high proportion of failing or succeeding invasions. Schroeder (1972) and Trepl (1984, 1990) have discussed in depth the role of stochastic events relative to invasion success: *Impatiens parviflora* would not have become the most successful invader in central European forests, had it not been dispersed incidentally by humans, and the non-arrival of most North American herbs to European forests may be explained by the absence of dispersal vectors. Conversely, the rapidly spreading *Prunus serotina* had been planted directly in forests (Starfinger 1990) and needed to bridge no gaps in reaching suitable sites for germination.

Finally, it is illustrated how the accessibility of safe sites may influence the length of lag phase. The reluctant spread of species such as *Vitis vinifera, Prunus persica, Prunus armeniaca*, and *Ficus carica* can be ascribed to climatic effects. Additionally, the popularization of these formally exotic fruits during the last decades has probably enlarged the dispersal of uneaten fragments of fruits to safe sites. *Prunus persica,* for example, has become common in a densely built-up area in Berlin (Böcker 1991), and fruit trees (*Prunus* spp.) conspicuously line some frequently used forest ways (W. Tigges, personal communication), as well as the railways of the urban transport system, when such space is not treated by herbicides. These sites had been available for more than 150 years, but they only became accessible through changes in dispersal conditions.

Conclusions

The historical reconstruction of invasion dynamics of woody species alien to Brandenburg revealed that less than 10% of the introduced species began to spread within the last 200 years. The number of newly initiated invasions has increased, especially since the 1850s. Only about 2% of the introduced species were able to establish self-sustaining populations, and less than half of this group became members of the natural vegetation, whereas 1% became extinct after invasion. On average, species began to spread only 147 years after their first introduction to Brandenburg. Successful invaders (in terms of naturalization or frequency) are not necessarily more rapid in starting invasions than less successful species. From the discussion of these results, the conclusions can be summarized under two headings: firstly, emphasizing the role of chance and timing in invasions and, secondly, discussing some implications for the 'exotic species model'.

Chance and timing in biological invasions

The first step to invasion may depend on deterministic factors such as the time needed for producing the first propagules. Considering that in many species, the length of lag phases is longer than the time expected to become feral, this stresses the role of other than intrinsic factors for the invasion success. The exponential population growth in *Ailanthus altissima, Acer negundo, Robinia pseudoacacia* or *Clematis vitalba* after the Second World War in Berlin cannot be explained sufficiently by the concept of minimum viable population. The key factor is considered to be the shift in open space available in bombed areas. Seen ecologically, this creation of safe sites resulted from stochastic events. The accessibility of safe sites may be also due to stochasticity, as shown by the example of exotic fruit trees compared to forest trees. There is some evidence however, that the opening of invasion windows does not necessarily result only from stochastic events (Johnstone 1986) but also from directional changes in environmental qualities. The invasion of several tree and shrub species has probably been enhanced by the trend to warmer temperatures since the 1850s, which moreover coincided in Berlin with an ameliorating urban climate. The invasion success of thermophilous species on bombed sites illustrates well the interaction of chance and timing in biological invasions. This has been emphasized recently by Crawley (1989). Supposing Berlin had been bombed 150 years before and the same pool of propagules was present, the colonization of the open space would differ conspicuously due to the less favourable climate. Who could have predicted in the last century the present-day success of species such as *Ailanthus*?

The risk of biological invasions and the 'exotic species' model

The invasion of alien species into areas in which they have not evolved can be used as a model for the release of genetically engineered organisms to the environment (Sharples 1982; Regal 1986; Kowarik and Sukopp 1986; Kowarik 1990c, 1992). The advantages and limitations of this 'exotic species' model have recently been discussed by Regal (1993) and Sukopp and Sukopp (1993). There are two main implications for its appliance which can be derived from the results of this study:
a. Because of the lag phases preceding the invasion of species introduced several

decades or centuries previously, the ratio between introduced and spreading species is not constant (Table 1). Even if no new species were introduced, the number of invasive species will increase through time. Consequently, the risk of invasions breaking out is higher than anticipated by such rules as Williamson's (1993) '10:10 rule' or even if refined, as proposed, as a '10:2:1' rule. 10:2:1 means that less than 10% of the species introduced to an area begin to spread, roughly 2% become established, and 1% is established in the natural vegetation. Considering the high probability of future break outs of new invasions, such rules have to be ascertained from time to time.

b. The analyses of invasion dynamics showed that the success of species may be delayed several decades and even centuries, considering both the beginning of invasion and a possible switch to a significantly higher rate of population growth. Being aware that the opening of invasion windows may result from deterministic as well as stochastic factors in concert with directional environmental changes, the failure of a new species under the current environmental conditions may hide its potential future success. Thus, risk assessments by short-termed experiments are completely insufficient in identifying the risk of future invasions. This is true for both currently successful and unsuccessful invaders. In this regard, there is no difference between introduced alien species and genetically engineered organisms.

Acknowledgments

My thanks are due to Karel Prach, Petr Pyšek, Marcel Rejmánek, Ulrike Sachse, Uwe Starfinger, Ludwig Trepl and Herbert Sukopp for their comments on the previous versions of the manuscript and to Timothy Doyle, Max Wade and Lois Child for improving my English. R. Böcker, A. Brande, and H.-D. Krausch supported the preceding work by their comments and by providing unpublished data on introductions and invasions. I also thank J. Meißner for his assistance and U. Jonczyk for preparing the figures. Parts of this study have been supported by a grant (03 19304 A) of the Federal Ministry for Research and Technology.

References

Baker, H.G. 1965. Characteristics and modes of origins of weeds. In: H.G. Baker and G.L. Stebbins (eds.), The Genetics of Colonizing Species, pp. 141-172. Academic Press, London.

Barrett, S.C.H. and Richardson, B.J. 1986. Genetic attributes of invading species. In: R.H. Groves and J.J. Burdon (eds.), Ecology of Biological Invasions: An Australian Perspective, pp. 22-33. Australian Academy of Science, Canberra.

Bartels, H., Bärtels, A., Schroeder, F.-G. and Seehann, G. (eds.) 1981. Erhebung über das Vorkommen winterharter Freilandgehölze. I. Die Gärten und Parks mit ihrem Gehölzbestand. Mitt. Deutsch. Dendr. Ges. 73: 1-468.

Bartels, H., Bärtels, A., Schroeder, F.-G. and Seehann, G. (eds.) 1982. Erhebung über das Vorkommen winterharter Freilandgehölze. II. Die Gehölze mit ihrer Verbreitung in den Gärten und Parks. Mitt. Deutsch. Dendr. Ges. 74: 1-377.

Bazzaz, F.A. 1986. Life history of colonizing plants: some demographic, genetic, and physiological features. In: H.A. Mooney and J.A. Drake (eds.), Ecology of Biological Invasions of North America and Hawai, pp. 96-110. Springer-Verlag, New York.

Böcker, R. 1991. Floreninformationssystem für den Bezirk Wedding von Berlin. Ermittlung des floristischen Kartieraufwandes für Biotopkartierungen in Städten. Habil. Thesis Technische Universität Berlin.

Bolle, C., 1887. Andeutungen über die Freiwillige Baum- und Strauchvegetation der Provinz Brandenburg. Verlag des Märkischen Provinzial-Museums, Berlin.

Boring, L.R. and Swank, W.T. 1984. The role of black locust (*Robinia pseudoacacia*) in forest succession. J. Ecol. 72: 749-766.
Brennenstuhl, G. 1990. Zur Verwilderung von *Platanus* x *hybrida* Brot. in Ost-Berlin und Dresden. Flor. Rundbr. 24: 99-103.
Bretschneider, 1898. History of European Botanical Discoveries in China. Vol. 2. London.
Bünger, E. 1884. Die Adventiv-Flora auf dem Bau-Terrain am Stadtbahnhof Bellevue in Berlin. Verh. Bot. Ver. Prov. Brandenburg 26: 203-210.
Burton, G.W. and Monson, W.G. 1978. Registration of Tifton 44 Bermudagrass (Reg. No. 10). Crop Sci 18: 911.
Burton, R.M. 1983. Flora of the London Area. London Natural History Society.
Büttner, R., 1883. Flora advena marchica. Verh. Bot. Ver. Prov. Brandenburg 25: 1-59.
Cocks, P.S. and Phillips, J.R. 1979. Evolution of subterranean clover in South Australia. I. The strains and their distribution. Austral. J. Agricult. Res. 30: 1035-1052.
Crawley, M.J. 1987. What makes a community invasible? In: M.J. Crawley, P.J. Edwards and A.J. Gray (eds.), Colonization, Succession and Stability, pp. 629-654. Blackwell Scientific Publ., Oxford.
Crawley, M.J. 1989. Chance and timing in biological invasions. In: J.A. Drake, H.A. Mooney, F. di Castri, R.H. Groves, F.J. Kruger, M. Rejmánek and M. Williamson (eds.), Biological Invasions: A Global Perspective, pp. 407-423. John Wiley and Sons, Chichester.
Davis, M.B. 1986. Climatic instability, time lags, and community disequilibrium. In: J. Diamond and T.J. Case (eds.), Community Ecology, pp. 269-284. Harper and Row, New York.
De Candolle, A. 1855. Géographie Botanique Raisonnée. Paris.
Di Castri, F. 1989. History of biological invasions with special emphasis on the old world. In: J.A. Drake, H.A. Mooney, F. di Castri, R.H. Groves, F.J. Kruger, M. Rejmánek and M. Williamson (eds.), Biological Invasions: A Global Perspective, pp. 1-30. John Wiley and Sons, Chichester.
Egler, F.E. 1961. The nature of naturalization. Recent advances in botany, pp. 1341-1345. University of Toronto Press, Toronto.
Ellenberg, H. Sr. 1988. Vegetation Ecology of Central Europe. 731 pp. Cambridge University Press, Cambridge.
Ellenberg H. Sr., Weber H.E., Düll, R., Wirth, V., Werner W. and Paulißen, D. 1991. Zeigerwerte von Pflanzen in Mitteleuropa. Scripta Geobot. 18: 1-248.
Ellenberg, H. Jr. 1985. Veränderungen der Flora Mitteleuropas unter dem Einfluß von Düngung und Deposition. Schweiz. Z. Forstwes. 136: 19-39.
Elton, C.S. 1958. The Ecology of Invasions by Animals and Plants. 181 pp. Methuen, London.
Erkamo, V. 1956. Untersuchungen über die pflanzenbiologischen und einige andere Folgeerscheinungen der neuzeitlichen Klimaschwankung in Finnland. Ann. Bot. Soc. Zool. Bot. Fenn. Vanamo 28(3): 1-290.
Feret, P.P. 1985. *Ailanthus*: variation, cultivation, and frustration. J. Arbor. 11: 361-368.
Feret, P.P. and Bryant, R.Y., 1974. Genetic differences between American and Chinese *Ailanthus* seedlings. Silvae Genetica 23(5): 144-148.
Fitter, R.S.R. 1946. London's Natural History. 282 pp. London.
Fowells, H.A. 1965. Silvics of Forest Trees of the United States. Agricult. Handbook No. 271. Forest Service USDA, Washington.
Fox, M.D. and Fox, B.J. 1986. The susceptibility of natural communities to invasions. In: R.H. Groves and J.J. Burdon (eds.), Ecology of Biological Invasions: An Australian Perspective, pp. 57-66. Australian Academy of Sciences, Canberra.
Franke, J. 1594. Hortus Lusatiae. Reprint in Johannes Franke "Hortus Lusatiae" Bautzen 1594 mit einer Biographie neu herausgegeben, gedeutet und erklärt. Edited by R. Zaunick, K. Wein and M. Militzer 1930. Bautzen.
Gianoni, G., Carraro, G. and Klötzli, F. 1988. Thermophile, anlaurophylle pflanzenartenreiche Waldgesellschaften im hyperinsubrischen Seenbereich des Tessins. Ber. Geobot. Inst. ETH, Stiftung Rübel 54: 164-180.
Goeze, E. 1916. Liste der seit dem 16. Jahrhundert bis auf die Gegenwart in die Gärten und Parks Europas eingeführten Bäume und Sträucher. Mitt. Deutsch. Dendr. Ges. 25: 129-201.
Guillerm, J.L., Le Floch, E., Maillet, J. and Boulet, C. 1990. The invading weeds within the western Mediterranean basin. In: F. di Castri, A.J. Hansen and M. Debussche (eds.), Biological Invasions in Europe and the Mediterranean Basin, pp. 61-84. Kluwer Academic Publ., Dordrecht.
Gutte, P., Klotz, S., Lahr, C., Trefflich, A. 1987. *Ailanthus altissima* (Mill. Swingle) - eine vergleichende pflanzengeographische Studie. Folia Geobot. Phytotax. 22: 241-262.
Harper, J.L. 1977. Population Biology of Plants. 892 pp. Academic Press, London.
Heimans, J. 1954. L'accessibilité, terme nouveau en phytogéographie. Vegetatio 5(6): 142-146.
Hellmann, G. 1910. Das Klima von Berlin. II. Lufttemperatur. Veröff. K. Preuss. Met. Inst. 221, 3(6): 1-108.
Hellmann, G. 1917. Über strenge Winter. Ber. K. Preuss. Akad. Wiss. 52: 738-759.
Hengeveld, R. 1987. Theories on biological invasions. In: I.A.W. Joenje, K. Bakker and L. Vlijm (eds.), The Ecology of Biological Invasions, pp. 45-49. Proceedings of the Royal Dutch Academy of Sciences, Series C, 90(1).

Hengeveld, R. 1989. Dynamics of Biological Invasions. 160 pp. Chapman and Hall, London.
Jäger, E. 1977. Veränderungen des Artenbestandes von Floren unter dem Einfluß des Menschen. Biol. Rundschau 15: 287-300.
Jäger, E. 1986. *Epilobium ciliatum* Raf. (*E. adenocaulon* Hausskn.) in Europa. Wiss. Z. Univ. Halle, Ser. Math.-Nat. 35: 122-134.
Jäger, E.J. 1988. Möglichkeiten der Prognose synanthroper Pflanzenausbreitungen. Flora 180: 101-131.
Johnstone, I.M. 1986. Plant invasion windows: a time-based classification of invasion potential. Biol. Rev. 61: 369-394.
Kiermeier, P. 1988. "Einen Garten ohne Exoten könnte man mit der Natur verwechseln" oder: Das Vordringen fremder Pflanzen in die Gärten des 19. Jahrhunderts. Das Gartenamt 37(6): 369-375.
Kohler, A. 1963. Zum pflanzengeographischen Verhalten der *Robinie* in Deutschland. Beitr. Naturk. Forsch. SW-Deutschl. 22(1): 3-18.
Kohler, A. and Sukopp, H. 1964. Über die Gehölzentwicklung auf Berliner Trümmerstandorten. Ber. Deutsch. Bot. Ges. 76(10): 389-406.
Kornas, J. 1990. Plant invasions in central Europe: historical and ecological aspects. In: F. di Castri, A.J. Hansen and M. Debussche (eds.), Biological Invasions in Europe and the Mediterranean Basin, pp. 19-36. Kluwer Academic Publ., Dordrecht.
Kowarik, I. 1983. Zur Einbürgerung und zum pflanzengeographischen Verhalten des Götterbaumes (*Ailanthus altissima* (Mill.) Swingle) im französischen Mittelmeergebiet (Bas-Languedoc). Phytocoenologia 11: 389-405.
Kowarik, I. 1984. *Platanus hybrida* Brot. und andere adventive Gehölze auf städtischen Standorten in Berlin (West). Göttinger Florist. Rundbr. 18(1/2): 7-17.
Kowarik, I. 1985. Die Zerreiche (*Quercus cerris* L.) und andere wärmeliebende Gehölze auf Berliner Bahnanlagen. Berliner Naturschutzbl. 29(3): 71-75.
Kowarik, I. 1990a. Zur Einführung und Ausbreitung der Robinie (*Robinia pseudoacacia* L.) in Brandenburg und zur Gehölzsukzession ruderaler Robinienbestände in Berlin. Verh. Berliner Bot. Ver. 8: 33-67.
Kowarik, I. 1990b. Some responses of flora and vegetation to urbanization in central Europe. In: H. Sukopp, S. Hejný and I. Kowarik (eds.), Urban Ecology, pp. 45-74. SPB Academic Publ., The Hague.
Kowarik, I. 1990c. Ecological consequences of the introduction and dissemination of new plant species. An analogy with the release of genetically engineered organisms. In: D. Leskien and J. Spangenberg (eds.), European Workshop on Law and Genetic Engineering, pp. 67-71. BBU-Verlag, Bonn.
Kowarik, I. 1992. Einführung und Ausbreitung nichteinheimischer Gehölzarten in Berlin und Brandenburg und ihre Folgen für Flora und Vegetation. Ein Modell für die Freisetzung gentechnisch veränderter Organismen. Verh. Bot. Ver. Berlin Brandenburg, Beiheft 3: 1-188.
Kowarik, I. 1995. On the role of alien species in urban flora and vegetation. In: P. Pyšek, K. Prach, M. Rejmánek and M. Wade (eds.), Plant Invasions - General Aspects and Special Problems, pp. 85-103. SPB Academic Publ., Amsterdam.
Kowarik, I. and Böcker, R. 1984. Zur Verbreitung, Vergesellschaftung und Einbürgerung des Götterbaumes (*Ailanthus altissima* (Mill.) Swingle) in Mitteleuropa. Tuexenia 4: 9-29.
Kowarik, I. and Sukopp, H. 1986. Ökologische Folgen der Einführung neuer Pflanzenarten. In: R. Kollek, B. Tappeser and G. Altner (eds.), Die ungeklärten Gefahrenpotentiale der Gentechnologie, pp. 111-135. Gentechnologie 10. J. Schweitzer, München.
Kreh, W. 1952. Der Fliederspeer (*Buddleja variabilis*) als Jüngsteinwanderer unserer Flora. Aus der Heimat (Öhringen) 60: 20-25.
Kunick, W. 1990. Spontaneous woody vegetation in cities. In: H. Sukopp, S. Hejný and I. Kowarik (eds.), Urban Ecology, pp. 167-174. SPB Academic Publ., The Hague.
LaMarche, V.C. 1973. Holocene climatic fluctuations inferred from treeline fluctuations in White Mountains, California. Quat. Res. 3: 632-660.
Landolt, E. 1991. Distribution patterns of flowering plants in the city of Zurich. In: G. Esser and D. Overdieck (eds.), Modern Ecology: Basic and Applied Aspects, pp. 807-822. Elsevier Science Publ., Amsterdam.
Lohmeyer, W. and Sukopp, H. 1992. Agriophyten in der Vegetation Mitteleuropas. Schr. R. Vegetationskd. 25: 1-185.
Mack, R.N. 1986. Alien plant invasion into the intermountain west. A case history. In: H.A. Mooney and J.A. Drake (eds.), Ecology of Biological Invasions of North America and Hawai, pp. 191-213. Springer-Verlag, New York.
Mooney, H.A. and Drake, J.A. 1989. Biological invasions: a SCOPE program overview. In: J.A. Drake, H.A. Mooney, F. di Castri, R.H. Groves, F.J. Kruger, M. Rejmánek and M. Williamson (eds.), Biological Invasions: A Global Perspective, pp. 491-506. John Wiley and Sons, Chichester.
Noble, I.R. 1989. Attributes of invaders and the invading process: terrestrial and vascular plants. In: J.A. Drake, H.A. Mooney, F. di Castri, R.H. Groves, F.J. Kruger, M. Rejmánek and M. Williamson (eds.), Biological Invasions: A Global Perspective, pp. 301-313. John Wiley and Sons, Chichester.

Orians, G.H. 1986. Site characteristics favoring invasions. In: H.A. Mooney and J.A. Drake (eds.), Ecology of Biological Invasions of North America and Hawai, pp. 133-148. Springer-Verlag, New York.
Pan, E. and Bassuk, N. 1986. Establishment and distribution of *Ailanthus altissima* in the urban environment. J. Environ. Hort. 4: 1-4.
Perrins, J., Williamson, M. and Fitter, A. 1992. A survey of differing views of weed classification: implications for regulation of introductions. Biol. Conserv. 60: 47-56.
Pyšek, P. 1991. *Heracleum mantegazzianum* in the Czech Republic: the dynamics of spreading from the historical perspective. Folia Geobot. Phytotax. 26: 439-454.
Pyšek, P. 1995. On the terminology used in plant invasion studies. In: P. Pyšek, K. Prach, M. Rejmánek and M. Wade (eds.), Plant Invasions - General Aspects and Special Problems, pp. 71-81. SPB Academic Publ., Amsterdam.
Pyšek, P. and Prach, K. 1993. Plant invasions and the role of riparian habitats: a comparison of four species alien to central Europe. J. Biogeogr. 20: 413-420.
Pyšek, P., Prach, K. and Šmilauer, P. 1995. Relating invasion success to plant traits: an analysis of the Czech alien flora. In: P. Pyšek, K. Prach, M. Rejmánek and M. Wade (eds.), Plant Invasions - General Aspects and Special Problems, pp. 39-60. SPB Academic Publ., Amsterdam.
Regal, P.J. 1986. Models of genetically engineered organisms. In: H.A. Mooney and J.A. Drake (eds.), Ecology of Biological Invasions of North America and Hawai, pp. 111-132. Springer-Verlag, New York.
Regal, P.J. 1993. The true meaning of 'exotic species' as a model for genetically engineered organisms. Experientia 49: 225-234.
Rejmánek, M. 1989. Invasibility of plant communities. In: J.A. Drake, H.A. Mooney, F. di Castri, R.H. Groves, F.J. Kruger, M. Rejmánek and M. Williamson (eds.), Biological Invasions: A Global Perspective, pp. 369-388. John Wiley and Sons, Chichester.
Rejmánek, M. 1995. What makes a species invasive? In: P. Pyšek, K. Prach, M. Rejmánek and M. Wade (eds.), Plant Invasions - General Aspects and Special Problems, pp. 3-13. SPB Academic Publ., Amsterdam.
Richter-Dyn, N. and Goel, N.S. 1982. On the extinction of a colonizing species. Theor. Pop. Biol. 3: 406-433.
Rothmaler, W. 1976. Exkursionsflora für die Gebiete der DDR und der BRD, Band 2 Gefäßpflanzen. Volk und Wissen Volkseigener Verlag, Berlin.
Roy, J. 1990. In search of the characteristics of plant invaders. In: F. di Castri, A.J. Hansen and M. Debussche (eds.), Biological Invasions in Europe and the Mediterranean basin, pp. 335-352. Kluwer Academic Publ., Dordrecht.
Sachse, U. 1989. Die anthropogene Ausbreitung von Berg- und Spitzahorn. Ökologische Voraussetzungen am Beispiel Berlins. Landschaftsentwickl. Umweltforsch. 63: 1-132.
Sachse, U. 1991. Die Populationsbiologie von *Acer negundo* L., einem aggressiven Neophyten in Eurasien. Post-doc. Paper SA 445/1-1 Deutsch. Forschungsgem., 111 pp. (unpubl.).
Sachse, U. 1992. Invasion patterns of boxelder on sites with different levels of disturbance. Verh. Ges. Ökol. 21: 103-111.
Salisbury, E.J. 1943. The flora of bombed areas. Proc. Roy. Inst. Great Britain 32: 435-455.
Sarisaalo-Taubert, A. 1963. Die Flora in ihrer Beziehung zur Siedlung und Siedlungsgeschichte in den südfinnischen Städten Provoo, Loviisa und Hamina. Ann. Bot. Soc. Zool. Bot. Fenn. Vanamo 35: 1-190.
Schlaak, P. 1982. Skizzen der Wetter- und Witterungsverhältnisse und ihre Auswirkungen auf Land, Leute und Wirtschaft zur Zeit des Aufstiegs Preussens (1640-1850). Beilage zur Berliner Wetterkarte 32/82, SO 4/82. pp. 1-32.
Schmitz, J. 1991. Vorkommen und Soziologie neophytischer Sträucher im Raum Aachen. Decheniana 144: 22-38.
Scholz, H. 1956. Die Ruderalvegetation Berlins (Artenbestand, Geschichte, Gesellschaften). PhD Thesis, FU Berlin.
Scholz, H. 1960. Die Veränderungen in der Berliner Ruderalflora. Ein Beitrag zur jüngsten Florengeschichte. Willdenowia 2: 379-397.
Scholz, H. (ed.) 1987. Botany in Berlin. Englera 7: 1-288.
Schroeder, F.-G. 1969. Zur Klassifizierung der Anthropochoren. Vegetatio 16: 225-238.
Schroeder, F.-G. 1972. *Amelanchier*-Arten als Neophyten in Europa. Abh. Naturw. Ver. Bremen 37(3): 287-419.
Sebold, O, Seybold, S. and G. Philippi. 1990. Die Farn- und Blütenpflanzen Baden-Württembergs. Vol. 1-3. Eugen Ulmer, Stuttgart.
Sharples, F.A. 1982. Spread of organisms with novel genotypes: thoughts from an ecological perspective. ORNL/TM-8473, Oak Ridge National Laboratory Environmental Sciences Division Publication No 2040.
Skeen, J. N. 1976. Regeneration and survival of woody species in a naturally created forest opening. Bull. Torrey Botany Club 103: 259-265.

Starfinger, U. 1990. Die Einbürgerung der Spätblühenden Traubenkirsche (*Prunus serotina* Ehrh.) in Mitteleuropa. Landschaftsentwickl. Umweltforsch. 69: 1-136.

Starfinger, U. 1991. Population biology of an invading tree species - *Prunus serotina*. In: A. Seitz and V. Loeschke (eds.), Species Conservation: A Population-Biological Approach, pp. 171-183. Birkhäuser, Basel.

Sukopp, H. 1971. Beiträge zur Ökologie von *Chenopodium botrys* L. 1. Verbreitung und Vergesellschaftung. Verh. Bot. Ver. Prov. Brandenburg 108: 3-25.

Sukopp, H. 1972. Wandel von Flora und Vegetation in Mitteleuropa unter dem Einfluß des Menschen. Ber. Landwirtsch. 50: 112-130.

Sukopp, H. 1976. Dynamik und Konstanz in der Flora der Bundesrepublik Deutschland. Schr. R. Vegetationskde. 10: 9-27.

Sukopp, H. 1987. On the history of plant geography and plant ecology in Berlin. Englera 7: 85-103.

Sukopp, H. (ed.) 1990. Stadtökologie. Das Beispiel Berlin. 455 pp. Reimer, Berlin.

Sukopp, H. and Sukopp, U. 1993. Das Modell der Einführung und Einbürgerung nicht einheimischer Arten. Gaia 2: 268-288.

Sukopp, H. and Trepl, L. 1987. Extinction and naturalization of plant species as related to ecosystem structure and function. Ecol. Studies 61: 245-276.

Thellung, A. 1912. La flore adventice de Montpellier. Mém. Soc. Sci. Nat. Cherbourg 38: 622-647.

Trepl, L. 1983. Zum Gebrauch von Pflanzenarten als Indikatoren der Umweltdynamik. SBer. Ges. Naturforsch. Freunde Berlin N.F. 23: 151-171.

Trepl, L. 1984. Über *Impatiens parviflora* DC. als Agriophyt in Mitteleuropa. Diss. Bot. 73: 1-400.

Trepl, L. 1990. Zum Problem der Resistenz von Pflanzengesellschaften gegen biologische Invasionen. Verh. Berliner Bot. Ver. 8: 195-230.

Trepl, L. and Sukopp, H. 1993. Zur Bedeutung der Introduktion und Naturalisation von Pflanzen und Tieren für die Zukunft der Artenvielfalt. In: Dynamik von Flora und Fauna - Artenvielfalt und ihre Erhaltung, pp. 127-142. Rundgespräche der Kommission für Ökologie der Bayrischen Akademie der Wissenschaften. Vol. 6. München.

Van Breemen, N. and Van Dijk, H.F.G. 1988. Ecosystem effects of atmospheric deposition of nitrogen in the Netherlands. Environ. Poll. 54: 249-274.

Von Burgsdorf, F.A.L. 1787. Ueber die in den Waldungen der Kurmark-Brandenburg befindlichen einheimischen und in etlichen Gegenden eingebrachten fremden Holzarten. Schr. Ges. Naturforsch. Freunde zu Berlin 7: 236-266.

Von Burgsdorf, F.A.L. 1806. Forsthandbuch. Allgemeiner theoretisch-praktischer Lehrbegriff sämtlicher Försterwissenschaften. Berlin.

Von Stülpnagel, A. 1987. Klimatische Veränderungen in Ballungsgebieten unter besonderer Berücksichtigung der Ausgleichswirkung von Grünflächen, dargestellt am Beispiel von Berlin (West). Thesis Tech. Univ. Berlin.

Von Stülpnagel, A., Horbert, M. and Sukopp, H. 1990. The importance of vegetation for the urban climate. In: H. Sukopp, S. Hejný and I. Kowarik (eds.), Urban Ecology, pp. 175-193. SPB Academic Publ., The Hague.

Webb, D.A. 1985. What are the criteria for presuming native status? Watsonia 15: 231-236.

Weeda, E.J. 1987. Invasions of vascular plants and mosses into the Netherlands. In: I.A.W. Joenje, K. Bakker and L. Vlijm (eds.), The Ecology of Biological Invasions, pp. 19-29. Proceedings of the Royal Dutch Academy of Sciences, Series C, 90(1).

Wein, K. 1930. Die erste Einführung nordamerikanischer Gehölze in Europa. I. Mitt. Deutsch. Dendr. Ges. 42: 137-163.

Wein, K. 1931. Die erste Einführung nordamerikanischer Gehölze in Europa. II. Mitt. Deutsch. Dendr. Ges. 43: 95-154.

Wein, K. 1939, 1940, 1942. Die älteste Einführungs- und Ausbreitungsgeschichte von *Acorus calamus*. 1-3. Hercynia 1: 367-450, 3: 72-128, 3: 214-291.

Werner, D.J., Rockenbach, T. and Hölscher, M.-L. 1991. Herkunft, Ausbreitung, Vergesellschaftung und Ökologie von *Senecio inaequidens* DC. unter besonderer Berücksichtigung des Köln-Aachener Raumes. Tuexenia 11: 73-107.

Willdenow, C.L., 1787. Florae Berolinensis Prodromus. Reprint Verh. Berliner Bot. Ver. Sonderband, Berlin 1987.

Williamson, M. 1993. Invaders, weeds and the risk from genetically manipulated organisms. Experientia 49: 219-224.

Williamson, M.H. and Brown, K.C. 1986. The analysis and modelling of British invasions. Phil. Trans. Roy. Soc. Lond. B 314: 505-522.

Woodwell, G.M. 1990. The earth unter stress: a transition to climatic instability raises questions about patterns of impoverishment. In: G.M. Woodwell (ed.), The Earth in Transition. Patterns and Processes of Biotic Impoverishment. Cambridge University Press, Cambridge.

RELATING INVASION SUCCESS TO PLANT TRAITS: AN ANALYSIS OF THE CZECH ALIEN FLORA

Petr Pyšek[1], Karel Prach[2] and Petr Šmilauer[2]
[1]*Institute of Applied Ecology, University of Agriculture Prague, CZ-281 63 Kostelec nad Černými lesy, Czech Republic;* [2]*Faculty of Biological Sciences, University of South Bohemia, Branišovská 31, CZ-370 01 České Budějovice, Czech Republic*

Abstract

Alien species introduced into the Czech Republic since 1492 (excluding rare and ephemeral introductions) are analysed and various species characteristics (geographical area of origin, taxonomic position, species height, life form, life strategy, pollination and dispersal agents, planting history, and ecological requirements expressed using the Ellenberg indicator values) are used to explain their invasion success. The species success in man-made and seminatural habitats was evaluated separately by using a semiquantitative three degree scale. Total success in both types of habitats was also assessed. The aliens were also compared with native flora for the particular characteristics. A total of 132 alien species were analysed of which about 20% may be considered successfully naturalized. Invasion success in seminatural habitats was found to be favoured by height, hemicryptophyte life form and C-strategy. The successful invaders into this habitat type also differed from unsuccessful ones in *(1)* requiring sites which are more moist and *(2)* more frequent planting in the past. Species successfully invading man-made habitats showed an increased representation of therophyte life form and of C- or CR-strategy. They were mostly introduced spontaneously and are able to grow successfully in drier sites. There is an increased representation of therophytes and members of *Asteraceae* family among the whole set of aliens when compared with the native flora. Further, species of North American and Asian origin and those confined to sites with higher nitrogen input were overrepresented. The study confirms the fact that there is no single characteristic which can reliably predict the success of any particular species as an invader. However, if a large data set is used, some differences between alien and native species and differences among the aliens invading contrasting habitat types can be revealed.

Introduction

Ecologists interested in biological invasions have consistently attempted to answer two main questions: *(a)* what are the traits of invasive species? and *(b)* what are the characteristics of those communities or ecosystems vulnerable to invasions? (Baker 1965, 1974; Barrett and Richardson 1986; Gray 1986; Hobbs 1989; Noble 1989; Rejmánek 1989; Roy 1990; Lodge 1993a).

So far, however, extensive sets of species or communities have only exceptionally been taken into account and analysed quantitatively with respect to species traits and invasion success (Newsome and Noble 1986). This is in part due to the fact that considering a large set of species, viewed at the landscape level, brings about inevitable limitations in the exactness and amount of detail of the analyses conducted.

The present study attempts to extend the usual coordinates of ecological studies (Pimm 1994) to an area of 72,000 km^2, for more than a hundred species, and over an historical time dimension. As there is an increasing agreement that it is probably impossible to identify simple, unequivocal plant traits with a high predictive power of invasion success (Lodge 1993a), it is becoming accepted that the performance of invading species should be assessed with respect to a particular ecological situation

Plant Invasions - General Aspects and Special Problems, pp. 39-60
edited by P. Pyšek, K. Prach, M. Rejmánek and M. Wade

(Crawley 1987; Noble 1989; Mooney and Drake 1989; Lodge 1993a). To take this into account in the present study, invaders into seminatural vegetation and those successful in man-made habitats were treated separately. The following questions were addressed: *(a)* Are there any plant traits on the basis of which the alien species can be distinguished from native ones? *(b)* Are there any particular traits favouring invasion success of certain alien species, and if so, do these differ in contrasting (*i.e.*, seminatural *versus* man-made) habitats?

Materials and methods

Species selection

Alien species introduced into Europe after the discovery of America in 1492 (*i.e.*, those classified as neophytes in central European phytogeographical terminology, see *e.g.*, Kornas 1990) were identified from the database covering the territory of former East Germany compiled by Frank and Klotz (1990). The list was revised with respect to the territory of the Czech Republic using local floras (Dostál 1982; Hejný and Slavík 1988-1992). Those species whose spontaneous occurrence had not been reported from this territory were excluded and those missing from the list of Frank and Klotz were added. Additionally, information from archaeological research (Opravil 1980) was considered in order to exclude those species introduced before 1492 (*i.e.*, archaeophytes). Only permanently established aliens were considered, those with only ephemeral occurrence were also excluded. An overview of alien species is given in Table 1 and Appendix 1.

Traits considered

Each species was characterized using the following traits:

a. Area of origin (extracted from Dostál 1958; Hejný and Slavík 1988-1992; Frank and Klotz 1990; Lohmeyer and Sukopp 1992).
b. Taxonomic position at the family level.
c. Plant stature expressed as the maximum height reached by a species (from Dostál 1958; Hejný and Slavík 1988-1992).
d. Life form according to Raunkiaer's scheme (see *e.g.*, Mueller-Dombois and Ellenberg 1974; taken from Frank and Klotz 1990; Ellenberg *et al.* 1991).
e. Life strategy according to Grime (1979), taken from Frank and Klotz (1990).
f. Pollination agents (Frank and Klotz 1990).
g. Dispersal agents (Frank and Klotz 1990).
h. Mode of spread, *i.e.*, by seed or vegetatively. A species was considered to possess the latter only if vegetative propagation was important for its spread at a site (taken from Dostál 1958; Hejný and Slavík 1988-1992).
i. Planting history, *i.e.*, whether the species had been planted in the territory of the Czech Republic (Dostál 1958, 1982; Hejný and Slavík 1988-1992).
j. Ecological requirements expressed by using Ellenberg indicator values for light, moisture, temperature and nitrogen (Ellenberg *et al.* 1991; Frank and Klotz 1990).

Assessment of invasion success

Evaluation of species invasion success was made semiquantitatively, based on the authors' personal experiences from the territory of the Czech Republic. The invasion success of each species was assessed separately for seminatural and man-made habitats using the following scale: 0 - absent, 1 - rare, 2 - scattered over the whole territory or locally abundant, 3 - common, *i.e.*, abundant within the whole territory. Man-made habitats included settlements, dumps in open landscape, arable land and various other disturbed areas. Meadows and grasslands, wetlands, water courses, shrubs and forests were considered as seminatural habitats and include rare remnants of natural communities in the territory under study, *i.e.*, natural habitats were not distinguished as a separate category (see *e.g.*, Kornas 1990) because of the problems of distinguishing between natural and seminatural habitats in the central European landscape.

Data treatment

A two-step comparison was used to analyse the data:
1. Comparison of characteristics of the total set of alien species with those of native flora. Since comparable concise information is not available for the flora of the Czech Republic, the characteristics of native flora were taken from Frank and Klotz (1990). This database, covering the territory of former East Germany, comprises 1,669 native species. Even if the floristic differences between East Germany and Czech Republic are taken into account, the database used undoubtedly provides a highly representative sample, characteristics of which may be considered as very similar to those of the Czech native flora.
2. Comparison within aliens for which three categories were distinguished: *(a)* species successful in seminatural habitats, *(b)* species successful in man-made habitats, and *(c)* unsuccessful species, *i.e.*, those successful in neither *(a)* nor *(b)* (see Table 1 for the list of successful species and Appendix 1 for the list of unsuccessful species). Species reaching the value of 0 or 1 in the above scale are termed as 'unsuccessful', those with scores 2 or 3 as 'successful'.*

Data were analysed using standard methods (Sokal and Rohlf 1981).

The most successful invaders, *i.e.*, those that attained score of total success at least 3 (obtained by summing the scores for seminatural and man-made habitats), were subjected to ordination to analyse the relationships between species traits and their success as invaders. (The only woody species among the most successful invaders, *Robinia pseudoacacia*, was excluded from the evaluation.) The technique of Canonical Correspondence Analysis (CCA; Ter Braak 1987) was used with the program CANOCO 3.12. Input data used were: success in seminatural habitats, success in man-made habitats, and total invasion success scores together with selected species traits. To stress the life history characteristics, *i.e.*, those supposed to be functionally related to a species success, only those traits listed under (c)-(h) were considered (*i.e.*, excluding the area of origin, taxonomic position and Ellenberg indicator values). The Monte-Carlo permutation test was used to evaluate whether the relationship between species traits and invasion success was significant.

*In the present paper, 'successful' and 'unsuccessful' are used as technical terms defined by a species' position at the 0-3 abundance scale.

Results

Overall characteristics of alien flora

In total 132 alien species appeared on the list, about 20% of which may be considered to be successfully naturalized. The proportion of successful invaders was similar in seminatural and man-made habitats (Table 2).

Table 1. Overview of successful aliens in the Czech flora. Species are divided according to their success in particular habitat categories: 0: absent; 1: rare; 2: scattered over the whole territory or locally abundant; 3: common. Total success is expressed as the sum of values in seminatural and man-made habitats. Origin is shown: N: north; E: east; W: west; S: south; C: central.

	Seminatural	Man-made	Total	Origin
Successful in both groups of habitats				
Epilobium adenocaulon	2	3	5	N America
Juncus tenuis	3	2	5	N America
Reynoutria japonica	2	3	5	E Asia
Solidago canadensis	2	3	5	N America
Bidens frondosa	2	2	4	N America
Heracleum mantegazzianum	2	2	4	W Asia
Robinia pseudoaccacia	2	2	4	N America
Trifolium hybridum	2	2	4	W Europe
Successful in seminatural habitats				
Elodea canadensis	3	0	3	N America
Impatiens parviflora	3	1	4	C Asia, Siberia
Solidago gigantea	3	1	4	N America
Acorus calamus	2	0	2	Asia
Impatiens glandulifera	2	1	3	S Asia
Lupinus polyphyllus	2	1	3	N America
Pinus nigra	2	0	2	S Europe
Pinus strobus	2	0	2	N America
Successful in man-made habitats				
Amaranthus retroflexus	0	3	3	N America
Cardaria draba	0	3	3	S Europe, C Asia
Chamomilla suaveolens	0	3	3	W Asia
Conyza canadensis	0	3	3	N America
Galinsoga ciliata	0	3	3	C,S America
Galinsoga parviflora	0	3	3	S America
Veronica persica	0	3	3	Asia
Amaranthus chlorostachys	0	2	2	tropical America
Bunias orientalis	0	2	2	S Europe
Lycium barbarum	0	2	2	S Europe
Medicago sativa	0	2	2	Asia

Table 2. Overall characteristics of the alien flora analysed. Only species introduced after 1492 are considered. Very rare, ephemeral and taxonomically problematic introductions were omitted.

	Number of species	Percentage
Aliens total	132	100.0
Successful in seminatural habitats	16	12.1
Successful in man-made habitats	19	14.4
Successful in both	8	6.1
Successful total	27	20.5
Unsuccessful total	105	79.5

Features distinguishing alien species from the others

Taxonomic position

The composition of the alien flora with respect to the proportion of species in particular families differed remarkably from that of the native flora (Fig. 1a). Among the families overrepresented in the alien flora were *Asteraceae* (20.5% in alien flora *vs.* 10.0% in native), *Brassicaceae* (8.3% *vs.* 4.0%), *Chenopodiaceae* (4.5% *vs.* 1.7%), *Onagraceae* (3.8% *vs.* 1.1%) and *Fabaceae* (7.6% *vs.* 3.9%). On the other hand, the proportion of *Poaceae* and *Rosaceae* was higher in the native flora (4.5%

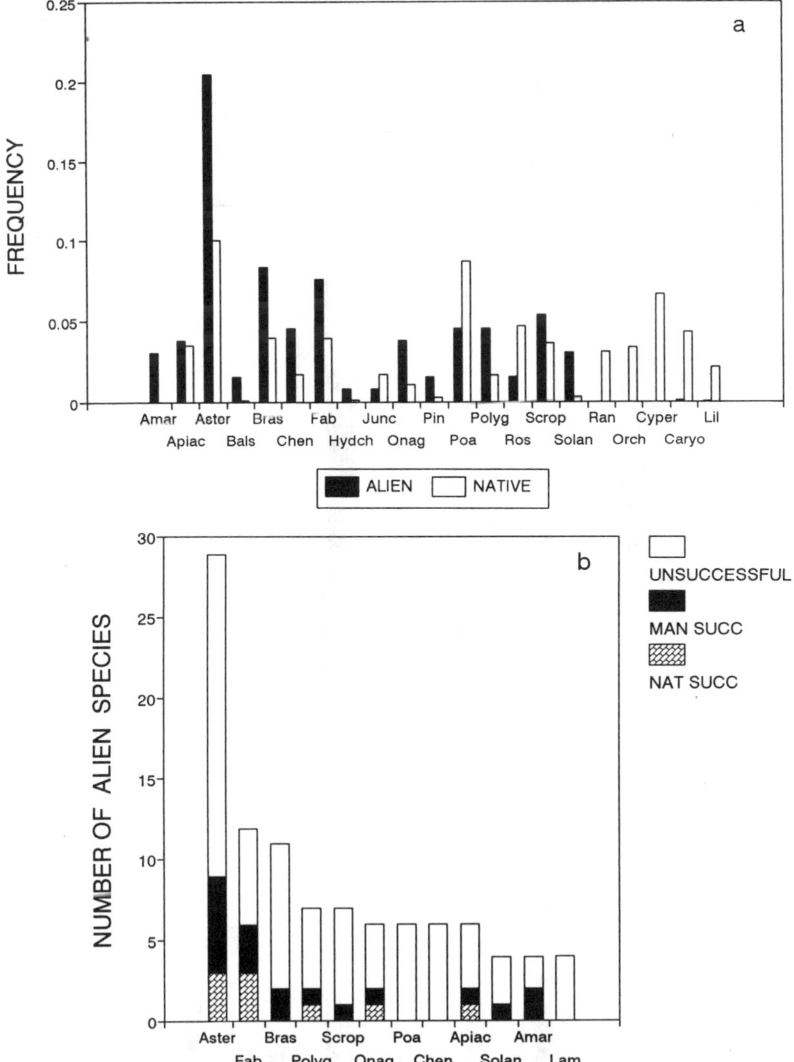

Fig. 1. Frequency of particular families among aliens in comparison with the native flora *(a)*, and differences in taxonomic position of alien species successful in man-made habitats (MAN SUCC), in seminatural habitats (NAT SUCC) and those relatively UNSUCCESSFUL *(b)*. Family codes: Amar: *Amaranthaceae*; Apiac: *Apiaceae*; Aster: *Asteraceae*; Bals: *Balsaminaceae*; Bras: *Brassicaceae*; Chen: *Chenopodiaceae*; Fab: *Fabaceae*; Hydch: *Hydrocharitaceae*; Junc: *Juncaceae*; Onag: *Onagraceae*; Pin: *Pinaceae*; Poa: *Poaceae*; Polyg: *Polygonaceae*; Ros: *Rosaceae*; Scrop: *Scrophulariaceae*; Solan: *Solanaceae*; Ran: *Ranunculaceae*; Orch: *Orchidaceae*; Cyper: *Cyperaceae*; Caryo: *Caryophyllaceae*; Lil: *Liliaceae*.

Table 3. Differences in plant traits between particular groups of aliens classified according to their success (I - successful in seminatural habitats, successful in man-made habitats, unsuccessful) and between the total set of alien species and native flora (II). The following null hypotheses were tested: I: Categories distinguished among aliens do not differ in the proportion of species with particular characters (chi^2 test on contingency tables); II: Alien species do not differ from native flora in the proportion of species with particular characters. Characteristics of native flora, if given (N.G.: not given), were extracted from Frank and Klotz (1990). N.T.: not tested as alien species differ in origin from others by definition. Test used was the chi^2 goodness-of-fit test, with expected value derived from the proportions of considered categories in the native flora. N.S. means that we cannot reject null hypothesis on significance level less than 0.05.

	I. Within aliens			II. Aliens *vs.* native flora		
	df	chi^2-value	*P*	df	chi^2-value	*P*
Origin	12	11.03	N.S.		N.T.	
Life form	10	25.56	<0.01	5	75.36	<0.001
Life strategy	12	14.11	N.S.	6	215.64	<0.001
Pollination	4	0.94	N.S.	2	2.42	N.S.
Dispersal	8	14.15	N.S.	4	163.47	<0.001
Mode of spread	4	1.50	N.S.		N.G.	
Planting	2	6.08	<0.05		N.G.	

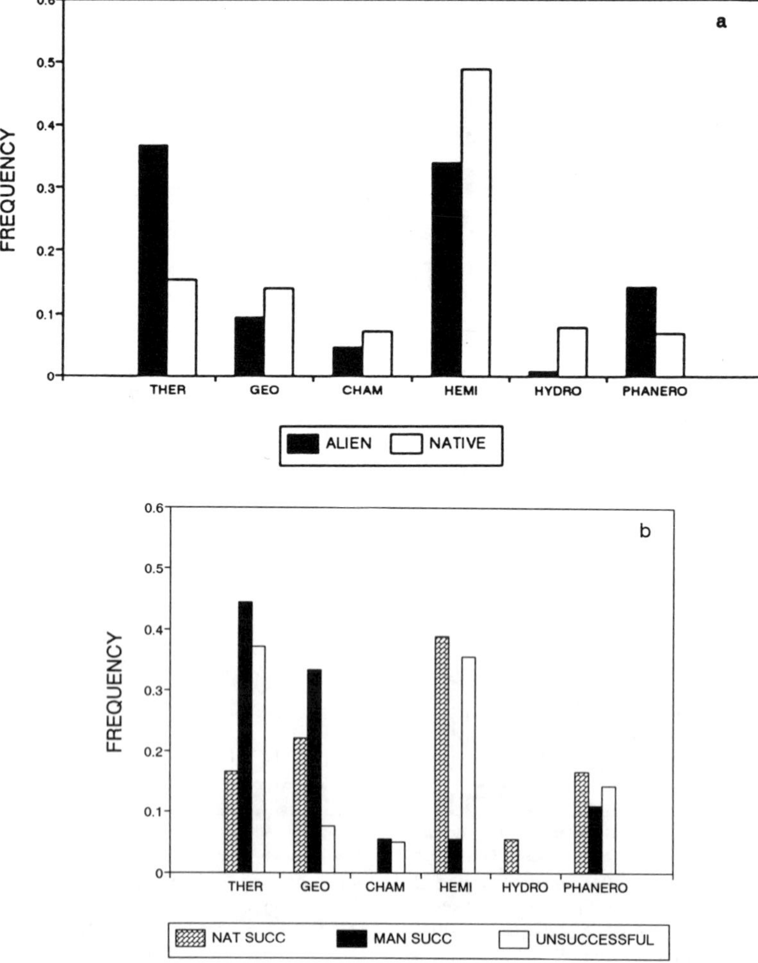

Fig. 2. Frequency of particular Raunkiaer's life forms among aliens and native flora *(a)*, and among the particular groups of alien species *(b)*. See Fig. 1 for codes of particular groups of aliens.

vs. 8.7% and 1.5% *vs.* 4.7%, respectively). No alien species occurred in the list among some of the families with relatively high representation in the native flora (*Ranunculaceae, Orchidaceae, Cyperaceae* and *Caryophyllaceae*, Fig. 1a).

Biological and ecological traits

The alien species differed significantly from the native flora in the frequency distribution of life forms and life strategies (Table 3). Therophytes and phanerophytes showed remarkably higher representation among aliens than among the native flora whereas hemicryptophytes and hydrophytes were underrepresented (Fig. 2a). There was a high representation of species with C- and CR-strategies among aliens. On the contrary, those species possessing a combination including S-strategy (CS, CSR, SR) contributed conspicuously more to the native flora (Fig. 3a).

No significant differences between aliens and native flora were found with respect to pollination (Fig. 4a). There was a significant difference in dispersal agents, with higher frequency of species dispersed by man among aliens (Fig. 5b, Table 3).

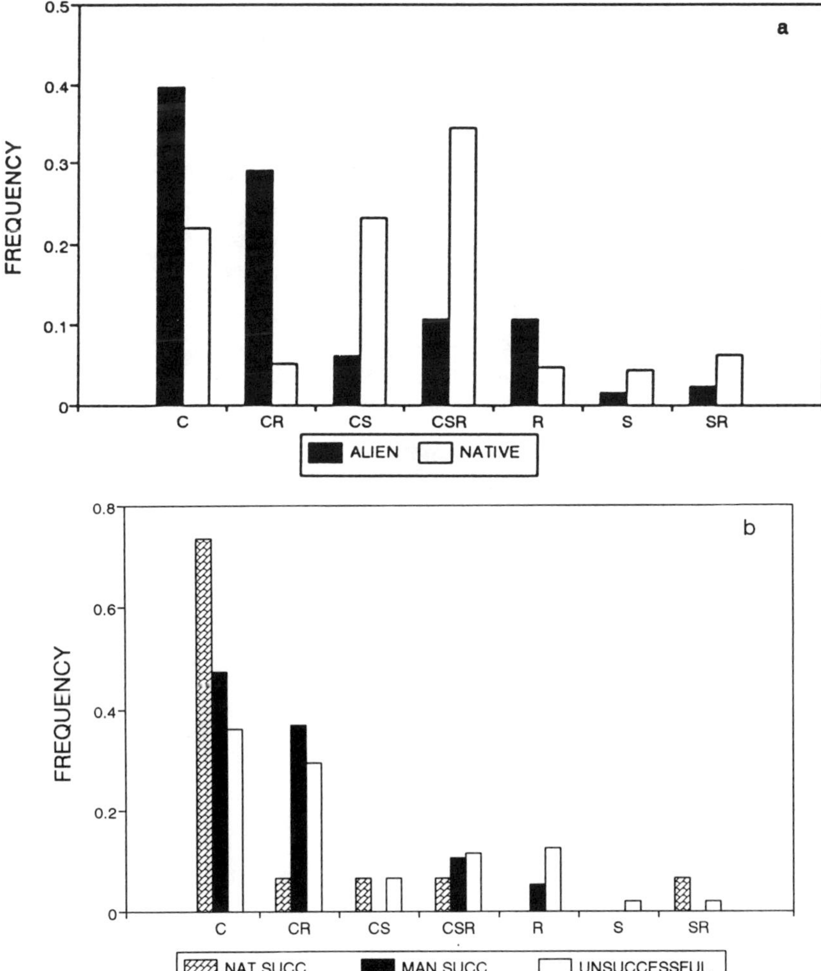

Fig. 3. Frequency of life strategies (Grime 1979) among aliens and native flora *(a)*, and among the particular groups of alien species *(b)*. See Fig. 1 for codes of particular groups of aliens.

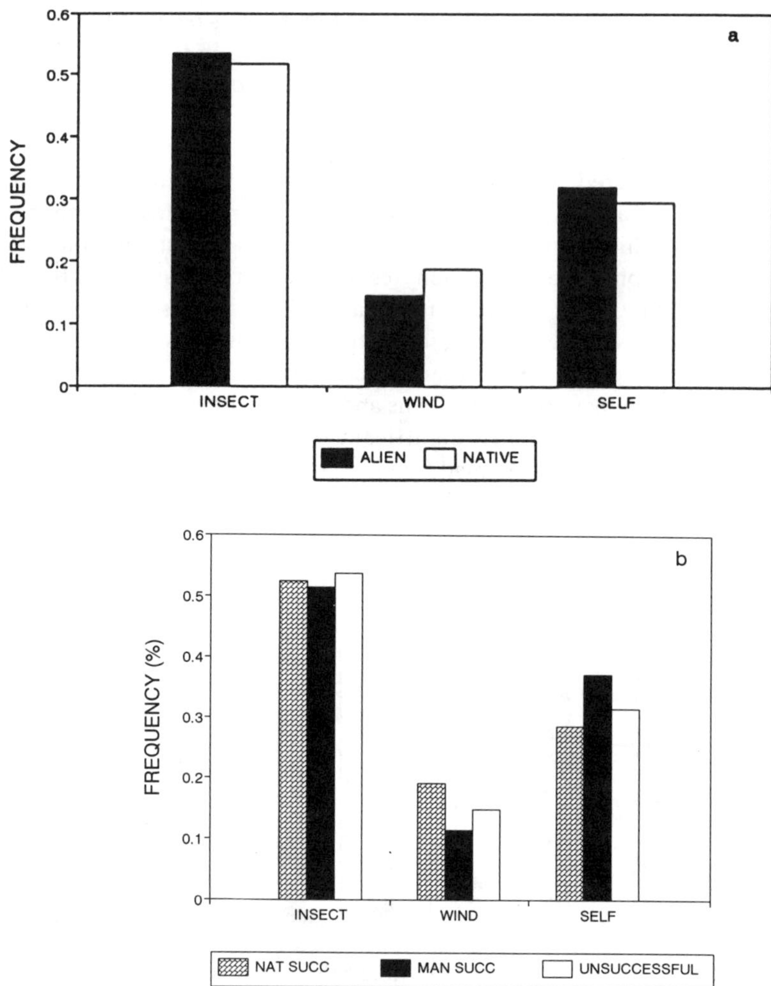

Fig. 4. Frequency of main pollination agents among aliens and native flora *(a)*, and among the particular groups of aliens *(b)*. See Fig. 1 for codes of particular groups of aliens.

Table 4. Comparison of ecological requirements of alien flora in the Czech Republic. Ellenberg indicator values are given (mean±S.D.) for each factor. Number of species for which the tabulated value was available is shown on the bottom line. Data characterizing native flora (n=1,669) were extracted from Frank and Klotz (1990). Kruskal-Wallis test was used to test for the differences. Means followed by the same letter row-wise were not significantly different. Level of significance of the difference in means for the comparison of aliens with native flora is indicated between the values (***: P<0.001).

	Successful in seminatural	Successful in man-made	Unsuccessful	Aliens		Native
Light	6.46±1.31[a] 16	6.87±1.23[a] 17	7.42±0.90[b] 75	7.38±1.22 95	***	6.76±1.52 1313
Moisture	6.57±2.49[a] 15	5.00±1.57[b] 18	4.67±1.49[b] 65	4.90±1.79 88	***	5.91±2.57 1239
Temperature	6.07±1.00[a] 14	6.47±1.45[ab] 16	6.98±1.38[b] 57	6.84±1.37 77	***	5.51±1.19 916
Nitrogen	6.75±1.12[a] 16	6.47±1.56[a] 19	6.45±1.47[a] 64	6.67±1.45 88	***	4.32±2.09 1311

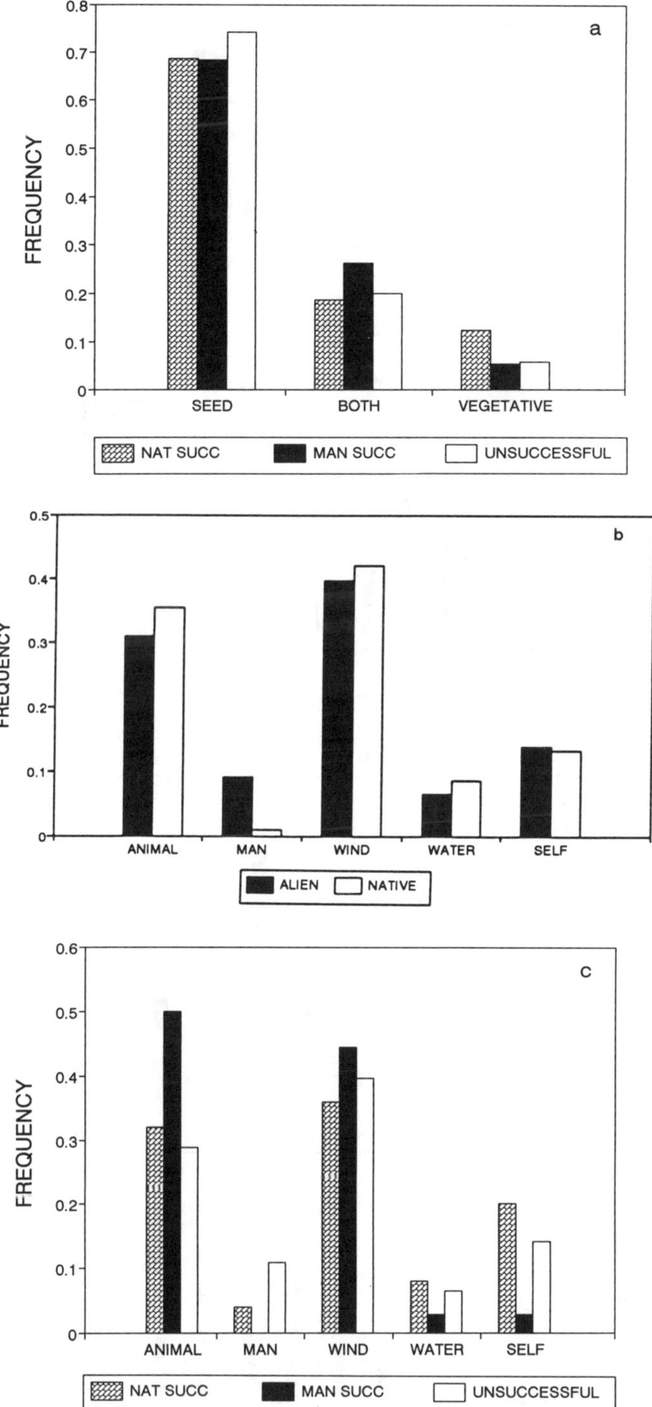

Fig. 5. Frequency of two main modes of spread (by seeds and vegetatively) among the particular groups of alien species *(a)*. Comparison of aliens with native flora is not given because the relevant data on native flora are not available in Frank and Klotz 1990. Frequency of main dispersal agents among aliens and native flora *(b)*, and among the particular groups of alien species *(c)*. See Fig. 1 for codes of particular groups of aliens.

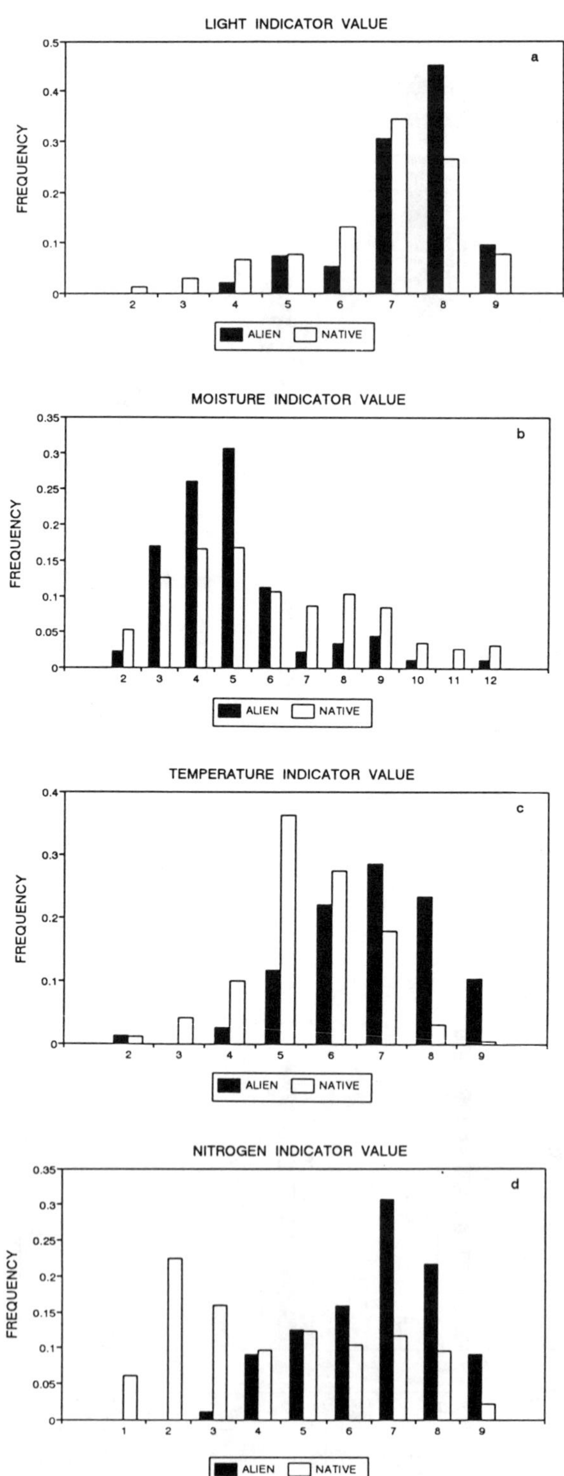

Fig. 6a-d. Frequency of Ellenberg's indicator values for light, moisture, temperature and nitrogen among aliens and native flora. See Fig. 1 for codes of particular groups of aliens.

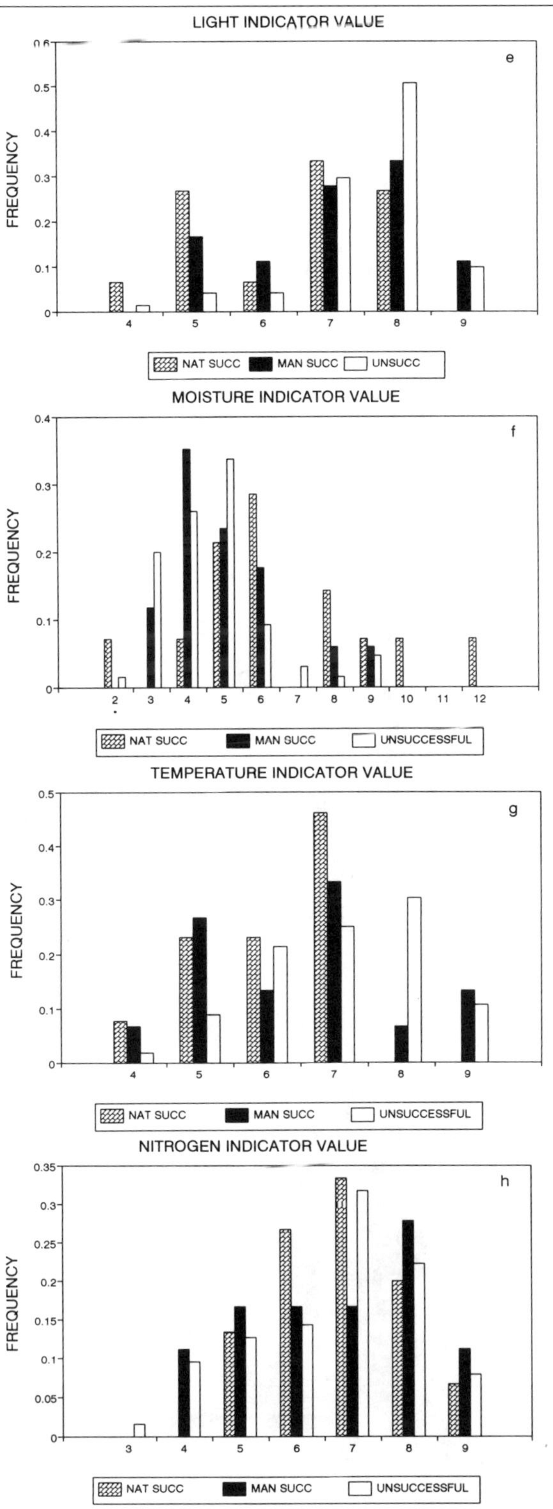

Fig. 6e-h. Frequency of Ellenberg's indicator values for light, moisture, temperature and nitrogen among the particular groups of alien species. See Fig. 1 for codes of particular groups of aliens.

Ecological requirements

Compared to the native flora, alien species showed higher mean indicator values for light, temperature and nitrogen (Fig. 6a, c,d) and these differences were highly significant (Kruskal-Wallis test, H=17.12, 66.18 and 95.19, respectively, $P<0.001$, Table 4). The native flora exhibited a significantly higher mean indicator value for moisture (H=11.73, $P<0.001$, Table 4, Fig. 6b).

Effect of plant traits on invasion success

Origin

There were apparent differences in the frequency distribution of areas of origin between particular groups of aliens classified according to invasion success, but these

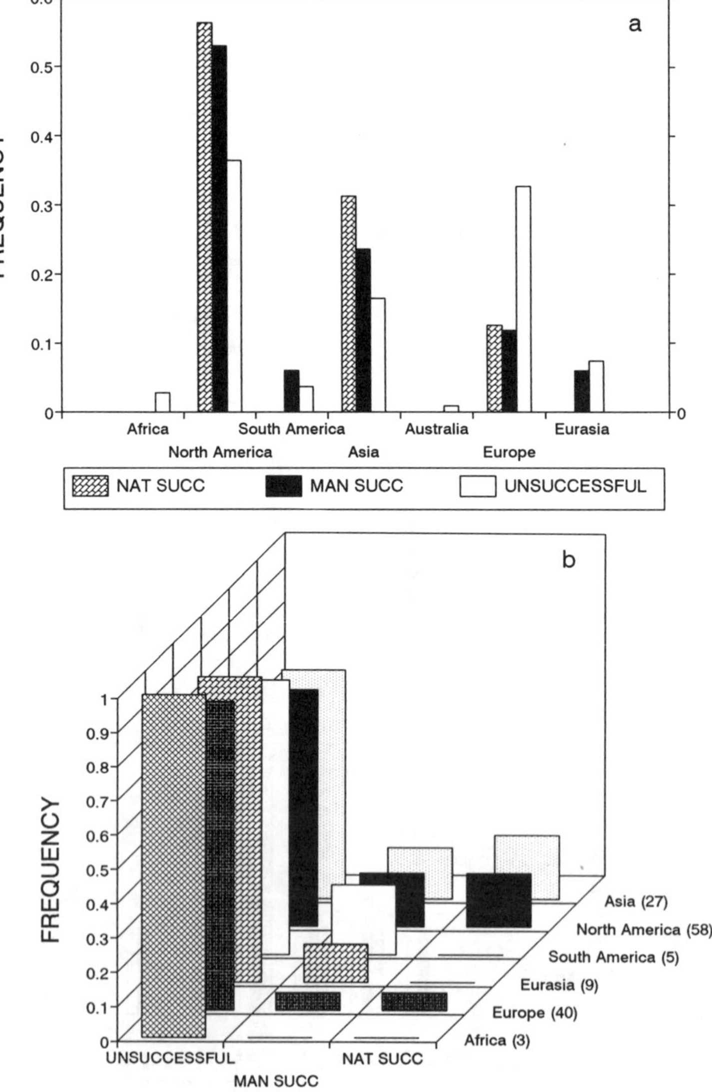

Fig. 7. Geographical origin of alien species distinguished according to their invasion success: frequency *(a)*, and relative frequency *(b)*. The numbers of species of respective origin are given with the names of the continents in *(b)*.

differences were not statistically significant (Table 3). The proportion of North American and Asian species was higher in aliens successful in both seminatural and man-made habitats than in those considered unsuccessful. In the latter, European species made a greater contribution (Fig. 7a). The relative contribution of successful and unsuccessful species for particular areas of origin is shown in Fig. 7b. In seminatural habitats, only North American, Asian and European species were among those successfully naturalized whereas in man-made habitats, naturalization also occurred in some species of South American and Eurasian origin. The percentage of successful invaders (for both seminatural and man-made habitats) decreased in the following order: Asia 33.3, North America 31.0, South America 20.0, Eurasia 11.1, Europe 10.0, Australia 0, Africa 0.

Taxonomic position
Asteraceae and *Fabaceae* contributed most to the number of successful aliens in both seminatural and man-made habitats. Species successful in both habitat groups were also found among *Polygonaceae, Onagraceae* and *Apiaceae.* The representatives of *Brassicaceae, Scrophulariaceae, Solanaceae* and *Amaranthaceae* were only successful in man-made habitats and the proportion of successful invaders was remarkably high in the latter family. No successful invaders were present among, *e.g.*, *Poaceae, Chenopodiaceae* and *Lamiaceae* (Fig. 1b).

Biological and ecological traits
There was a significant effect on species invasion success of plant stature (expressed as its maximum height) (ANOVA, $P<0.01$), those considered successful in seminatural habitats were taller. The effect was still significant ($P<0.05$) when shrubs and trees were excluded from the analysis. There was no difference in height between successful and unsuccessful species in man-made habitats (Table 5).

Table 5. Summary of ANOVAs showing an effect of plant height on invasion success of a species. Heights are given (m). Log-transformation of the data was used to achieve normality.

	Seminatural habitats				Man-made habitats			
	Mean±S.E.	df	F	*P*	Mean±S.E.	df	F	*P*
Aliens total								
Successful	5.82±2.38	1,129	11.18	0.0011	2.31±1.01	1,129	0.019	0.8910
Unsuccessful	1.84±0.37				2.33±0.49			
Forbs only								
Successful	1.78±0.31	1,121	4.73	0.036	1.32±0.27	1,121	0.063	0.8045
Unsuccessful	1.18±0.09				1.23±0.09			

Particular categories of invasion success differed in the frequency distribution of life forms and in planting history (Table 3).

Therophytes and geophytes were conspicuously overrepresented among aliens successful in man-made habitats compared to those successfully invading seminatural vegetation. The reverse was the situation in hemicryptophytes. Unsuccessful species were mostly classified as therophytes or hemicryptophytes (Fig. 2b).

The proportion of species which had been planted in the territory of the Czech Republic was remarkably lower among the aliens successful in man-made habitats

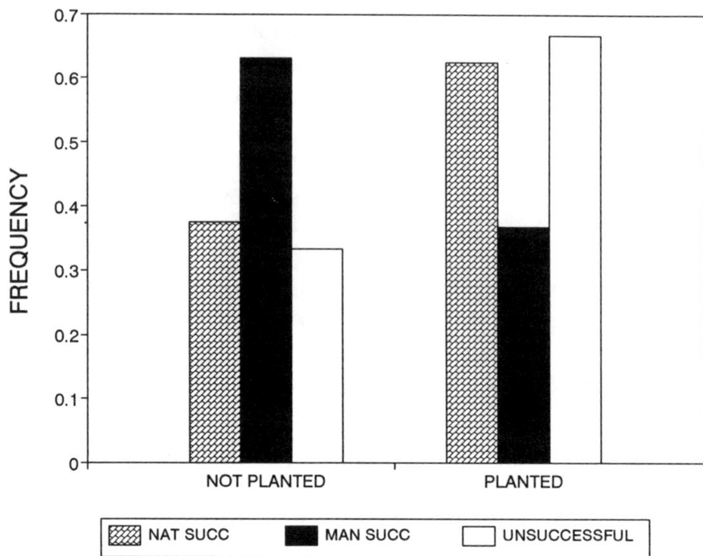

Fig. 8. Planting history, *i.e.*, whether or not species had been planted in the territory of the Czech Republic, distinguished according to invasion success.

than in those successfully invading seminatural sites (Fig. 8).

Only species possessing C-strategy were successful in seminatural habitats (the representation of other strategies was very low) whereas in man-made sites, CR-strategists were equally important (Fig. 3b). These differences, however, were not significant (Table 3).

Similarly, no significant differences between particular categories of invasion success were found with respect to pollination (Fig. 4b), dispersal agents (Fig. 5c), and mode of spread (Fig. 5a, Table 3).

Ecological requirements

Successful invaders showed lower requirements for light than unsuccessful species (Table 4). Species successfully invading vegetation of man-made sites had a higher mean indicator value for light than those in seminatural habitats. This difference was, however, not significant (Kruskal-Wallis test; Fig. 6e).

Invaders into seminatural vegetation exhibited a higher mean indicator value for moisture than either unsuccessful species or those successful in man-made habitats (Table 4, Fig. 6f).

Fig. 9. (a) CCA ordination biplot representing the relationship of alien species to scores describing their invasion success in man-made habitats (SuccMM), seminatural habitats (SuccNat), and total success as a sum of both values (SuccTot). Abbreviations of species names are composed from the first four letters of genera and species names which are listed in Table 1. Only the most successful species, *i.e.*, those which attained the score of total success at least 3 were considered. *(b)* CCA ordination biplot of species traits (full arrows) related to species invasion success (dashed arrows). Explanations of species traits - Life forms according to Raunkiaer: therophytes (Ther), geophytes (Geoph), hemicryptophytes (Hemicr), chamaephytes (Cham) (phanerophytes were excluded from this evaluation); Life strategies according to Grime: C,S,R; Maximum height reached by a species (Height); Pollination agents: wind (windP), insect (insectP), self-pollinated (selfP); Dispersal agents: wind (Dwind), animals (Danim), water (Dwater), man (Dman), and self-dispersed mechanisms (Dself), vegetatively dispersed species (Veget), and seed dispersed species (Seeds); Planted species (Planted).

SuccMM
SuccTot
SuccNat
SoliGiga
SoliCana
ReynJapo
HeraMant
EpilAden
CardDrab
VeroPers
ConyCana
GaliCili+GaliParv
ChamSuav
BideFron
AmarRetr
TrifHybr
ImpaGlan
ElodCana
LupiPoly
JuncTenu
ImpaParv
+1.0
-1.0
-1.0
+1.0

Fig. 9a.

SuccMM
SuccTot
SuccNat
Danim
selfP
insectP
C
Height
Geoph
Veget
Dwater
Hemicr
Dself
S
Planted
Ther
Dwind
windP
Seeds
Dman
Cham
R
+0.5
-0.3
-0.4
+0.5

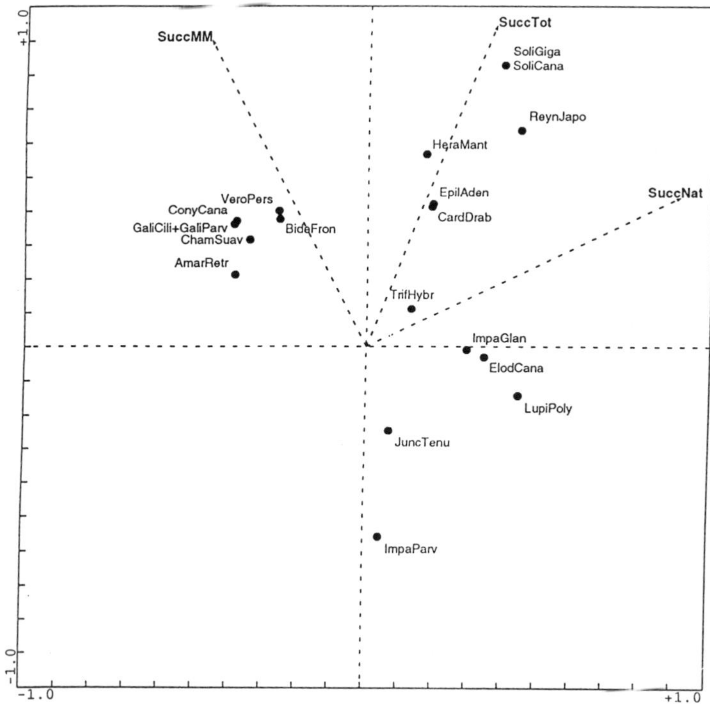

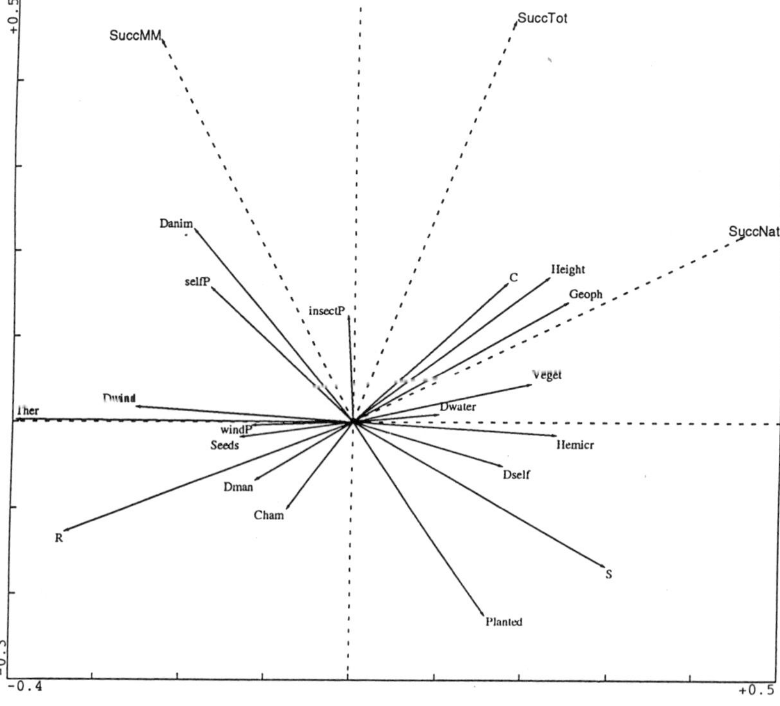

Fig. 9b.

Requirements for temperature increased from the invaders of seminatural habitats category through those invading man-made sites to the unsuccessful aliens category (Table 4, Fig. 6g).

No differences between invasion success categories were found with respect to nitrogen indicator values (Table 4, Fig. 6h).

Relations between species traits and species success evaluated by CCA

The distribution of species in the ordination biplot reveals several groups of species distinguishable in relation to their success in various habitats (Fig. 9a). Arable weeds are clustered along the arrow symbolizing success in man-made habitats; strong competitors such as both *Solidago* species, *Reynoutria japonica*, and *Heracleum mantegazzianum* were distinquished as the most successful in both types of habitats. The group of species more successful in seminatural habitats is less compact. The position of an outlier *Impatiens parviflora* reflects the fact that it is the only SR-strategist.

The core information obtained by the ordination is displayed in Fig. 9b in which the species traits are related to the invasion success. Remember that the longer the arrow the higher the importance of a trait and the more acute the angle between a 'trait' arrow and a 'success' arrow the closer the positive relation between both (see Jongman *et al.* 1987). The height of a species, vegetative spreading, geophyte life form and C-strategy as a general characteristic appear to favour the species success in seminatural habitats, whereas zoochory as well as self- and insect pollination are the traits best related to its success in man-made habitats. On the other hand, R-strategy, therophyte life form, and anemochory are negatively correlated with success in seminatural habitats. Further, those species that were introduced intentionally *via* cultivation and those possessing S-strategy are not successful as invaders in man-made habitats. The Monte-Carlo permutation test showed a significant correspondence ($P<0.01$) between species traits and their success as invaders.

Discussion

The results of the present study indicate that there are *(1)* some significant differences between alien and native flora and *(2)* some plant traits which can favour the invasion success of a species (Table 6). Among the main features distinguishing the alien species from the others is an increased proportion of therophytes, especially those adopting CR-strategy (*sensu* Grime 1979). The C-strategy is also overrepresented in aliens which, further, appear to be very weak stress-tolerators (see Noble 1989). Ecological requirements of alien species are noticeably different from those of the local central European flora. They have higher demands for light, nitrogen and temperature and are confined to drier sites than local species. These conclusions correspond to the areas of origin which are, in the vast majority of cases, warmer compared to central European conditions. The correspondence between the performance of aliens in the territory under study and climatic conditions in the areas of their native distribution indicates the importance of homoclimatic analysis for understanding the pattern of plant invasions (Kruger *et al.* 1989; Lodge 1993b). Capability of growing in nitrogen-rich sites seems to support successful performance in many habitats of nutrient-overdosed Czech landscape.

Table 6. A simplified generalized scheme of the features of 'an average successful invader' and comparison of 'an average alien' with native flora. Traits in which the successful invader of seminatural habitats differs unambiguously from that of man-made sites are in bold font. N.A.: not analysed.

	Alien compared to native flora	Successful invader	
		Seminatural habitats	Man-made habitats
Taxonomy	prev. *Asteraceae*, some families not represented	*Asteraceae, Fabaceae*	*Asteraceae, Fabaceae, Brassicaceae, Amaranthaceae*
Origin	outside central Europe	prev. N America, Asia (3 areas in total)	prev. N America, Asia (5 areas in total)
Stature	N.A.	tall	normal
Life form	therophytes overrepresented	hemicryptophyte	therophyte, geophyte
Life strategy	more C, CR, less S	C	C, CR
Dispersal	no difference	wind, animal, self	animal, wind
Pollination	no difference	insect	insect
Planting	N.A.	escaped from cultivation	introduced spontaneously
Mode of spread	N.A.	seed	seed
Ecology	drier, warmer, nutrient-rich open sites	moister sites with higher nitrogen input	drier, warmer sites with higher nitrogen input

The traits conditioning successful invasion differ with respect to habitat type. An 'ideal' successful invader into seminatural vegetation appears to be a geophyte or hemicryptophyte, perennial forb or woody plant, with C-strategy, the spread of which has been assisted by cultivation as an ornamental plant. These features are associated with a high competitive ability which allows the species to compete successfully with native flora even in moist sites. Given that the plant size is related to competitive ability (Grime 1979; Keddy 1989), the taller stature of aliens successful in seminatural habitats may be taken as another fact supporting this view.

On the other hand, a species successfully naturalized in man-made sites is likely to be a therophyte (annual) using CR-strategy (note that R-strategy itself is not sufficient - see Kornas 1990) or a geophyte using C-strategy, introduced unintentionally rather than *via* escape from cultivation. The difference in planting history can reflect the fact that *(a)* species introduced unintentionally by humans are more confined to man-made habitats even in the area of their origin (so that they have a better chance to be transported by activities linked with trade, traffic *etc.* - Sýkora 1990), and *(b)* their ability to spread in man-made habitats is higher (which is evidenced by the simple fact of them having been transported from the area of origin) than that of those escaped from cultivation (which were transported intentionally, often with less consideration of their ecological demands). Consequently the chance of naturalization in the disturbed vegetation of man-made sites is higher for those species unintentionally introduced than for those escaped from cultivation. Higher temperature requirements, capability of growing in drier sites and greater importance of spread by animals are among further features distinguishing the invader naturalized in man-made sites from the species successful in seminatural vegetation.

Invaders into both habitat types come mostly from North America and Asia and are concentrated in the *Asteraceae* family (Kornas 1990; Sýkora 1990). The high number of aliens among *Asteraceae* (and among other families rich in species and therefore contributing remarkably to local flora) in itself would not be surprising because many of the characteristics conditioning their evolutionary success has also contributed to their success as invaders (Heywood 1989). However, the relative representation of aliens among this family compared to that of native species reveals that the former is really noticeably high. Surprisingly, no clear difference was found with respect to the mode of spread, suggesting some kind of trade-off between *(a)* taking advantage of numerous seeds produced with good long-distance dispersal, and *(b)* efficient use of the space once occupied through vegetative spread (see Grime 1979; Noble 1989; Di Castri 1990; Roy 1990). Generally, there is no simple biological predictor of a successful invasion (Noble 1989; Lodge 1993a; *etc.*). However, promising results emerge from analyses using species' native geographical ranges as predictors of their invasiveness (Rejmánek 1995).

Community-centered interpretation of our results suggests that communities in nutrient-rich sites in warmer parts of the country appear to be more susceptible to invasions as the invaders often require such conditions. Higher requirement for light among invaders suggests that more open, *i.e.*, early successional stages or disturbed habitats also tend to be more vulnerable to invasions (see Rejmánek 1989). However, these hypotheses are not easy to verify and the data set analysed in the present study is insufficient to do that.

In searching for general characteristics conditiong invasion success, mutual comparison of a large set of species is the most promising and, in fact, the only possible method. However, when carrying out such an analysis, one meets a number of methodological problems among which incomplete data is the main one. These problems are faced at various stages of the research:

1. Compiling the list of aliens for a given territory is rather difficult. Sources differ in their approaches to the alien status of a species which is, to a certain extent, due to phytogeographical reasons. Species considered in more northern parts of central Europe as aliens can reach the southern part of the Czech Republic so that they are native in the territory covered by this study. Moreover, increasing knowledge arising from developments within the field changes the view and some species having been considered as aliens in the past may be proved to be native. Similarly, it may be revealed that those considered as neophytes have occurred in the territory before 1492. Furthermore, lists including species only temporarily introduced or rarely escaping from cultivation can hardly ever be complete. Similar problems to those associated with the alien status of a species concern the area of its origin; especially in widespread species, it is sometimes difficult to tell the range of their original distribution.

   For these reasons, the set of alien species analysed in the present paper was restricted to those permanently established. Based on our personal view, it certainly lacks some of the species which other workers would include. More precise information about both native and alien flora is needed to assess the proportion of invaders in the Czech flora and to evaluate the probability that, once introduced, an alien species will become successfully established (see Kowarik 1995).
2. Species traits for which data are available are necessarily selective and we do not have enough quantitative data for those attributes which are extremely important for the process of plant invasions (*i.e.,* seed set, germination conditions, growth rate, regeneration ability, competitive ability, physiological features, effect of pests and pathogens *etc.* - Noble 1989; Roy 1990). The present study shows that schemes summarizing biological and ecological traits (namely life forms and Grime's strategies) are among the characteristics best correlated with the species invasion success. The separate analysis of particular traits would provide us with

a more detailed understanding of the invasion process.

3. Data necessary to describe invasion success quantitatively are not available for large sets of species (Lodge 1993a). The simple semiquantitative assessment of species abundance in the landscape which was adopted in the present study must be considered as the first step to the detailed study of the issue. Certainly, lack of a detailed knowledge of species' invasion history makes the analysis more difficult. For example, the comparison of successful and unsuccessful invaders is necessarily biased by the unavailability of information about the date of species' introduction into the area. It is clear that not all the species have had the same historical chance to reach the invasion success simply because some of them were introduced later - hence their time still may come (consider the duration of lag phases reported in *e.g.*, Pyšek and Prach 1993; Kowarik 1995). However, studies on invasion dynamics based on historical reconstruction indicate that in central Europe, there is a great potential for describing the invasion process of particular species in detail (Pyšek 1991; Pyšek and Prach 1993).

Conclusions

The results obtained by our study confirm that there is no simple biological predictor of invasion success (see also Di Castri 1990; Roy 1990; Lodge 1993a). Nevertheless, some species traits appear to occur more frequently in either alien or native flora, and particular species characteristics also appear to determine a species' invasion success or failure in particular habitat types.

Despite a certain scepticism associated with the correlative approach used in the present study, we believe this is the first step which should precede any detailed analysis of functional relations between the species traits and their invasion success. Considering the knowledge accumulated so far in the field of plant invasions, it seems probable that the invasion success of any particular species can be predicted only if both the potential invader and its target community are intensively studied (Lodge 1993a).

Acknowledgments

We thank Marcel Rejmánek, Davis, for comments on the manuscript, and Lois Child and Max Wade, Loughborough, for improving our English. The study was supported by a grant of the Grant Agency of the Czech Republic no. 204/93/2440.

References

Baker, H.G. 1965. Characteristics and modes of origin of weeds. In: H.G. Baker and C.L. Stebbins (eds.), The Genetics of Colonizing Species, pp. 147-169. Academic Press, New York.

Baker, H.G. 1974. The evolution of weeds. Ann. Rev. Ecol. Syst. 5: 1-24.

Barrett, S.C.H. and Richardson, B.J. 1986. Genetic attributes of invading species. In: R.H. Groves and J.J. Burdon (eds.), Ecology of Biological Invasions: An Australian Perspective., pp. 21-33. Australian Academy of Sciences, Canberra.

Di Castri, F. 1990. On invading species and invaded ecosystems: the interplay of historical chance and biological necessity. In: F. di Castri, A.J. Hansen and M. Debussche (eds.), Biological Invasions in Europe and the Mediterranean Basin, pp. 3-16. Kluwer Academic Publ., Dordrecht.

Crawley, M.J. 1987. What makes a community invasible? In: A.J. Gray, M.J. Crawley and P.J. Edwards (eds.), Colonization, Succession and Stability, pp. 429-454. Blackwell Scientific Publ., Oxford.

Dostál, J. 1958. Key to the Flora of Czechoslovakia. 2nd Ed. Academia, Praha (in Czech).

Dostál, J. 1982. List of vascular plants of Czechoslovak flora. Trója Botanical Garden, Praha (in Czech).

Ellenberg, H. *et al.* 1991. Zeigerwerte von Pflanzen in Mitteleuropa. Scripta Geobot. 18: 1-248.

Frank, D. and Klotz, S. 1990. Biologish-ökologische Daten zur Flora der DDR. Wiss. Beitr. Martin Luther Univ. Halle-Wittenberg 32: 1-167.

Gray, A.J. 1986. Do invading species have definable genetic characteristics? Phil. Trans. Roy. Soc. Lond. B 314: 675-693.

Grime, J.P. 1979. Plant Strategies and Vegetation Processes. 222 pp. John Wiley and Sons, Chichester.

Hejný, S. and B. Slavík (eds.) 1988-1992. Flora of the Czech Republic. Vol. 1-3. Academia, Praha (in Czech).

Heywood, V.H. 1989. Patterns, extents and modes of invasions by terrestrial plants. In: J.A. Drake, H.A. Mooney, F. di Castri, R.H. Groves, F.J. Kruger, M. Rejmánek and M. Williamson (eds.), Biological Invasions: A Global Perspective, pp. 31-60. John Wiley and Sons, Chichester.

Jongman, R.H., Ter Braak, C.J.F. and Van Tongeren, O.F.R. 1987. Data Analysis in Community and Landscape Ecology. 299 pp. Pudoc, Wageningen.

Keddy, P.A. 1989. Competition. Chapman and Hall, London.

Kornas, J. 1990. Plant invasions in Central Europe: historical and ecological aspects. In: F. di Castri, A.J. Hansen and M. Debussche (eds.), Biological Invasions in Europe and the Mediterranean Basin, pp. 19-36. Kluwer Academic Publ., Dordrecht.

Kowarik, I. 1995. Time lags in biological invasions with regard to the success and failure of alien species. In: P. Pyšek, K. Prach, M. Rejmánek and M. Wade (eds.), Plant Invasions - General Aspects and Special Problems, pp. 15-38. SPB Academic Publ., Amsterdam.

Kruger, F.J., Breytenbach G.J., Macdonald, I.A.W. and Richardson, D.M. 1989. The characteristics of invaded mediterranean-climate regions. In: J.A. Drake, H.A. Mooney, F. di Castri, R.H. Groves, F.J. Kruger, M. Rejmánek and M. Williamson (eds.), Biological Invasions: A Global Perspective, pp. 181-213. John Wiley and Sons, Chichester.

Lodge, D.M. 1993a. Biological invasions: lessons for ecology. Trends Ecol. Evolut. 8: 133-137.

Lodge, D.M. 1993b. Species invasions and deletions - community effects and responses to climate and habitat change. In: P.M. Kareiva, J.G. Kingsolver and R.B. Huey (eds.), Biotic Interactions and Global Change, pp. 367-387. Sinauer Publ.

Lohmeyer, W. and Sukopp H. 1992. Agriophyten in der Vegetation Mitteleuropas. Schr. R. Vegetationskd. 25: 1-185.

Mooney, H.A. and Drake, J.A. 1989. Biological invasions: a SCOPE program overview. In: J.A. Drake, H.A. Mooney, F. di Castri, R.H. Groves, F.J. Kruger, M. Rejmánek and M. Williamson (eds.), Biological Invasions: A Global Perspective, pp. 491-506. John Wiley and Sons, Chichester.

Mueller-Dombois, D. and Ellenberg, H. 1974. Aims and Methods of Vegetation Ecology. John Wiley and Sons, Chichester.

Newsome, A.E. and Noble, I.R. 1986. Ecological and physiological characters of invading species. In: R.H. Groves and J.J. Burdon (eds.), Ecology of Biological Invasions: An Australian Perspective, pp. 1-20. Australian Academy of Sciences, Canberra.

Noble, I.R. 1989. Attributes of invaders and the invading process: terrestrial and vascular plants. In: J.A. Drake, H.A. Mooney, F. di Castri, R.H. Groves, F.J. Kruger, M. Rejmánek and M. Williamson (eds.), Biological Invasions: A Global Perspective, pp. 301-313. John Wiley and Sons, Chichester.

Opravil, E. 1980. On the history of synanthropic vegetation 1-3. Živa 28 (66): 4-5, 53-55, 88-90 (in Czech).

Pimm, S.L. 1994. Importance of watching birds from airplanes. Trends Ecol. Evolut. 9: 41-43.

Pyšek, P. 1991. *Heracleum mantegazzianum* in the Czech Republic: the dynamics of spreading from the historical perspective. Folia Geobot. Phytotax. 26: 439-454.

Pyšek, P. and Prach, K. 1993. Plant invasions and the role of riparian habitats: a comparison of four species alien to central Europe. J. Biogeogr. 20: 413-420.

Rejmánek, M. 1989. Invasibility of plant communities. In: J.A. Drake, H.A. Mooney, F. di Castri, R.H. Groves, F.J. Kruger, M. Rejmánek and M. Williamson (eds.), Biological Invasions: A Global Perspective, pp. 369-388. John Wiley and Sons, Chichester.

Rejmánek, M. 1995. What makes a species invasive? In: P. Pyšek, K. Prach, M. Rejmánek and M. Wade (eds.), Plant Invasions - General Aspects and Special Problems, pp. 3-13. SPB Academic Publ., Amsterdam.

Roy, J. 1990. In search of the characteristics of plant invaders. In: F. di Castri, A.J. Hansen and M. Debussche (eds.), Biological Invasions in Europe and the Mediterranean Basin, pp. 335-352. Kluwer Academic Publ., Dordrecht.

Sokal, R.P. and Rohlf, F.J. 1981. Biometry. W.H. Freeman, San Francisco.

Sýkora, K.V. 1990. History of the impact of man on the distribution of plant species. In: F. di Castri, A.J. Hansen and M. Debussche (eds.), Biological Invasions in Europe and the Mediterranean Basin, pp. 37-50. Kluwer Academic Publ., Dordrecht.
Ter Braak, C.J.F. 1987. CANOCO - a Fortran program for canonical community ordination by (partial) (detrended) (canonical) correspondence analysis, principal components analysis and redundancy analysis. 90 pp. ITI-TNO, Wageningen.

Appendix 1

List of unsuccessful invaders, *i.e.*, those receiving a higher score than 1 in neither seminatural nor man-made habitats. For a complete picture of the alien flora analysed in this paper, see also Table 1.

Acer negundo, Ailanthus altissima, Allium paradoxum, Amaranthus albus, Amaranthus blitoides, Ambrosia artemisiifolia, Ambrosia trifida, Amorpha fruticosa, Anethum graveolens, Angelica archangelica, Antirrhinum majus, Artemisia annua, Artemisia verlottorum, Asclepias syriaca, Aster lanceolatus, Aster novi-belgii, Atriplex hortensis, Bryonia alba, Calystegia pulchra, Cannabis sativa, Centaurea solstitialis, Cerastium tomentosum, Chenopodium botrys, Chenopodium foliosum, Chenopodium pumilio, Consolida orientalis, Cornus alba, Coronopus didymus, Corydalis lutea, Cymbalaria muralis, Digitalis purpurea, Echinocystis lobata, Elaeagnus angustifolia, Elsholtzia ciliata, Erigeron annuus, Erigeron strigosus, Fallopia aubertii, Galega officinalis, Geranium pyrenaicum, Guizotia abyssinica, Helianthus tuberosus, Hirschfeldia incana, Hordeum jubatum, Inula helenium, Isatis tinctoria, Iva xanthiifolia, Kochia densiflora, Kochia scoparia, Laburnum anagyroides, Lactuca tatarica, Lepidium densiflorum, Lepidium virginicum, Lolium multiflorum, Lonicera tatarica, Lycopersicon esculentum, Mahonia aquifolium, Malva mauritiana, Malva verticillata, Mentha spicata, Mimulus guttatus, Mimulus moschatus, Myrrhis odorata, Nicandra physalodes, Oenothera biennis, Oenothera erythrosepala, Oenothera parviflora, Oenothera renneri, Ornithogalum nutans, Oxalis corniculata, Oxalis dillenii, Panicum capillare, Parthenocissus inserta, Phacelia tanacetifolia, Phalaris canariensis, Physocarpus opulifolius, Potentilla norvegica, Quercus rubra, Reynoutria sachalinensis, Rhus typhina, Rudbeckia laciniata, Rumex patientia, Rumex scutatus, Rumex triangulivalvis, Salvia officinalis, Salvia verticillata, Sedum spurium, Sempervivum tectorum, Senecio vernalis, Silybum marianum, Sinapis alba, Sisymbrium irio, Sisymbrium volgense, Sisyrinchium angustifolium, Smyrnium perfoliatum, Solanum tuberosum, Sorghum halepense, Syringa vulgaris, Tanacetum parthenium, Telekia speciosa, Trifolium resupinatum, Veronica filiformis, Vicia lutea, Vicia sativa, Xanthium spinosum, Zea mays.

COMPARISON OF DISPERSAL STRATEGIES OF ALIEN AND NATIVE SPECIES IN THE DANISH FLORA

Ulla Vogt Andersen
Royal Veterinary and Agricultural University, Botanical Section, Rolighedsvej 23, DK-1958 Frederiksberg C, Denmark

Abstract

Dispersal strategies were determined and dispersal spectra constructed for three groups of Danish flora: *(1)* native species in disturbed habitats, *(2)* alien species in disturbed habitats, and *(3)* alien species invading seminatural habitats. Diaspore morphology was used as evidence for the mode of dispersal. The dispersal spectra of the three groups were significantly different. Among the native species of disturbed habitats, epizoic and 'no special device' dispersal were the most important strategies. Wind dispersal and endozoic dispersal were most important among aliens invading seminatural vegetation. In contrast, there was a high percentage of censers among the aliens of disturbed habitats. Although the seed dispersal is important, the dispersal strategy on its own does not determine the invasiveness of a species. Vegetative propagation plays an important role, being the only means of dispersal in several species invading seminatural vegetation.

Introduction

Dispersal of diaspores is necessary for maintaining the existence of plant communities and the development of new ones (Welch *et al.* 1990). Seed dispersal is a way of not only colonizing new habitats but also maintaining genetic variation between existing populations. This paper focuses on dispersal in space, not in time, though the role of seed banks as a means of dispersal in time is very important to both native and alien species. The purpose of the present study is to investigate the dispersal strategies of aliens in Denmark, in relation to their introduction and subsequent spread in the Danish landscape.

A species is classified as an alien if it has been introduced to areas outside its natural distribution ranges (Pyšek 1995). Although a large proportion of aliens fail to establish viable populations, some may become invasive and sucessfully enter local vegetation. It has been estimated, however, that a very low proportion of introduced plant species eventually spread into natural plant communities (Di Castri 1989). Most invasive aliens are introduced from distant areas and therefore arrive without their co-evolved predators and parasites (Macdonald *et al.* 1989), which provides them with competitive advantage over the native flora. There are many examples of invasive species that have been introduced intentionally, *e.g.*, garden ornamentals (Bazzaz 1986).

Successful invasions often take place in environments highly disturbed by man. New species in the Danish flora are recruited from urban areas, dumps, roadsides, and railway lines (Jensen 1989). Cities are important immigration areas for alien plants introduced either with or without the help of humans (Sukopp and Werner 1983). Invasible communities are often characterized by young age, a low plant cover and are subject to frequent disturbance (Crawley 1987; Cook 1990). This does not

Plant Invasions - General Aspects and Special Problems, pp. 61-70
edited by P. Pyšek, K. Prach, M. Rejmánek and M. Wade

mean that natural communities comprising of a number of coexisting species are safe from invasions (Rejmánek 1989), even if communities with a high diversity are supposed to be less susceptible to invasions (Fox and Fox 1986). Counts from England indicate that aliens make up more than 50% of the flora on urban waste ground, but usually less than 5% in unmanaged, native woodlands (Crawley 1987).

Methods

Three groups of species were compared in the present study: *(1)* The first group consisted of 40 species native to Denmark, preferably those which occur in arable fields or other disturbed, man-made habitats (Jessen and Lind 1923; Sepstrup 1974; Andreasen 1990). *(2)* The second group consisted of 61 alien species bound to arable fields and other disturbed habitats; these species do not invade seminatural or natural vegetation in Denmark (Hansen 1981; Madsen and Lyck 1991). *(3)* The third group consisted of 32 aliens which are naturalized in Denmark and considered to be invasive in semi-natural or natural habitats (Hansen 1981; Madsen and Lyck 1991). They are thus able to invade mature communities and eliminate the resident species.

These three groups were compared with respect to dispersal strategies. Knowledge of diaspore morphology, habit, and natural history of the species were used to determine dispersal strategy. Ripe diaspores from the herbarium at the Royal Veterinary and Agricultural University were studied under a dissection microscope and dispersal strategies of the species were classified according to the diaspore morphology. When no diaspore samples of a species were available, the following references were used in order to determine the morphological adaptations: Sernander (1901); Ridley (1930); Ross-Craig (1957-59); Kiffmann (1960); Clapham *et al.* (1962); Van der Pijl (1969); Christiansen (1970); Staniforth and Cavers (1976); Hansen (1981); Beerling and Perrins (1993), and Andersen (1993).

Dispersal spectra were constructed according to Willson *et al.* (1990), and the species were ranked into the following eight dispersal strategy categories, on the basis of their morphological dispersal devices: *(1)* the species dispersed by wind: bristles, awns, plumed seeds, pappuses, inflated perianth, wings, or samaras; *(2)* water: air trapped in seed or fruit under perianth or in corky tissue; *(3)* epizoic: sticky diaspores, burrs, spines, or hooks; *(4)* endozoic: drupes, berries or similar fleshy diaspores; *(5)* ants: diaspores with eleiosome; *(6)* ballists: explosive or dramatically dehiscent fruits; *(7)* censers: diaspores kept in, for example, a poricidal capsule or a capitulum, follicles on lignified stems; *(8)* no special device: diaspores without any evident morphological specialization associated with dispersal.

Small diaspores are often classified as adapted to wind dispersal (Van der Pijl 1969; Salisbury 1975). However, here they are classified in the 'no special device' category, as there is no basis for choosing a specific size below which species would be treated as having a wind dispersal strategy. Species with diaspores equipped for two dispersal strategies contributed to two different dispersal categories.

The dispersal spectra of different groups of species were compared using a likelihood-ratio chi-square test. When expected values were too small, the smallest classes of explanatory variables, *i.e.,* dispersal categories were aggregated as described by Jongmann *et al.* (1987).

Results

As to the life form, 86 of the 133 species recorded in total were classified as annuals or biennials and 47 as perennials. Their dispersal spectra were compared (Fig. 1) and although there was a slight overrepresentation of perennials among wind, endozoic, and no special device dispersers, the test turned out to be non-significant (G^2=6.204, DF=6, *P*=0.401).

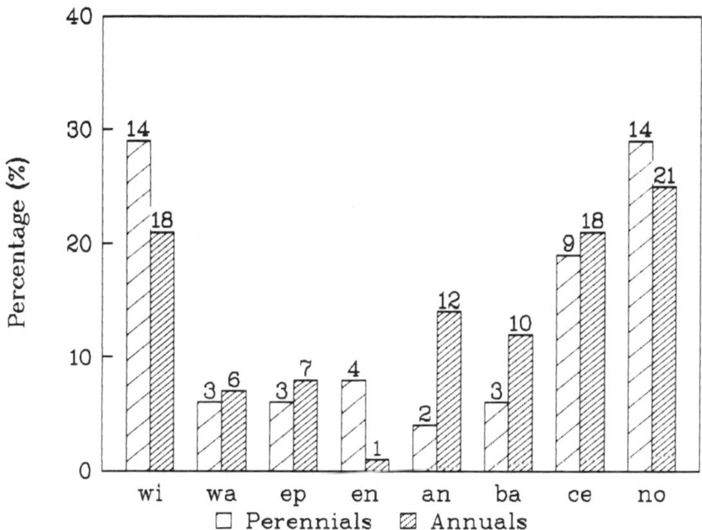

Fig. 1. Comparison between 48 perennial and 85 annual species. Numbers on top of bars indicate absolute number of species in each category. One species can be included in two categories according to polymorphic dispersal structures. wi: wind dispersal; wa: water dispersal; ep: epizoic, dispersal by adhesion; en: endozoic, dispersal through digestive tracts of animals; an: ant dispersal; ba: dispersal by ballistic mechanisms; ce: dispersal by censor mechanisms; no: dispersal by no special device. The following data are grouped in the statistical treatment: ep + en. G^2=6.204, DF=6, *P*>0.05.

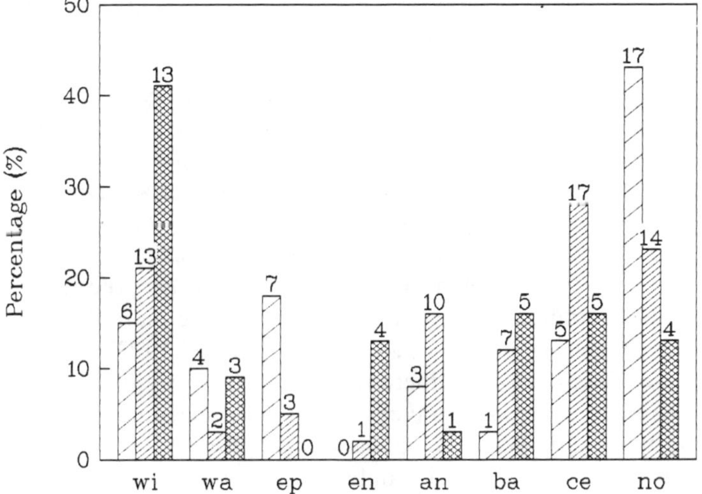

Fig. 2. Comparison between 40 species native in disturbed habitats (hatched), 61 invaders of disturbed habitats (densely hatched), and 32 invaders of seminatural vegetation (cross-hatched). The following data are grouped in the statistical treatment: wi + wa, ep + en, and ba + an. G^2=21.099, DF=8, *P*>0.01. For legend see Fig. 1.

The dispersal spectra of native species of disturbed habitats, invaders to disturbed habitats and invaders to seminatural habitats were significantly different (G^2=21.099, DF=8, *P*=0.007). Among the species invasive in seminatural habitats, more have adaptations for wind, endozoic, and ballistic dispersal, whereas a larger percentage of the native species posses adaptations for epizoic or water dispersal. The aliens occurring in disturbed habitats have the highest percentage of ant and censer dispersed species. The group 'no special device' contributed 43% to the native species, 23% to the aliens of disturbed habitats, and 13% to the invaders to seminatural vegetation (Fig. 2).

A reduced model was applied to test the difference between aliens of disturbed habitats and aliens invading seminatural vegetation. Their dispersal spectra appeared to be significantly different (G^2=12.502, DF=4, *P*=0.0014). A comparison was also made between native species of disturbed habitats and those aliens of disturbed habitats which have arrived since the early middle ages and a significant difference in dispersal spectra was found (G^2=12.426, DF=4, *P*=0.014). Water, epizoic, and no special device dispersal were the prevalent strategies among the native species. The aliens were overrepresented among wind, ant, ballistic, and censer dispersers.

Discussion

Many different dispersal strategies are found among aliens in the Danish flora. According to Crawley (1987) wind and 'no special device' dispersal appear to predominate among aliens. In the present study wind, endozoic, and ballistic dispersal were most widespread among invaders to seminatural vegetation, while ant and censer dispersers were more abundant among the aliens invading disturbed habitats. Eleven of 32 are woody species having either endozoic, ant/ballistic, or wind dispersal. Relatively few herbaceous plants disperse their diaspores as fleshy fruits; this dispersal mode is mostly restricted to woody plants (Sorensen 1986; Willson *et al.* 1990). Only one species (*Solanum nigrum*) among the herbaceous species in this study has this dispersal strategy.

Wind dispersal is traditionally regarded as being a feature of pioneer plants such as species occurring in arable fields. In the present study, few species occurring in arable fields (be it aliens or not) are adapted to wind dispersal. In contrast, 41% of the aliens invasive in seminatural habitats have a wind dispersal strategy, indicating that possession of adaptations to wind dispersal is an important feature to a potentially invasive species. According to Sheldon and Burrows (1973) wind dispersed diaspores from low-growing species seldom travel long distances, mainly because the height of release is not sufficient to allow them to be airborne for long periods of time. Other studies indicate, however, that wind is the main dispersal agent to plant immigration in windy habitats (Dale 1989; Andersen 1993).

Epizoic dispersed diaspores can adhere to and be carried by their dispersal agents indefinitely, having the potential to travel very long distances (Sorensen 1985, 1986; Welch *et al.* 1990). Epizoic dispersal appears to be quite widespread among the native species of the present study. This could be related to necessity for a species to disperse efficiently by several mechanisms given that abiotic agents are ineffective (Willson *et al.* 1990). In Denmark epizoic dispersal is very common among herbaceous species occurring in the understory of the later tree-dominated stages of suc-

Table 1. List of native and alien species, commonly occurring in agricultural fields or other disturbed habitats in Denmark compiled from Sepstrup (1974) and Andreasen (1990). Dispersal strategies are listed according to the morphological adaptations *e.g.*, all pappus possessing species are listed as wind dispersed, even though they can be dispersed by other means such as water. If not given, the species has no obvious morphological adaptations to dispersal (corresponds to the 'no special device' group), and no strategy is indicated. Species marked * are native to Denmark. Nomenclature follows Hansen (1981).

Species	Adaptation	Strategy
*Achillea millefolium**	lignified stem	censer
Aethusa cynapium	lignified stem	censer
Anagallis arvensis	-	-
Anchusa arvensis	eleiosome	ants
Anthemis arvensis	lignified stem	censer
Apera spica-venti	awn	wind
Aphanes arvensis	-	-
*Arabidopsis thaliana**	-	-
*Arenaria serpyllifolia**	-	-
*Artemisia vulgaris**	lignified stem	censer
*Atriplex patula**	air trapped under perianth	water
Avena fatua	awn	wind
Barbarea vulgaris ssp. *vulgaris*	lignified stem	censer
*Bidens triparitus**	achene with spines	epizoic
Brassica campestris	lignified stem	censer
*Bromus hordeaceus**	-	-
Capsella bursa-pastoris	lignified stem	censer
Carduus crispus	pappus/eleiosome	wind/ants
Centaurea cyanus	pappus/eleiosome	wind/ants
Cerastium fontanum ssp. *triviale**	-	-
Chaenorrhinum minus	-	-
Chamomilla recutita	lignified stem	censer
Chamomilla suaveolens	lignified stem	censer
*Chenopodium album**	-	-
Chrysanthemum segetum	lignified stem	censer
*Cirsium arvense**	pappus/eleiosome	wind/ants
Convolvulus arvensis	-	-
Crepis capillaris	pappus	wind
Crepis tectorum	pappus	wind
Daucus carota	schizocarp with burrs	epizoic
Descurainia sophia	lignified stem	censer
*Elytrigia repens**	-	-
*Equisetum arvense**	-	-
Erodium cicutarium	hairy beek	epizoic
*Erophila verna**	-	-
Erysimum cheiranthoides	lignified stem	censer
Euphorbia exigua	explosive capsule/eleiosome	ballist/ants
Euphorbia helioscopia	explosive capsule/eleiosome	ballist/ants
Euphorbia peplus	explosive capsule/eleiosome	ballist/ants
Fumaria officinalis	-	-
*Galeopsis speciosa**	-	-
Galeopsis tetrahit	-	-
Galinsoga ciliata	scale-like pappus	wind
Galinsoga parviflora	scale-like pappus	wind
*Galium aparine**	fruits with burrs	epizoic
Geranium dissectum	explosive capsule	ballist
Geranium molle	explosive capsule	ballist
Geranium pusillum	explosive capsule	ballist
*Gnaphalium uliginosum**	pappus	wind
*Juncus bufonius**	sticky seeds	epizoic
Lamium amplexicaule	eleiosome	ants
Lamium purpureum	eleiosome	ants
Lapsana communis	lignified stem	censer
*Leucanthemum vulgare**	lignified stem/sticky achene	censer/epizoic
*Lolium perenne**	-	-

Table 1. Continuation.

Species	Adaptation	Strategy
Lolium multiflorum	awn	wind
Medicago lupulina	-	-
*Mentha arvensis**	-	-
*Myosotis arvensis**	hook-shaped hairs on calyx	epizoic
Myosotis discolor	-	-
Papaver dubium	lignified stem	censer
Papaver rhoeas	lignified stem	censer
*Plantago lanceolata**	sticky seeds	epizoic
*Plantago major**	sticky seeds	epizoic
Poa annua	-	-
*Poa trivialis**	-	-
*Polygonum aviculare**	air trapped under perianth	water
Polygonum convolvulus	air trapped under perianth	water
*Polygonum lapathifolium**	air trapped under perianth	water
Polygonum persicaria	air trapped under perianth	water
*Ranunculus repens**	-	-
Raphanus raphanistrum	-	-
*Rumex acetosella**	lignified stem	censer
*Rumex crispus**	fruits with spongy appendix	water
*Sagina procumbens**	-	-
Scleranthus annuus ssp. *annuus*	-	-
Senecio vernalis	pappus	wind
Senecio viscosus	pappus	wind
*Senecio vulgaris**	pappus	wind
Sherardia arvensis	fruits with spines and burrs	epizoic
Silene noctiflora	lignified stem	censer
Sinapis arvensis	lignified stem	censer
Sisymbrium officinale	lignified stem	censer
Solanum nigrum	berry	endozoic
*Sonchus arvensis**	pappus	wind
Sonchus asper	pappus	wind
Sonchus oleraceus	pappus	wind
Spergula arvensis	-	-
*Stellaria media**	-	-
Taraxacum sp.*	pappus	wind
Thlaspi arvense	lignified stem	censer
*Tripleurospermum inodorum**	lignified stem	censer
*Tussilago farfara**	pappus	wind
*Urtica urens**	-	-
Veronica agrestis	eleiosome	ants
Veronica arvensis	-	-
*Veronica hederifolia**	eleiosome	ants
Veronica persica	-	-
*Vicia hirsuta**	-	-
Viola arvensis	explosive capsule/eleiosome	ballist/ants
*Viola tricolor**	explosive capsule/eleiosome	ballist/ants

cession, and some of these species are obviously able to become weedy. The dispersal spectrum of the aliens of disturbed habitats reflects the distribution of dispersal strategies in open, disturbed, treeless, pioneer vegetation throughout the world, where the wind is the primary dispersal agent, and the species therefore have adaptations for wind or censer dispersal (Willson *et al.* 1990).

Water dispersal is widespread among native species occurring in arable fields and much less important among aliens occupying the same habitat. This may be due to their origin from seashore communities, where water dispersal is one of the main strategies (Andersen 1993). These communities are particularly poor in aliens which

Table 2. List of alien species naturalized in seminatural vegetation in Denmark. See Table 1 for details. [1]indicates that vegetative propagation is more important than seed propagation for the species. [2] indicates that the species is available for sale from plant nurseries in Denmark.

Species	Adaptation	Strategy
Acer pseudoplatanus[2]	samara	wind
Aegopodium podagraria[1]	-	-
Campanula rapunculoides[1]	lignified stem	censer
Clematis vitalba[2]	hairy style	wind
Conyza canadensis	pappus	wind
Elodea canadensis[1]	air-filled 'berry'	water
Epilobium adnatum	plumed seeds	wind
Epilobium ciliatum	plumed seeds	wind
Geranium pyrenaicum	explosive fruit	ballist
Heracleum mantegazzianum	winged fruits/spongy tissue	wind/water
Impatiens glandulifera	explosive capsule	ballist
Impatiens parviflora	explosive capsule	ballist
Lysimachia punctata[1,2]	lignified stem	censer
Lycium barbarum	berry	endozoic
Mimulus guttatus[1,2]	-	-
Oxalis europaea[1]	explosive capsule	ballist
Pastinaca sativa	winged fruits	wind
Petasites hybridus[1,2]	pappus	wind
Pinus mugo[2]	winged seeds	wind
Prunus serotina[2]	drupe	endozoic
Reynoutria japonica[1,2]	nuts with winged perianth	wind
Reynoutria sachalinensis[1]	nuts with winged perianth	wind
Rosa rugosa[2]	air-filled, fleshy 'hip'	water/endozoic
Solidago canadensis[2]	pappus	wind
Solidago gigantea	pappus	wind
Spartina alterniflora x *maritima*[1]	-	-
Spiraea alba[1,2]	follicles on lignified stems	censer
Spiraea douglasii[1,2]	follicles on lignified stems	censer
Spiraea tomentosa[1,2]	follicles on lignified stems	censer
Symphoricarpus rivularis[2]	drupe	endozoic
Ulex europaea	dehiscent legume/eleiosome	ballist/ants
Veronica filiformis[1]	-	-

might reflect the fact that most maritime plants are very widespread simply because dispersal by sea is so effective, and has already introduced most potential colonists to most suitable habitats before man became the major dispersal agent (Crawley 1987).

The majority of species occupying arable fields (both native and aliens) analysed in this study were found to lack adaptations to long distance dispersal. As a consequence they are directly dependent on agricultural practice and many species are sown inadvertantly with crop seed. Many species adapted to censer dispersal are harvested together with the crops as the fruiting of capsules, siliquas, or capitulums are synchronized with the crops, and the seeds are dispersed over considerable distances by the combine harvesters. There are many additional methods of dispersal associated with agricultural operations and machinery (Marshall and Hopkins 1990). For instance, many 'no special device' dispersers have small diaspores which can stick to the muddy tyres of farm machinery. Most of the species listed in Table 1 can only survive by the help of humans, creating sites in which they are released from competition. It is likely that many weedy species have evolved rather recently under cultivation pressure (Holzner 1977).

For 56% of the invaders to seminatural habitats vegetative reproduction is more important than sexual reproduction (see Table 2), and for some species, *e.g., Elodea canadensis* and *Reynoutria* spp. it is the only way of propagating as they never set seeds in Denmark. Their distribution is a result of the dispersal of vegetative fragments by humans, machinery, or water. The absence of seed production is not a disadvantage to the dispersal of these species (Arthington and Mitchell 1986; Hodgson and Grime 1990). Of the species invading seminatural vegetation, 28% are related to riparian or seashore habitats, but only 9% have diaspores adapted to water dispersal, as water dispersal is achieved mostly by vegetative fragments. Dispersal must be considered in connection with other aspects of the ecology of a species, and it must be emphasized that, irrespective of the dispersal strategy a species will be successful only if it is capable of exploiting the habitat to be colonized. Many unexploited niches probably still remain in the range of productive habitats created by modern land use (Hodgson and Grime 1990). This may explain the success of aliens such as *Reynoutria japonica* whose tall, dense leaf canopy and vigorous lateral vegetative growth appear to make the species more competitive than most native species exploiting similar undisturbed habitats (Hodgson and Grime 1990).

There is not a single collection of eco-physiological characters possessed by successful invaders, but rather a number of characters depending on the specialization of the invader. These include both *(1)* plant traits such as mode of reproduction, reproductive capacity, tolerance to stress, rate of vegetative growth, and unpalatability to grazing animals; and *(2)* environmental factors such as disturbance, nutrient availability, competition from the native flora, and the absence of natural enemies (Shaw 1984; Arthington and Mitchell 1986). In the future the introduction of alien species will no doubt continue as a result of increased international transport (Di Castri 1989; Marshall and Hopkins 1990). It has been now recognized that successful invasion by an alien often facilitates further invasions, that is, the process is self-reinforcing (Macdonald *et al.* 1989).

Conclusion

The present study identifies human activities as the most important factor in the introduction of potentially invasive species. Human beings are likewise the most important vector for the dispersal of arable weeds, and they play a very important role in the dispersal of those vegetatively propagated invaders into seminatural habitats by moving topsoil containing rhizomes or root fragments, and by the deliberate breeding and distribution from plant nurseries (Table 2).

Most invasive aliens possess a well developed adaptation to one or more long distance dispersal strategies. When it comes to invasive sexually reproducing species, other agents such as wind, water, and animals are responsible for the further spread of most of the invasive aliens in Denmark once they have been introduced into the country. The present study indicates that a special attention should be given to wind dispersed species, as these have a large potential for becoming invasive in Danish seminatural vegetation.

Acknowledgments

I would like to thank P. Pyšek, C. Bay, and J. Fredshavn for critical reading and constructive comments on the manuscript and the National Forest and Nature Agency for providing information on the sale and distribution of invasive species. Max Wade and Lois Child kindly improved my English. The study was financially supported by the Agricultural Student Foundation.

References

Andersen, U.V. 1993. Dispersal strategies of Danish seashore plants. Ecography 16: 289- 298.
Andreasen, C. 1990. The Occurrence of Weed Species in Danish Arable Fields. 53 pp. Thesis. Royal Veterinary and Agricultural University, Copenhagen.
Arthington, A.H. and Mitchell, D.S. 1986. Aquatic invading species. In: R.H. Groves and J.J. Burdon (eds.), Ecology of Biological Invasions, pp. 34-53. Cambridge University Press, Cambridge.
Bazzaz, F.A. 1986. Life history of colonizing plants: some demographic, genetic, and physiological features. In: H.A. Mooney and J.A. Drake (eds.), Ecology of Biological Invasions of North America and Hawaii, pp. 98-110. Springer-Verlag, New York.
Beerling, D.J. and Perrins, J.M. 1993. Biological flora of British Isles: *Impatiens glandulifera* Royle (*Impatiens roylei* Walp.) J. Ecol. 81: 367-382.
Clapham, A.R., Tutin, T.G. and Warburg, E.T. 1962. Flora of the British Isles. 1269 pp. Cambridge University Press, Cambridge.
Christiansen, M.S. 1970. Danmarks Vilde Planter. Vol. 1-3. 768 pp. Politikens Forlag, Copenhagen.
Cook, C.D.K. 1990. Origin, autecology, and spread of some of the worlds most troublesome aquatic weeds. In: A.H. Pieterse and K.J. Murphy (eds.), The Ecology and Management of Nuisance Aquatic Vegetation, pp. 31-38. Oxford University Press, Oxford.
Crawley, M.J. 1987. What makes a community invasible? In: A.J. Gray, M.J. Crawley and P.J. Edwards (eds.), Colonization, Succession and Stability, pp. 429-451. Blackwell Scientific Publ., Oxford.
Dale, V.H. 1989. Wind dispersed seeds and plant recovery on the Mount St. Helens debris avalanche. Can. J. Botany 67: 1434-1441.
Di Castri, F. 1989. History of biological invasions with special emphasis on the Old World. In: J.A. Drake, H.A. Mooney, F. di Castri, R.H. Groves, F.J. Kruger, M. Rejmánek and M. Williamson (eds.), Biological Invasions: A Global Perspective, pp. 1-30. John Wiley and Sons, Chichester.
Fox, M.D. and Fox, B.J. 1986. The susceptibility of natural communities to invasion. In: R.H. Groves and J.J. Burdon (eds.). Ecology of Biological Invasions, pp. 57-66. Cambridge University Press, Cambridge.
Hansen, K. 1981. Dansk Feltflora. 757 pp. Gyldendal, Copenhagen.
Hodgson, J.G. and Grime, J.P. 1990. The role of dispersal mechanisms, regenerative strategies and seed banks in the vegetation dynamics of the British landscape. In: R.G.H. Bunce and D.C. Howard (eds.), Species Dispersal in Agricultural Habitats, pp. 64-81. Belhaven Press, London.
Holzner, W. 1977. Weed species and weed communities. Vegetatio 38: 13-20.
Jensen, J. 1989. Ruderal sites and other urban areas. In: P. Vestergaard and K. Hansen (eds.), Distribution of Vascular Plants in Denmark. Botanica 96: 69-72.
Jessen, K. and Lind, J. 1923. Det danske markukrudts historie. 496 pp. Det Kgl. Danske Vidensk. Selsk. Skrifter.
Jongmann, R.H.G., Ter Braak, C.J.F. and Van Tongeren, O.F.R. 1987. Data Analysis in Community and Landscape Ecology. 299 pp. Pudoc, Wageningen.
Kiffmann, R. 1960. Bestimmungsatlas für Sämenreien der Wiesen- und Weidepflanzen. 202 pp. Freising-Weihenstephan.
Macdonald I.W.A., Loope, L.L., Usher, M.B. and Hamann, O. 1989. Wildlife conservation and the invasion of nature reserves by introduced species: a global perspective. In: J.A. Drake, H.A. Mooney, F. di Castri, R.H. Groves, F.J. Kruger, M. Rejmánek and M. Williamson (eds.), Biological Invasions: A Global Perspective, pp. 215-255. John Wiley and Sons, Chichester.
Madsen, H.E.S. and Lyck, G. 1991. Introducerede Planter. 183 pp. Institut for Okologisk Botanik, Kobenhavns Universitet, Skov- og Naturstyrelsen.
Marshall, E.J.P. and Hopkins, A. 1990. Plant species composition and dispersal in agricultural land. In: R.G.H. Bunce and D.C. Howard (eds.), Species Dispersal in Agricultural Habitats, pp. 98-115. Belhaven Press, London.

Pyšek, P. 1995. On the terminology used in plant invasion studies. In: P. Pyšek, K. Prach, M. Rejmánek and M. Wade (eds.), Plant Invasions - General Aspects and Special Problems, pp. 71-81. SPB Academic Publ., Amsterdam.

Rejmánek, M. 1989. Invasibility of plant communities. In: J.A. Drake, H.A. Mooney, F. di Castri, R.H. Groves, F.J. Kruger, M. Rejmánek and M. Williamson (eds.), Biological Invasions: A Global Perspective, pp. 369-388. John Wiley and Sons, Chichester.

Ridley, H.N. 1930. The Dispersal of Plants throughout the World. 744 pp. L. Reeve and Co., Ltd., Ashford.

Ross-Craig, S. 1957-59. Drawing of British Plants. 1317 pp. Bell and Sons Ltd., London.

Salisbury, E.J. 1975. The survival value of modes of dispersal. Proc. Roy. Soc. London., B 188: 183-188.

Salisbury, E.J. 1961. Weeds and Aliens. 388 pp. Collins, London.

Sepstrup, P. 1974. Markskelvegetationens Sammensaetning, Økologi og Spredningsevne. 103 pp. Thesis. Royal Veterinary and Agricultural University, Copenhagen.

Sernander, R. 1901. Den Skandinaviska Vegetationens Spridningsbiologi. 459 pp. Uppsala.

Shaw, M.W. 1984. *Rhododendron ponticum* - ecological reasons for the success of an alien species in Britain and features that may assist in its control. In: Weed Control and Vegetation Management in Forests and Amenity Areas. Aspects Appl. Biol. 5: 231-243.

Sheldon, J.C. and Burrows, F.M. 1973. The dispersal effectiveness of achene pappus units of selected compositae in steady wind convection. New Phytol. 72: 665-675.

Sorensen, A.E. 1985. Seed dispersal and the spread of weeds. In: E.S. Delfosse (ed.), Proceedings of the VIth International Symposium on the Biological Control of Weeds, 19-25 August 1984, pp. 121-126. Vancouver.

Sorensen, A.E. 1986. Seed dispersal by adhesion. Ann. Rev. Ecol. Syst. 17: 443-463.

Staniforth, R.J. and Cavers, P.B. 1976. An experimental study of water dispersal in *Polygonum* spp. Can. J. Botany 54: 2587-2596.

Sukopp, H. and Werner, P. 1983. Urban environment and vegetation. In: W. Holzner, M.J.A. Werger and I. Ikusima (eds.), Man's Impact on Vegetation, pp. 247-260. Dr. W. Junk Publ., The Hague.

Van der Pijl, L. 1969. Principles of Dispersal in Higher Plants. 214 pp. Springer-Verlag, Berlin.

Welch, D., Miller, G.R. and Legg, C.J. 1990. Plants dispersal in moorlands and heathlands in Britain. In: R.G.H. Bunce and D.C. Howard (eds.), Species Dispersal in Agricultural Habitats, pp. 117-131. Belhaven Press, London.

Willson, M.F., Rice, B.L. and Westoby, M. 1990. Seed dispersal spectra: a comparison of temperate communities. J. Veget. Sci. 1: 547-560.

ON THE TERMINOLOGY USED IN PLANT INVASION STUDIES

Petr Pyšek
Institute of Applied Ecology, University of Agriculture Prague, CZ-281 63 Kostelec nad Černými lesy, Czech Republic

Abstract

An attempt to clarify the meaning of terms used in studies on plant invasions is presented. An alien species is understood as one which reached the area as a consequence of the activities of neolithic or post-neolithic man or of his domestic animals. There are at least 14 terms used to describe the alien status of a species. Of the 1172 studies analysed, 60.6% give an indication in their title that the paper is focused upon plant invasions and their proportion has been increasing over time. The term 'invasive' has been used most frequently (37.1% of studies that indicate focus on invasions) and its use increased dramatically in the mid-eighties. Some terms which have ambiguous meaning if related to plant invasions are still widely used (*e.g.*, weed, 22.9%). Most papers do not give explicite definitions of the terms used. Comparison of some definitions of the term 'invasive' (including 'invasion' and 'invader') has shown that some of them do not consider aliens established in man-made habitats and some have applied the term to native species. It is proposed to use the term 'invasive' for an alien whose distribution and/or abundance in the area is increasing, *i.e.*, for the one that can be considered as a successful alien. For a native species increasing its range, 'expanding' appears to be a convenient term. Comparison of European classification of man-accompanying plants with the present terminology used in plant invasion studies is given and difficulties associated with deciding about a species' native *vs.* alien status are discussed.

Introduction

Plant invasions are receiving more and more attention and in the last few years as a conservative estimate at least hundred new studies have appeared annually (Pyšek 1995). Although the first attempts to classify man-accompanying plants go back to as early as the second half of the last century (De Candolle 1855; Ascherson 1883), it is the recent burst of studies on plant invasions which brought about some attempts to clarify the terminology being used in the field (Prach and Wade 1992; Binggeli 1994; Rejmánek 1995). Plants occurring in the region where they are not native have been termed aliens, invaders, exotics, introduced, translocated, neophytes, adventive, weeds, newcomers, naturalized, colonizers, non-native, non-indigeneous, and immigrants. In many studies, these terms are being used without precise definitions, sometimes freely synonymized throughout a paper. Furthermore, some terms have been questioned recently as having anthropocentric implications (Barber 1987; Garthwaite 1993; see discussion in Binggeli 1994). Moreover, as pointed out by Binggeli (1994), only a few terms are truly needed.

In this paper, I will briefly discuss criteria for distinguishing between a species' native and alien status, provide a quantitative insight into how frequently the particular terms have been used and investigate in detail how the term 'invasive' is being interpreted by people involved in studies on plant invasions. For the purpose of the present paper, plant invasion studies are considered as those that deal with any aspect of the biology and ecology of an alien species in an area of its secondary distribution, *i.e.*, in an area to which it is not native.

Plant Invasions - General Aspects and Special Problems, pp. 71-81
edited by P. Pyšek, K. Prach, M. Rejmánek and M. Wade

Alien *versus* native species status

To distinguish between the native and alien status of a species is the first problem one must face when studying plant invasions (Sukopp 1972; Heywood 1989; Pyšek *et al.* 1995). Ecologists and invasion biologists, especially when analysing whole floras, are largely dependent on taxonomists and on data available in local floras. Although most floras make an attempt to consider whether the species is native or introduced, the decisions are often based on inappropriate criteria (uncritical acceptance of earlier opinions, misinterpretation of fossil records) or essentially intuititive grounds, often biased by irrelevant emotional views (Webb 1985). Webb (1985) has proposed eight useful criteria for deciding about native status of which, however, only fossil records (see *e.g.,* Byrne and McAndrews 1975; Betancourt *et al.* 1984) and historical evidence can prove the status (the former one the native, the latter the alien, but neither can prove the reverse) whereas the others (habitat, geographical distribution, ease of known naturalization elsewhere, genetic diversity, reproductive pattern, and supposed means of introduction) can only provide an indication. The problem has been also discussed by Preston (1986) and Smith (1986).

The view has been generally accepted and used as a basic criterion that a species can only be regarded as native to a given area if its occurrence is independent of human activities. However, another important limitation must be added: those species which arrived before the beginning of the neolithic period (ca. 5-6000 years B.C.) should be also considered as native, even if introduced by man. Until that time, man was a part of Nature and his influence on species dispersal was equivalent to that of animals (Webb 1985). Also to the frequently used criterion that native species are those that evolved *in situ* two important points should be added. First, if a species occurred in the area before or during the last Ice Age, it was not under present conditions as the climate was different from today (Webb 1985). Species that became extinct during the last glacial and were reintroduced by man, cannot therefore be regarded as native (Kowarik 1995). Second, the species which arrived in the area in recent times by means independent of human activity, should be also regarded as native (Webb 1985). The latter point implies an interesting theoretical problem: what if a plant species is dispersed into an area in which it has never occurred by a wild living animal which is an alien to that area? Clearly, activities of domestic animals must be included into human activities in a broad sense (see Webb 1985). Following strictly the definition, such a species must be treated as an alien since had it not been for the activity of humans, it would never have reached the area under question.

Comparison of classification systems

More than a century of effort put into the classification of man-accompanying plants by European botanists has yielded a number of classification systems (De Candolle 1855; Ascherson 1883; Rikli 1903; Thellung 1922; Schroeder 1969; Holub and Jirásek 1967; Kornas 1990). These are mostly based on *(1)* time of immigration of a species into the region, *(2)* the means of introduction by humans, *i.e.,* intentionally or unintentionally, and *(3)* degree of its naturalization, *i.e.,* its ability to become established under local conditions (see discussion in Trepl 1990). Some of these systems are rather complicated and have led to the creation of an extensive jargon

(Binggeli 1994). Their complexity is probably the main reason why they did not come into wider use. Some of them (*e.g.*, Kornas 1990), by defining a number of categories often based on vague criteria, impose a number of practical problems that emerge if one is trying to use them, *e.g.*, how to distinguish seminatural and natural habitats or when is an alien already permanently established and when not.

Clearly, the use of these systems has been mostly restricted to continental (or even central) Europe and they have never reached a wide attention in Anglosaxon countries where the plant invasions are most intensively studied (Pyšek 1995). Table 1 represents an attempt to relate the classification system proposed by Holub and Jirásek (1967), in my opinion one of the most carefully elaborated, to the meaning of terms used worldwide in the contemporary literature on plant invasions. The main difference appears to be that species introduced before 1500 are usually not the subject of studies on plant invasions and as such they are often being deliberately dismissed (see *e.g.*, Weeda 1987; Pyšek *et al.* 1995). There are two reasons for that. The first reflects the difficulties associated with deciding about their alien or native status (many authors tend to treat these species as equivalent to natives, see Webb 1985) and the impossibility to follow their introduction into the area. The second is that, being a field with important practical implication, invasion ecology is simply not much concerned with these species because they usually neither cause management problems nor eliminate local flora or change ecosystem properties.

Terms used in studies on plant invasions

To obtain an insight into the use of various terms relating to a species alien status, 1172 papers dealing with various aspects of plant invasions worldwide (except those focused exclusively upon control) were analysed. The database used does not cover all the studies published on the topic but may be considered as a reasonably repre-

Table 1. Comparison of the phytogeographical terminology proposed for classification of human-accompanying plants (Holub and Jirásek 1967) with the meaning (as suggested in the present paper) of terms presently used in studies on plant invasions. Natural vegetation includes also semi-natural vegetation types (*i.e.*, those close to the natural ones); 'not considered' means that the category is usually not a subject of studies on plant invasions.

Term	Definition	Meaning
A. Apophytes[1]	native species occurring in man-made habitats	native
B. Anthropophytes	species introduced by man	
I. Hemerophytes	introduced intentionally[2]	
II. Xenophytes	introduced unintentionally[2]	
1. Archaeophytes	introduced before 1500	not considered alien
2. Neophytes	introduced after 1500	
a. Ephemerophytes	temporary occurrence, only in man-made habitats	not invasive
b. Epekophytes	established in man-made habitats	invasive in man-made habitats
c. Neoindigenophytes[3]	penetrating to natural habitats	invasive in natural habitats

[1]English transcription of terms which have been introduced in German is given.
[2]While no terms are being used in plant invasion studies to distinguish between these two categories, Frank and McCoy (1990) do so for animal invasions. Among non-native species (which they term 'adventive'), they proposed to distinguish those introduced deliberately by man (termed 'introduced') from those that arrived from elsewhere by their own volition (termed 'immigrants').
[3]Corresponds to the term 'agriophytes' *sensu* Schroeder 1969 which is also being frequently used (see *e.g.*, Lohmeyer and Sukopp 1992).

sentative sample.

First, I analysed how was that the study deals with alien species reflected in its title (Fig. 1a). Of the total number of studies, 475 (40.5%) explicitly used some of the terms to describe the alien status and 486 (41.5%) gave the name of the species studied in the title which may also be considered, at least in some cases, as an indication of the focus upon alien flora. Only 211 studies (18.0%) gave no indication, neither by using any of the terms nor by giving the species name. Undoubtedly, explicit formulation of the focus of the study improves the communication between people working in the field. However, to recognize a study on invasion only by the name of the species studied, the area to which the study is related must be indicated as well and one must know that the species under question is alien to that area. If we dismiss those rather ambiguous terms such as 'weed', we are left with 564 studies (48.1%) more or less clearly indicating by the title that they deal with alien species.

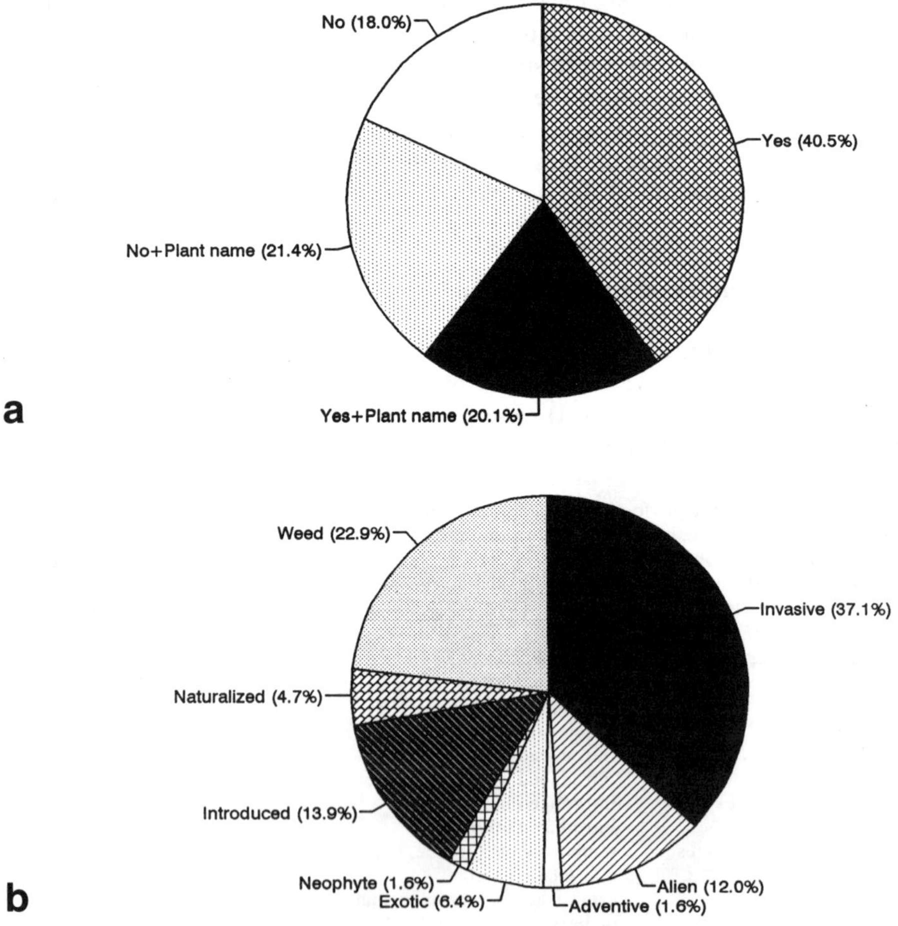

Fig. 1. (a) Analysis of 1172 studies showing whether the fact that they deal with plant invasions is reflected in their title (Yes - a term describing alien status is used in the title, No - no indication is given) or whether the name of the species studied is given or not. *(b)* Frequency of use of particular terms shown for those studies the focus of which is indicated by their title (n=710). Modifications of the terms (*e.g.*, invader, invasion, introduction, naturalization) were also considered.

One must bear in mind, that only titles were analysed here; had the keywords been included, the number would probably increase, but it is the title which is often used as a basis for a reader's decision whether to be interested in the paper or not. Hence, the present analysis can be taken as reasonably showing the potential for communication, *e.g.,* via some abstracting journals, databases or by scanning the references in papers.

There is a consistent increase in the proportion of papers indicating a focus on invasions from the beginning of the eighties (Fig. 2). This can be partly related to the increasing awareness among workers that the status of the species they study is an important aspect of their work.

Fig. 1b shows how frequently the particular terms are used. Of the total number of studies in which the status was indicated (710), there were 124 (17.5%) whose title contained more than one term (*i.e.,* invasive alien, introduced weed, naturalized exotic *etc.*). The terms* 'invasive' (309), 'weed' (191) and 'introduced' (116) were used most frequently. There is also a noticeable difference in temporal trends in using particular terms (Fig. 3a). An increase in the cumulative number of papers in whose titles the term was used appears to be be fairly constant for all but 'invasive', the use of which accelerated remarkably in the mid eighties (Fig. 3a, b). This was probably caused by the launch of the international SCOPE project on biological invasions in 1982 (Drake *et al.* 1989). The term 'weed' has been widely used since the beginning of the period analysed (Fig. 3a) but its proportional contribution is decreasing on the long-term scale (Fig. 3b).

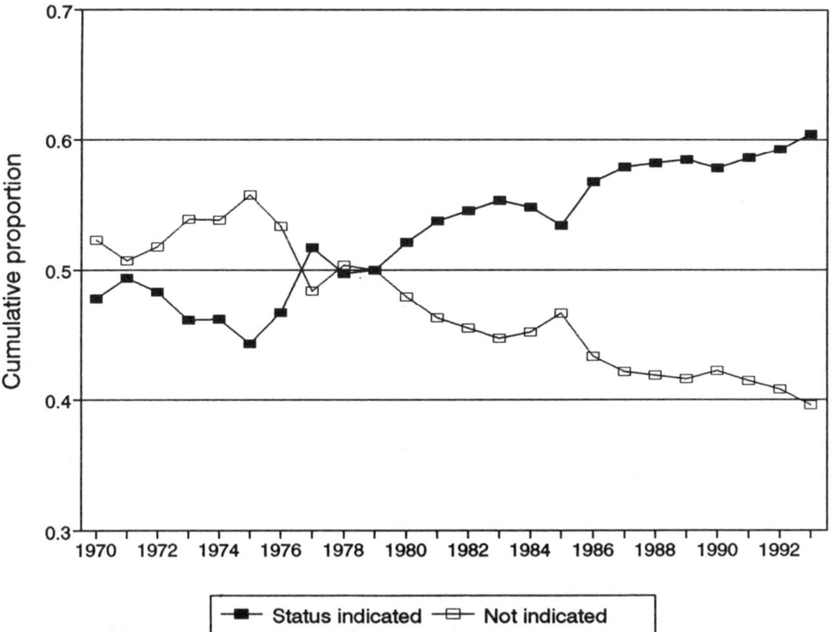

Fig. 2. Temporal trends in the proportion of studies with the focus on plant invasions *(a)* indicated and *(b)* not indicated by their title. Cumulative proportions are shown.

*In the present paper, when speaking about particular terms, terms with the same origin are also considered, *i.e.,* the number of records given for 'invasive' also includes 'invader' or 'invasion'; similarly for the other terms analysed, *e.g.,* 'introduction', 'naturalization'.

The term 'weed' is probably the best example of a term that is rather confusing when related to plant invasion studies as it refers to the anthropocentric viewpoint (Rejmánek 1995). It has been repeatedly pointed out that the term weed implies an

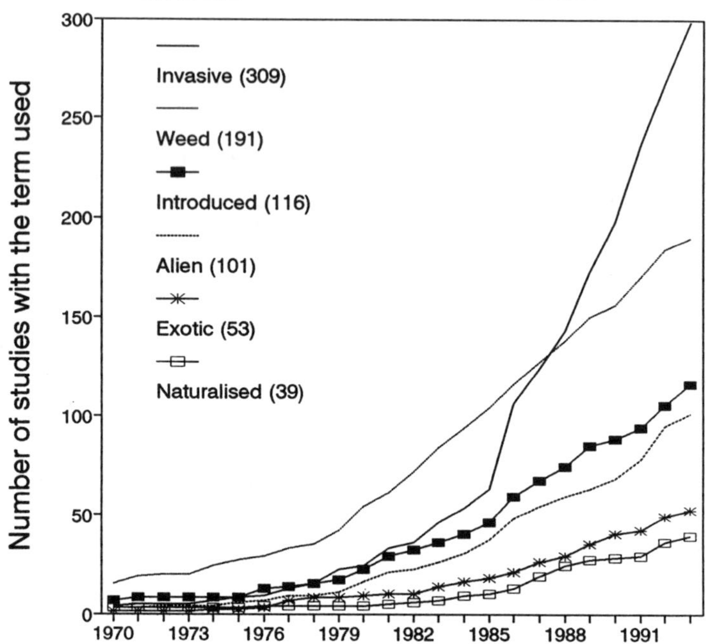

Fig. 3a. Increase in the cumulative number of studies using particular terms in their titles. The graph shows the situation up to 1993; total number of studies up to present (*i.e.*, those that were published in the first half of 1994) is given in parenthesis.

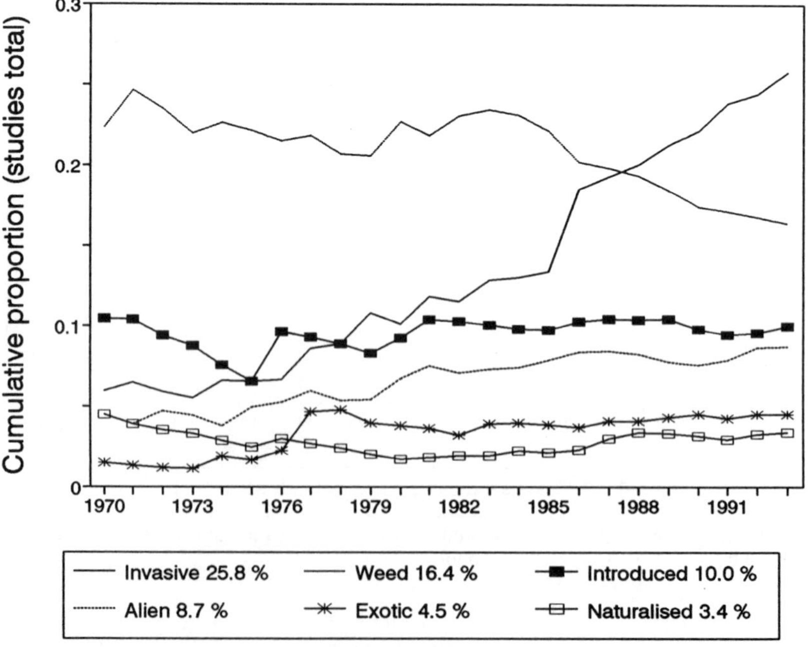

Fig. 3b. Changes in the cumulative contribution of particular terms to the total number of studies analysed (n=1172). Percentage contribution of particular terms is given.

interference with objectives and requirements of people (Binggeli 1994; Rejmánek 1995). Although the biological invasions are not the sole domain of biogeography, it is the biogeographical viewpoint from which the terminology should stem. As pointed out by Roy (1990) " ... the process of invasion brings an organism to an environment in which it did not evolve ... it is this evolutionary aspect which is unique to invaders and not implied by the concept of weeds, colonists or successional species."

How is the term 'invasive' being understood?

The term 'invasive' is not being used only in a strictly biogeographical sense, *i.e.*, referring to alien species. I have analysed the entries that appeared in Ecological Abstracts between 1987-1992 (excluding animal invasions) and found that out of 98, there were 14 cases (14.3%) of use in a different sense. Generally, the term is being used in *(a)* ecological papers, including both theoretical and field studies, to describe colonization of a community by a newly arrived species (*e.g.*, Sloan Wilson and Turelli 1986; Van Hulst 1987; Silvertown *et al.* 1994) or in *(b)* palaeontological studies referring to species migrations (*e.g.*, Davis 1987; Bennett 1987; Coope 1987). The former case also regards an interesting history of *Dittrichia viscosa*, a species native to the Mediterranean basin where it has been recently extending its geographical range by entering man-made habitats (Wacquant 1990).

Surprisingly, not many definitions of invasion or invaders are available and in the majority of studies, the term is used without explicit definition. To see how the term 'invasive' is being understood, I compared several definitions recently published in the literature with 5 theoretical situations: *(1)* a native species that is not increasing (*i.e.*, its geographical range is not extending and/or its abundance in the area under question is not increasing), *(2)* increasing native species, *(3)* not increasing alien species, *(4)* alien species increasing in man-made habitats, and *(5)* alien species increasing in natural habitats (Table 2). All the definitions in Table 2 match the category of an increasing alien. However, some of them do not include an alien increasing in man-made habitats and some also match other situations. If the logical sense of particular definitions is strictly followed, out of the total of 13, 7 do not exclude native species; of these 4 explicitly by definition (Nos. 3,5,8,9, *i.e.*, those that adopt an ecological rather than biogeographical view) and 3 (Nos. 2,6,7) by not taking into account the possibility of recent introduction independent of human activities. Three definitions (Nos. 2,8,13) do not exclude non increasing aliens (*i.e.*, group 3) and two exclude aliens increasing in man-made habitats (Nos. 1,4; the latter, however, purposely, as it applies to nature reserves). The remaining three (Nos. 10, 11, 12) consider as an invasive species an alien established in the wild in the area of introduction regardless of the habitat, which is the meaning in which, I suggest, the term should be used.

Even if the term 'invader' is used in biogeographical sense, the main problem associated with definitions given in Table 2 remains the spatial scale implied by various terms. There is a mention of 'area', 'geographical area', 'geographical range', 'region' or 'territory', mostly without an indication of how these terms are understood. Obviously, no definition of an invasive species would be absolutely satisfactory; however, a reasonable consensus about its meaning would be most useful.

Table 2. Comparison of some definitions of the term 'invasive' ('invader', 'invasion') in literature and their interpretation with respect to possible situations. 'Native' and 'alien' without specification mean any native or any alien species. When assessing the meaning, the terms 'spreading', 'extending range', 'colonizing' were assumed to indicate a species' invasion success, as opposite to 'entering' or 'occurring' which need not to be related to a successful alien but may also include the one that has failed to establish. Natural habitats also include semi-natural ones.

Author	Definition	Matches the situation				
		Native	Increasing native	Alien	Alien increasing in	
					man-made habitats	semi-natural habitats
1. Stirton 1979	(invaders) are alien plants that invade and oust native vegetation	No	No	No	No	Yes
2. Mack 1985	(invader) any taxon entering a territory in which it has never occurred before, regardless of circumstances (even if it fails to establish)	Yes	Yes	Yes	Yes	Yes
3. Joenje 1987	(invasion) the influx of numerous individuals of a species or the sudden increase of a founder population in an area	No	Yes	No	Yes	Yes
4. Macdonald *et al.* 1989	(invasive) the introduced species capable of establishing self-sustaining populations in area of natural or seminatural vegetation*	No	No	No	No	Yes
5. Mooney and Drake 1989	(invader) when it colonizes and persists in an ecosystem in which it has never been before	No	Yes	No	Yes	Yes
6. Di Castri 1990	(invader) a species which, most usually transported inadvertedly or intentionally by man, colonizes and spreads into new territories some distance from its home territory	No	Yes	No	Yes	Yes
7. Roy 1990	(invasion) the entering of a species into a territory in which it has never before occurred, followed by an extension of the range of that species	No	Yes	No	Yes	Yes
8. Gouyon 1990	(invader) it occurs in a kind of habitat where it was not present before and/or the number of its individuals in a place it was before is abnormally increasing	Yes	Yes	Yes	Yes	Yes
9. Le Floch *et al.* 1990	(invading) species having an expanding status either in terms of geographical area or in terms of increasing frequency and density	No	Yes	No	Yes	Yes
10. Prach and Wade 1992	(invasion) rapid increase of an alien species in a region	No	No	No	Yes	Yes
11. Rejmánek 1995	(invaders) are spreading into areas where they are not native	No	No	No	Yes	Yes
12. Binggeli 1994	(invasive) the establishment of self-regenerating, usually expanding, populations of an introduced species in a free-living state in the wild	No	No	No	Yes	Yes
13. Kowarik 1995	(invasion) the whole process of range extension of an alien species ... including its very beginning	No	No	Yes	Yes	Yes

*Applies to nature reserves.

In fact, the strict separation of the term 'alien' from 'invasive alien' is difficult. I understand the term 'invasive' as describing an alien which has become successful in the area into which it was introduced (some definitions mention as an important feature that the species is creating self-regenerating populations and is capable of further spread without direct support of humans, see *e.g.,* Binggeli 1994). Theoretically, an alien species can be considered as becoming invasive when it enters the exponential phase of spread. However, the studies measuring the rate of invasion in quantitative terms are rather rare (Connolly 1977; Trepl 1984; Trewick and Wade 1986; Pyšek and Prach 1993; Perrins *et al.* 1993) and this information is certainly not available for the vast majority of plant invasions. There is evidence in the literature database analysed in the present paper that at least some authors explicitly distinguish between 'alien' and 'invasive alien': 68 studies (*i.e.,* 9.6% of those which indicate the status) speak about 'invasive' alien (exotic, introduced, adventive, neophyte) species. These cases suggest the view that not every alien must be necessarily invasive. However, if we take into account the difficulties associated with deciding what is 'successfully established' and when, using the term 'invasive' as synonymous to 'alien' (*e.g.,* Mack 1985; Kowarik 1995) seems unavoidable and acceptable.

Conclusions - Suggestions of terminology

For the purpose of the vast majority of studies, we need to describe four basic situations with respect to a species' status (whether it is native or alien to the area) and the dynamics of its behaviour (whether it is spreading or not). There is no need to create another set of definitions since the terms needed have been properly defined elsewhere. Here are the definitions from the available literature and the suggested meaning of the terms:

- *Native* (indigenous) species is one which evolved in the area or which arrived there by one means or another before the beginning of the neolithic period or which arrived there since that time by a method entirely independent of human activity (Webb 1985);
- *Alien* (introduced, exotic, adventive) species is one which reached the area as a consequence of the activities of neolithic or post-neolithic man or of his domestic animals (Webb 1985);
- *Invasive* (naturalized) species is an alien the distribution and/or abundance of which in the wild is in the process of increasing regardless of habitat (see *e.g.,* Prach and Wade 1992; Binggeli 1994).

To avoid confusion, we also need a term describing the range extension and/or increase in abundance of a species native to the region (*e.g.,* the above mentioned example of *Dittrichia viscosa,* see Wacquant 1990). Here I refer to the terminology proposed by Prach and Wade (1992) who term a native species exhibiting such behaviour as *expanding* rather than invading.

Acknowledgments

I thank David J. Beerling, Sheffield, Pierre Binggeli, Ulster, and Karel Prach, Třeboň, for helpful comments on the manuscript. Roger L. Hall, Oxford, and Max Wade, Loughborough, kindly improved my English. My thanks also to the Department of Plant Sciences, University of Oxford for space and the possibility to use their facilities. The study was made possible by funding from The Leverhulme Trust Foundation; thanks are due to Dana Pavelčíková, British Council Prague, for her kind help. The study was partly supported by the Grant Agency of the Czech Republic (grant no. 204/93/2440).

References

Ascherson, P. 1883. Einfluß des Menschen auf Vegetation. In: J. Leunis (ed.), Synopsis der Pflanzenkunde, pp. 791-795. Hannover.

Barber, F. 1987. Sycamore - noxious weed or valuable tree? North West Naturalist, 1987, pp. 14-18.

Bennett, K.D. 1987. The rate of spread and population increase of forest trees during the postglacial. In: H. Kornberg and M.H. Williamson (eds.), Quantitative Aspects of the Ecology of Biological Invasions, pp. 523-531. Royal Society, London.

Betancourt, J.L., Long, A., Donahue, D.J., Jull, A.J.T. and Zabel, T.H. 1984. Pre-Columbian age for North American *Corispermum* L. (*Chenopodiaceae*) confirmed by accelerator radiocarbon dating. Nature 311: 653-655.

Binggeli, P. 1994. The misuse of terminology and anthropometric concepts in the description of introduced species. Bull. Brit. Ecol. Soc. 25(1): 10-13.

Byrne, R. and McAndrews, J.H. 1975. Pre-Columbian purslane (*Portulaca oleracea* L) in the New World. Nature 253: 276-277.

Connolly, A.P. 1977. The distribution and history in the British Isles of some alien species of *Polygonum* and *Reynoutria*. Watsonia 11: 291-311.

Coope, G.R. 1987. The invasion and colonization of the North Atlantic islands: a palaeological solution to a biogeographic problem. In: H. Kornberg and M.H. Williamson (eds.), Quantitative Aspects of the Ecology of Biological Invasions, pp. 619-635. Royal Society, London.

Davis, M.B. 1987. Invasion of forest communities during the Holocene: beech and hemlock in the Great Lakes region. In: A.J. Gray, M.J. Crawley and P.J. Edwards (eds.), Colonization, Succession and Stability, pp. 373-393. Blackwell Scientific Publ., Oxford.

De Candolle, A. 1855. Géographie Botanique Raisonée I, II. Masson, Paris.

Di Castri, F. 1990. On invading species and invaded ecosystems: the interplay of historical chance and biological necessity. In: F. di Castri, A.J. Hansen and M. Debussche (eds.), Biological Invasions in Europe and Mediterranean Basin, pp. 3-16. Kluwer Academic Publ., Dordrecht.

Drake, J.A., Mooney, H.A., Di Castri, F., Groves, R.H., Kruger, F.J., Rejmánek, M. and Williamson, M. (eds.) 1989. Biological Invasions: A Global Perspective. 525 pp. John Wiley and Sons, Chichester.

Frank, J.H. and McCoy, E.D. 1990. Endemics and epidemics of shibbolets and other things causing chaos. Florida Entomol. 73: 1-8.

Garthwaite, P.E. 1993. End to term 'native'. Quart. J. Forestry 87: 59-60.

Gouyon, P.H. 1990. Invaders and disequilibrium. In: F. di Castri, A.J. Hansen and M. Debussche (eds.), Biological Invasions in Europe and Mediterranean Basin, pp. 365-369. Kluwer Academic Publ., Dordrecht.

Heywood, V.H. 1989. Patterns, extents and modes of invasion by terrestrial plants. In: J.A. Drake, H.A. Mooney, F. di Castri, R.H. Groves, F.J. Kruger, M. Rejmánek and M. Williamson (eds.), Biological Invasions: A Global Perspective, pp. 31-55. John Wiley and Sons, Chichester.

Holub, J. and Jirásek, V. 1967. Zur Vereinheitlichung der Terminologie in der Phytogeographie. Folia Geobot. Phytotax. 2: 69-113.

Jarvis, P. (ed.) 1987-1992. Ecological Abstracts, Vol. 14-19. Elsevier Science Publ., Amsterdam.

Joenje, W. 1987. Remarks on biological invasions. In: W. Joenje, K. Bakker and L. Vlijm (eds.), The Ecology of Biological Invasions, pp. 15-18. Proceedings of the Royal Dutch Academy of Sciences, Series C, 90(1).

Kornas, J. 1990. Plant invasions in Central Europe: historical and ecological aspects. In: F. di Castri, A.J. Hansen and M. Debussche (eds.), Biological Invasions in Europe and Mediterranean Basin, pp. 19-36. Kluwer Academic Publ., Dordrecht.

Kowarik, I. 1995. Time lags in biological invasions with regard to the success and failure of alien species. In: P. Pyšek, K. Prach, M. Rejmánek and M. Wade (eds.), Plant Invasions - General Aspects and Special Problems, pp. 15-38. SPB Academic Publ., Amsterdam.

Le Floch E., Le Houerou, H.N. and Mathez, J. 1990. History and patterns of plant invasion in northern Africa. In: F. di Castri, A.J. Hansen and M. Debussche (eds.), Biological Invasions in Europe and Mediterranean Basin, pp. 105-133. Kluwer Academic Publ., Dordrecht.

Lohmeyer, W. and Sukopp H. 1992. Agriophyten in der Vegetation Mitteleuropas. Schr. R. Vegetationskd. 25: 1-185.

Macdonald, I.A.W., Loope L.L., Usher, M.B. and Hamann, O. 1989. Wildlife conservation and the invasion of nature reserves by introduced species: a global perspective. In: J.A. Drake, H.A. Mooney, F. di Castri, R.H. Groves, F.J. Kruger, M. Rejmánek and M. Williamson (eds.), Biological Invasions: A Global Perspective, pp. 215-255. John Wiley and Sons, Chichester.

Mack, R.M. 1985. Invading plants: their potential contribution to population biology. In: J. White (ed.), Studies on Plant Demography, pp. 127-142. Academic Press, London.

Mooney, H.A. and Drake, J.A. 1989. Biological invasions: a SCOPE program overview. In: J.A. Drake, H.A. Mooney, F. di Castri, R.H. Groves, F.J. Kruger, M. Rejmánek and M. Williamson (eds.), Biological Invasions: A Global Perspective, pp. 491-508. John Wiley and Sons, Chichester.

Perrins, J., Fitter, A. and Williamson, H. 1993. Population biology and rates of invasion of three introduced *Impatiens* species in the British Isles. J. Biogeogr. 20: 33-44.

Prach, K. and Wade, P.M. 1992. Population characteristics of expansive perennial herbs. Preslia 64: 45-51.

Preston, C.D. 1986. An additional criterion for assessing native status. Watsonia 16: 83.

Pyšek, P. 1995. Recent trends in studies on plant invasions (1974-1993). In: P. Pyšek, K. Prach, M. Rejmánek and M. Wade (eds.), Plant Invasions - General Aspects and Special Problems, pp. 223-236. SPB Academic Publ., Amsterdam.

Pyšek, P. and Prach, K. 1993. Plant invasions and the role of riparian habitats: a comparison of four species alien to central Europe. J. Biogeogr. 20: 413-420.

Pyšek, P., Prach, K. and Šmilauer, P. 1995. Relating invasion success to plant traits: an analysis of the Czech alien flora. In: P. Pyšek, K. Prach, M. Rejmánek and M. Wade (eds.), Plant Invasions - General Aspects and Special Problems, pp. 39-60. SPB Academic Publ., Amsterdam.

Rejmánek, M. 1995. What makes a species invasive? In: P. Pyšek, K. Prach, M. Rejmánek and M. Wade (eds.), Plant Invasions - General Aspects and Special Problems, pp. 3-13. SPB Academic Publ., Amsterdam.

Rikli, M. 1903. Die Anthropochoren und der Formenkreis der *Nasturtium palustre* DC. Ber. Zürich Bot. Ges. 13: 71-82.

Roy, J. 1990. In search of the characteristics of plant invaders. In: F. di Castri, A.J. Hansen and M. Debussche (eds.), Biological Invasions in Europe and Mediterranean Basin, pp. 335-352. Kluwer Academic Publ., Dordrecht.

Schroeder, F.G. 1969. Zur Klassifizierung der Anthropochoren. Vegetatio 16: 225-238.

Silvertown, J., Lines, C.E.M. and Dale, M.P. 1994. Spatial competition between grasses - rates of mutual invasion between four species and the interaction with grazing. J. Ecol. 82: 31-38.

Sloan Wilson, D. and Turelli, M. 1986. Stable underdominance and the evolutionary invasion of empty niches. Am. Natur. 127: 835-850.

Smith, P.M. 1986 Native or introduced? Problems in the taxonomy of some widely introduced annual brome grasses. Proc. Roy. Soc. Edinburgh, Sect. B, 273-281.

Stirton, C.H. 1979. Taxonomic problems associated with invasive alien trees and shrubs in South Africa. In: Proceedings of the 9th Plenary Meeting AETFAT, pp. 218-219.

Sukopp, H. 1972. Wandel von Flora und Vegetation in Mitteleuropa unter dem Einfluß des Menschen. Ber. Landwirtsch. 50: 112-130.

Thellung, A. 1922. Zur Terminologie der Adventiv- und Ruderalfloristik. Allg. Bot. Z., Karlsruhe, 1918/1919, 24/25: 36-42.

Trepl, L. 1984. Über *Impatiens parviflora* DC. als Agriophyt in Mitteleuropa. Diss. Bot. 73: 1-399.

Trepl, L. 1990. Research on the anthropogenic migration of plants and naturalisation: its history and current state of development. In: H. Sukopp, S. Hejný and I. Kowarik (eds.), Urban Ecology: Plants and Plant Communities in Urban Environments, pp. 75-97. SPB Academic Publ., The Hague.

Trewick, S. and Wade, P.M. 1986. The distributon and dispersal of two alien species of *Impatiens*, waterway weeds in the British Isles. In: Proceedings of the EWRS/AAB 7th Symposium on Aquatic Weeds, pp. 351-356.

Van Hulst, R. 1987. Invasion models of vegetation dynamics. Vegetatio 69: 123-131.

Wacquant, J.P. 1990. Biogeographical and physiological aspects of the invasion of *Dittrichia* (ex: *Inula*) *viscosa*, a ruderal species in the Mediterranean basin. In: F. di Castri, A.J. Hansen and M. Debussche (eds.), Biological Invasions in Europe and Mediterranean Basin, pp. 353-364. Kluwer Academic Publ., Dordrecht.

Webb, D.A. 1985 What are the criteria for presuming native status? Watsonia 15: 231-236.

Weeda, E.J. 1987: Invasions of vascular plants and mosses into the Netherlands. In: I.A.W. Joenje, K. Bakker and L. Vlijm (eds.), The Ecology of Biological Invasions, pp. 19-29. Proceedings of the Royal Dutch Academy of Sciences, Series C, 90(1).

II

INVASIVE SPECIES IN PLANT COMMUNITIES

ON THE ROLE OF ALIEN SPECIES IN URBAN FLORA AND VEGETATION

Ingo Kowarik
University of Hannover, Department of Landscape Architecture and Environmental Studies, Plant Ecology, Herrenhäuser Straße 2, D 30419 Hannover, Germany

Abstract

The role of alien species in urban vegetation is reviewed. Representation of aliens is compared with that of native species by taking into account both historical and spatial aspects. On a long-term time scale (more than a century), the importance of aliens, especially neophytes, is increasing. The representation of alien species, if expressed in quantitative terms, shows a close relationship to the spatial structure of the city, and decreases from the city centre towards the outskirts. In many cases, this trend is still valid at the level of particular species, as exemplified by an alien tree *Ailanthus altissima*. The relationship between species richness and the level of man-induced disturbance supports the intermediate disturbance theory only if the native species are considered; aliens, both archaeophytes and neophytes, are encouraged in sites subjected to high disturbance levels. At the level of phytosociological alliances, the proportion of alien species was closely correlated with the disturbance level, although the relationship was rather variable between particular alliances. As regards the role of alien species in succession on urban waste land, it appears that alien species may persist as dominants in succession for a long period. The following factors promoting the success of alien species in urban environment are discussed: availability of specific urban niches, high level of disturbance in urban environment, and the minor isolation of seed sources.

Introduction

In one of the first studies of rural-urban gradients, the Finnish botanist Linkola (1916) proved with quantitative data that the number of alien species increases from natural forests through meadows to arable land and human settlements. Later, comparisons between some European settlements have shown a close relationship between the presence of alien species and the size of settlements (Falinski 1971; Sukopp and Werner 1982; Pyšek 1989). This is usually explained by the considerable habitat heterogeneity, the role of big cities as centres of species' immigration and the better adaptation of alien species to man-made perturbations (Sukopp 1981; Sukopp and Trepl 1987; Kowarik 1990a, 1991; Pyšek 1993).

In this paper, the role of alien, compared to native, species in the urban flora and vegetation will be illustrated by addressing the following questions:

1. What is the proportion of alien compared to native species in the urban flora and do both groups differ in frequency patterns?
2. Do special patterns exist in the spatial distribution of alien *versus* native species?
3. Are alien species better adapted to man-made disturbance than natives?
4. What role do alien species play during the course of succession in urban vegetation?

Plant Invasions - General Aspects and Special Problems, pp. 85-103
edited by P. Pyšek, K. Prach, M. Rejmánek and M. Wade

Data sources and definitions

This paper refers mainly to Berlin, Germany, as a city with an extensive background of research in urban ecology during the last 40 years (Sukopp 1990, and papers in Sukopp *et al.* 1990). In order to elucidate some general trends in the performance of alien species in the urban environment, two extensive data sets are considered: the work of Kunick (1974, 1982a) who analysed spatial patterns in the occurence of species, and a dataset with 5136 vegetation relevés including most of the plant communities currently existing in Berlin (Kowarik 1988). Additionally, quantitative analysis on alien species in the woody vegetation of derelict sites (Kowarik 1992a) are used. Taxonomy and species' status (native/alien) is in accordance with the list in Sukopp *et al.* 1982.

Because of the long tradition of differently used terms some definitions are necessary: 'Alien' (= non-native, non-indigenous) species are those occurring in an area in which they have not evolved since the last Ice Age and whose introduction or immigration was assisted deliberately or involuntarily by human activities. This definition of alien species is in the tradition of Thellung (1912, 1918/19, see Trepl 1990a). It also follows the definition by Roy (1990), but enlarges it by two additions. First, the reference to the last Ice Age must be made; second by the connection to the role of humans assisting the introduction of alien species. The former is necessary in order to exclude species as natives which had formerly evolved in the area but became extinct during the colder periods. Re-introduced they should be treated as aliens because if they occurred in the area before or during the last Ice Age, it was not under present conditions as the climate was different from today (Webb 1985), and the chance of coevolution with other organisms was limited. The latter reference is necessary because most native species did not evolve *in situ* either, but arrived in the period since the last Ice Age, as Egler (1961) stated. Compared with alien species, however, the appearance of natives has not necessarily been supported by humans (see also Pyšek 1995).

Refering to Thellung, but accepting the terminological changes by Schroeder (1969), in the central European tradition alien species are usually divided in two groups according to the time of their introduction or immigration: archaeophytes brought in up to 1500 AD, and neophytes brought in after this date.

For studying the urban flora and vegetation, it is advantageous to consider the total area within the political borders of a city (Klotz 1990; Kowarik 1992b). Although this border is not defined ecologically, this approach is quite appropriate to the transition character of various habitats on the urban-rural gradient (*e.g.*, urban forests - 'natural' forests, urban settlements - rural settlements) which may be discussed in terms of the ecotone concept (Pyšek 1992) and which offers 'unexploited opportunities' for ecological research (McDonnel and Pickett 1990).

Species' responses to increasing disturbance have been studied by using the hemeroby-approach (Jalas 1955; Sukopp 1972, 1976; Kowarik 1988), which refers explicitly to the man-made components of disturbance (details in Kowarik 1990a).

Alien species in the urban flora

Species numbers and frequency

Dividing Berlin's present day flora into native and alien species reveals a large proportion of aliens: 593 alien species (41%), which include 167 archaeophytes and 426 neophytes, as opposed to 839 native species (59%) (Kowarik 1988). This differs from Falinski (1971) who assumed a proportion of aliens between 50-70% in big cities. A limit of roughly 50% aliens seems to exist in central Europe. Even from highly industrialized cities in Northrhine-Westphalia, only between 42 and 55% aliens are reported in the urban flora (Reidl and Dettmar 1993).

In Fig. 1, native and alien species of Berlin's flora are grouped in 8 classes according to their current frequency. In contrast to urban floras from the 19th century, most species are currently rare. The proportion between native and alien species changes directionally with increasing frequency: 48% of the rarest species are aliens, but only 18% of the most common species are aliens.

Some historical analyses have shown that, mainly due to industrialisation, a wave of newly introduced species reached central Europe in the second part of the 19th century (Scholz 1960; Sukopp 1976; Jäger 1977, but see Kowarik 1995 on effects of the climatic warming). Beyond changes in the species number, shifts in species' frequency have to be considered. A comparison of species' current frequency in Berlin with data from the 19th century (Ascherson 1864) reveals evident differences between native species and both groups of alien species (Table 1).

Two thirds of the species which became more frequent during the last 120 years

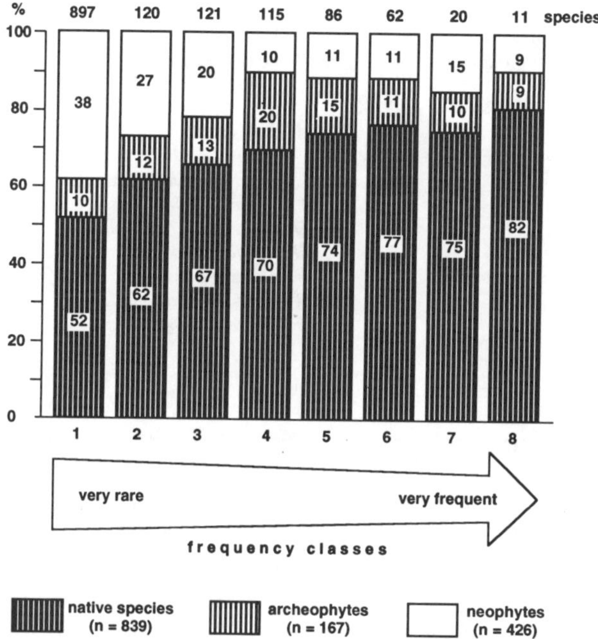

Fig. 1. Division of Berlin's flora according to frequency classes (based on the occurrence in 5136 phytosociological relevés made in the territory of the city). Proportion of native and alien species (the latter divided into archaeophytes and neophytes) is shown and absolute species numbers in each category are indicated for each frequency class on the top of the bar (after Kowarik 1988).

Table 1. Changes in frequency of native and alien species in the flora of Berlin during the last 120 years. Aliens are divided in archaeophytes and neophytes. Species with constant frequency are not shown. From Kowarik 1992b after Kutschkau 1982 *(a, b)* and Böcker *et al.* 1991 *(c)*. Frequency values in % are given.

	Native	Archaeophytes	Neophytes
a. Species with increasing frequency (100% = 361 species)	26.9	8.0	65.8
b. Species with decreasing frequency (100% = 821 species)	78.0	13.8	8.3
c. Extinct and endangered species (100% = 525 species)	80.2	11.2	8.6

Table 2. The currently most frequent alien species (archaeophytes and neophytes) in the flora of Berlin (frequency classes decreasing from 8 to 5, see Fig. 1, Kowarik 1988). Striking changes in frequency compared with Ascherson's flora from 1864 (after Kutschkau 1982) are indicated. Species with frequency having increased by +: 2-3 frequency classes; ++: 4-5 frequency classes; +++: 6 frequency classes. Species with frequency having decreased by -: 2-3 frequency classes; --: 4-5 frequency classes; --- 6 frequency classes. No symbol is given for species with more or less constant frequency. For native species, only the number of species belonging to a given frequency class is shown.

	Archaeophytes		Neophytes	Native
Frequency class 8				
	Poa annua	+++	*Solidago canadensis*	9 species
Frequency class 7				
	Capsella bursa-pastoris		*Conyza canadensis*	
+	*Plantago major*	+	*Impatiens parviflora*	
		++	*Prunus serotina*	15 species
Frequency class 6				
-	*Apera spica-venti*	+	*Arrhenatherum elatius*	
	Berteroa incana	++	*Bidens frondosa*	
	Convolvulus arvensis		*Galinsoga parviflora*	
	Fallopia convolvulus		*Oenothera biennis*	
	Plantago lanceolata	+	*Poa compressa*	
	Tripleurospermum inodorum	+++	*Rumex thyrsiflorus*	
	Veronica arvensis	+++	*Sisymbrium loeselii*	48 species
Frequency class 5				
-	*Ballota nigra*	++	*Acer negundo*	
	Bromus tectorum	++	*Clematis vitalba*	
+	*Crepis capillaris*	-	*Euphorbia cyparissias*	
	Euphorbia peplus	++	*Matricaria discoidea*	
-	*Hordeum murinum*	+	*Oenothera chicaginensis*	
-	*Lamium purpureum*	?	*Oxalis fontana*	
-	*Polygonum lapathifolium*	++	*Parietaria pensylvanica*	
	P. persicaria	++	*Robinia pseudoacacia*	
-	*Setaria viridis*	?	*Solidago gigantea*	
	Solanum nigrum			
	Sonchus oleraceus			
	Urtica urens			
	Viola arvensis			64 species

are neophytes. In contrast, most of the decreasing species are natives (78%) which are mainly listed in Red Data Books: in Berlin, 51% of the native species, 37% of the archaeophytes, but only 20% of the neophytes are endangered (Böcker *et al.* 1991). Obviously, unlike the neophytes, the archaeophytes which are mostly confined to agricultural disturbance regimes, are not encouraged by the effects of urbanization. Similar trends have been found in Halle, Warsaw, and Poznan (Klotz 1984; Sudnik-Wójcikowska 1987; Jackowiak 1989).

Table 2, showing the most common alien species in Berlin, also indicates conspicuous shifts in species' frequency. Half of the 20 most frequent neophytes have had a shift through more than three frequency classes during the last 120 years. In contrast, most of the common archaeophytes were already frequent, or more frequent, during the last century. The same is true for most native species. From these data, the increasing importance of alien species, mainly neophytes, in the urban environment can be predicted.

Spatial patterns

Abiotic environmental conditions usually change on rural-urban gradients from the centre to the outskirts of cities (Kuttler 1988; McDonnel and Pickett 1990; Bullock and Gregory 1991; Pyšek 1992). Due to the polycentric structure of many European cities, these changes are spatially diversified. There is a considerable evidence that the distribution patterns of plants and animals correspond significantly with the spatial structure of cities (for a review see Wittig *et al.* 1993).

In Berlin, Kunick (1974, 1982a) worked out a city zonation using floristic data. Table 3 shows that the proportion of natives and aliens changes directionally on the urban-rural gradient. In the city centre (zone 1), the proportion of natives is about 50%, increasing to more than 70% in the outskirts (zone 4). Data from surrounding rural districts of Brandenburg show that in less urbanized areas, the proportion of natives may increase to about 80% of the flora. A higher percentage of archaeophytes compared to neophytes is typical of rural landscapes (Pyšek and Pyšek 1988, 1990). City centres, however, may be characterized mainly by neophytes. One of them is *Ailanthus altissima*, native to China, which is virtually confined to urban-industrial areas in most parts of central Europe (Kowarik and Böcker 1984; Gutte *et al.* 1987; Müller 1987; Landolt 1991). Results of a grid mapping incorporated in Table 3 show a close relationship between the distribution of this woody species, and Kunick's city zonation of Berlin. Similar results have been found in Vienna (Kugler cited by Punz 1993).

Apart from the general trend on the urban-rural-gradient (increasing numbers of aliens from the outskirts to the centre), variations due to the specific habitat structure have to be considered. High numbers of aliens are usually found on industrial and

Table 3. Percentage of native and alien species (divided in archaeophytes and neophytes) on an urban-rural gradient. Shown are data from Berlin (city zonation according to Kunick 1982), from a grid mapping of *Ailanthus altissima* (Böcker and Kowarik 1982) and data from surrounding rural districts of Brandenburg (Klemm 1975). Data in %.

	Natives	Archaeophytes	Neophytes	*Ailanthus altissima**
Berlin				
Zone 1 (centre)	50.2	15.2	34.6	92.2
Zone 2	53.1	14.1	32.8	46.1
Zone 3	56.6	14.5	28.9	24.8
Zone 4 (outskirts)	71.5	10.2	18.3	3.2
Brandenburg districts				
Spremberg	74.8	8.1	17.1	-
Ruppiner Land	75.7	8.9	15.4	-
Priegnitz	77.5	9.1	13.4	-
Dahme	78.4	10.6	11.0	-
Spreewald	79.3	10.4	10.3	-

*Percentage of occuppied squares mapped in a grid in zones 1-4

railway sites as well as in built-up areas (Brandes 1983; Kowarik 1986; Rebele 1986; Wittig *et al.* 1989; Aey 1990; Dettmar 1991; Reidl and Dettmar 1993; Reidl 1993). The presence of safe sites for the germination and establishment of alien species may be due to the history of a city. Presumably, closely built-up city centres are less rich in alien species than cities including ruins and derelict areas. Differences between particular cities can be explained either by varying abiotic, mainly climatic, conditions or by methodological problems concerning the selected areas and species groups (Kunick 1982b; Brandes 1989; Wittig 1989; Pyšek 1993).

Responses to man-made disturbance

General pattern

The hypothesis that disturbance promotes the establishment and spread of alien species is accepted in most studies on biological invasions (*e.g.*, Trepl 1983, 1990b; Orians 1986; Crawley 1987; Sukopp and Trepl 1987; Hobbs 1989; Rejmánek 1989). No ecosystem is absolutely free of disturbance (Pickett and White 1985), but in the urban environment, it is mainly the man-made components of disturbance which have to be considered as driving forces affecting the species composition.

Referring to a data set of 5136 vegetation relevés which had each been assigned to one of nine levels of man-made disturbance (hemeroby, see Kowarik 1988, 1990a), the relationship between increasing disturbance and species richness in natives, archaeophytes and neophytes has been analysed. At first glance, the results shown in Fig. 2 confirm the intermediate disturbance hypothesis which means that species richness is highest at an intermediate level of disturbance and lowest under conditions of both high and low disturbance (Grime 1979; Connell 1979). But this is only

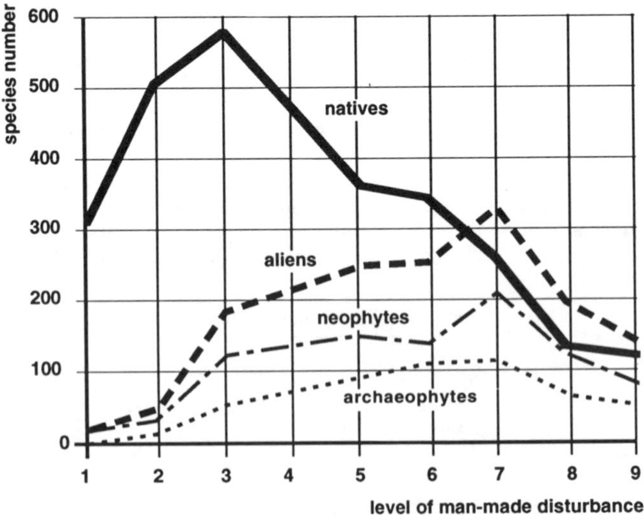

Fig. 2. The relationship between species richness and the level of man-made disturbance. Each of 5136 vegetation relevés made in the Berlin vegetation was classified using a 9-degree hemeroby scale (expressing the level of man-induced disturbance, see Kowarik 1990a). The total number of native and alien species (the latter divided into archaeophytes and neophytes) was then calculated for relevés belonging to the same hemeroby class.

true for native species. Regarding the alien species, an opposite trend is revealed: archaeophytes and neophytes are generally enhanced on sites subjected to a higher level of disturbance. These results may help to refine the intermediate disturbance hypothesis by differentiating according to species' origin. Furthermore, they support the theory on the positive relationship between disturbance and the promotion of alien species.

Differences among vegetation types

By grouping the 5136 relevés in 52 phytosociological alliances of the Braun-Blanquet system, which represent virtually the total vegetation of Berlin (according to the list of Sukopp 1979), the relationship between the average proportion of alien species and the average disturbance level which had been calculated for each of these 52 vegetation units, can be tested (data in Fig. 4). For each vegetation relevé the proportion of alien species (% of total species number) and the level of man-made disturbance was calculated, the latter by using species' indicator values for hemeroby from Kowarik (1988). In Fig. 3 average data for 45 vegetation units are shown. There is a significant correlation ($r = 0.95$, $P<0.001$) between the level of man-made disturbance and the proportion of alien species. Because of the varying ratio between archaeophytes and neophytes in several vegetation units (see Fig. 4), the correlation

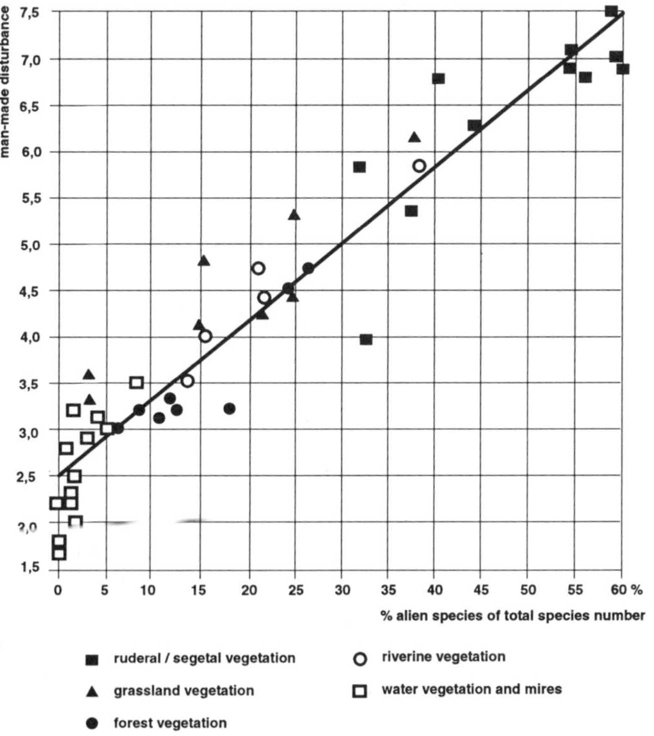

Fig. 3. The relationship between the level of man-made disturbance and the proportion of alien species in various vegetation units (phytosociological alliances representing virtually the total vegetation of Berlin). The average proportion of alien species and average disturbance level (expressed by a 9-degree hemeroby scale, see Kowarik 1990a) were calculated for each vegetation unit. Particular vegetation types are distinguished by using different symbols.

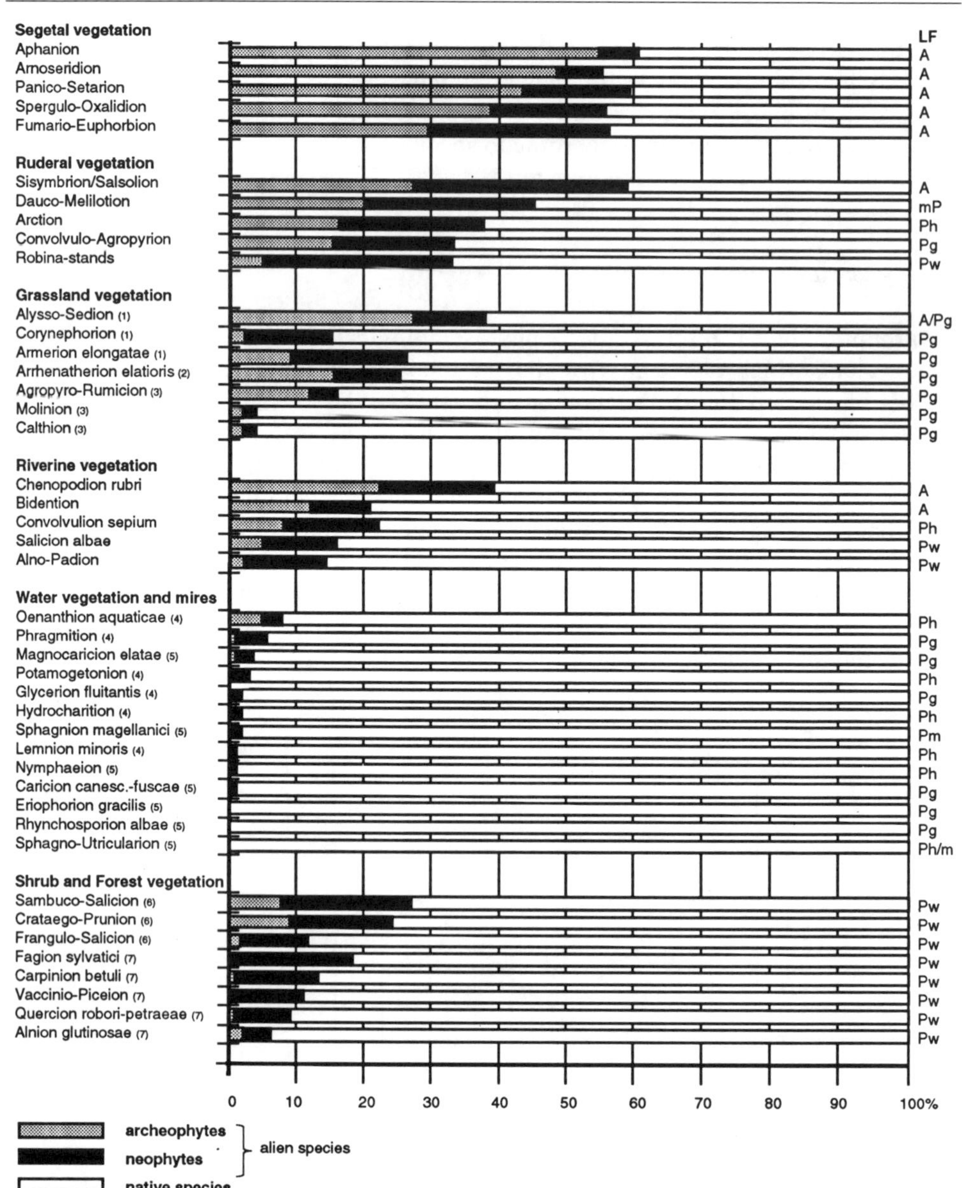

Fig. 4. The representation of native and alien species (the latter divided into archaeophytes and neophytes) in the vegetation of Berlin. The percentage contribution of each category to the total number of species is shown for particular phytosociological alliances (after Kowarik 1990a). The vegetation units are roughly arranged according to successional age and additional information on the vegetation type is indicated in brackets following the name of the alliance: 1: dry grassland; 2: mesic grassland; 3: wet grassland; 4: water vegetation; 5: mire vegetation; 6: shrub vegetation; 7: forest vegetation. Dominant life form (LF) is indicated: A: annual; mP: monocarpic perennial; Ph: perennial, mainly herbs; Pg: perennial, mainly grasses; Pm: perennial, mainly mosses; Pw: perennial, mainly woody species.

between disturbance and these groups is less strong (r = 0.89 and 0.68, respectively) but still significant ($P<0.001$).

Analysis of communities belonging to the same vegetation unit (at the level of

Table 4. Relationship between the proportion of archaeophytes (A), neophytes (N) and alien species in total (A+N) and the level of man-made disturbance in the vegetation of Berlin. Calculations were based on vegetation relevés assigned to particular phytosociological alliances (after Kowarik 1988). The significance level of the correlation coefficient (*t*-test) is shown: * $P<0.05$, ** $P<0.01$, *** $P<0.001$, NS: non significant; if not given, the respective unit was free of alien species. The following groups of vegetation are indicated: Pr: ruderal pioneer vegetation; Ps: segetal pioneer vegetation; R: riverine vegetation (RF: riverine forests); H: vegetation dominated by perennial herbs; G: grassland vegetation (Gd: dry, Gm: mesic, Gw: wet); W: water vegetation and mires; F: forest vegetation (Fr: ruderal forests, RF: riverine forests).

Alliance	A+N	A	N	Vegetation
Dauco-Melilotion	***	NS	***	Pr
Sisymbrion	***	***	***	Pr
Arnoseridion	NS	NS	NS	Ps
Sperqulo-Oxalidion	NS	NS	NS	Ps
Panico-Setarion	***	**	NS	Ps
Fumario-Euphorbion	***	***	NS	Ps
Aphanion	**	NS	*	Ps
Chenopodion rubri	*	**	NS	P/R
Bidention	***	***	*	R
Convolvulion	***	***	*	R
Agropyro-Rumicion	***	***	NS	R
Geo-Alliarion	***	***	**	H
Arction	***	***	NS	H
Aegopodion	*	***	NS	H
Convolvulo-Agropyrion	***	***	NS	H
Trifolion medii	NS	NS	NS	H
Molinion	***	**	***	Gw
Calthion	***	***	NS	Gw
Arrhenatherion	***	***	***	Gm
Polygonion avicularis	***	***	***	Gm
Koelerion	**	NS	**	Gd
Alysso-Sedion	***	***	NS	Gd
Corynephorion	**	***	*	Gd
Armerion	***	***	NS	Gd
Oenanthion	***	***	NS	W
Glycerion	NS	***	NS	W
Nanocyperion	*	**	NS	W
Phragmition	***	***	NS	W
Filipendulion	NS	**	NS	W
Magnocaricion	***	***	*	W
Caricion canescenti-fuscae	NS	NS	NS	W
Sphagnion magellanici	NS	NS	NS	W
Lemnion	NS	NS	NS	W
Hydrocharition	NS	NS	NS	W
Potamogetonion	NS	NS	NS	W
Nymphaeion	NS	NS	NS	W
Eriophorion	-	-	-	W
Sphagno-Utricularion	-	-	-	W
Alno-Padion	NS	***	**	RF
Salicion albae	NS	***	NS	RF
Crataego-Prunion	***	***	***	F
Frangulo-Salicion	**	***	**	F
Alnion glutinosae	***	***	***	F
Rubo-Salicion	*	*	NS	F
Robinia stands	NS	***	NS	Fr
Quercion robori-petraeae	***	***	***	F
Fagion	***	NS	***	F
Carpinion	**	**	*	F
Vaccinio-Piceion	NS	NS	NS	F

phytosociological alliances) but subjected to different levels of human impact, reveals less clearcut trends (Table 4). When the relationship between the proportion of aliens and level of man-made disturbance was tested (calculation based on single vegetation relevés available for a given alliance), a positive significant correlation between both variables was found in 32 alliances (65.3%) whereas in 17 alliances (34.7%) the relationship was not significant. A similar result was obtained for archaeophytes treated separately (significant correlation between both variables in 33 alliances, *i.e.*, 67.3%); for neophytes the respective figure was 19 (38.7%).

In detail, the following conclusions may be drawn from these results (Table 4):

a. Archaeophytes show a closer relationship to disturbance than neophytes: in 25 alliances, the significance level of the correlation coefficient was higher for archaeophytes than neophytes whilst in only 3 cases, neophytes were more closely correlated with the level of disturbance than archaeophytes.
b. In most vegetation units of wet sites (water and mire vegetation), no significant correlation between the representation of alien species and the level of disturbance was found.
c. The relationship between the representation of alien species and the level of disturbance was not significant in several units of segetal pioneer vegetation. This may be due to the constantly high proportion of aliens in this vegetation type.
d. With few exceptions, there was a close relationship between the proportion of aliens and the level of disturbance in vegetation units dominated by perennial grasses or herbs. This holds true for dry, mesic, and wet sites.

Alien species in the urban vegetation

Representation of alien species in particular vegetation types

From urban environments, plant communities dominated by alien species were mainly reported from railway areas and from heavily disturbed urban-industrial sites. In Essen, for example, Reidl and Dettmar (1993) found 55 communities (of a total of 262) which are characterized by alien species, among them 27 belonging to the annual and perennial ruderal vegetation (*e.g.*, communities with *Inula graveolens* and *Chenopodium botrys*, Dettmar and Sukopp 1991).

Considering the data set representing the vegetation of Berlin, comparable information can be derived on the role of alien species in various urban, seminatural and natural vegetation units. 5136 vegetation relevés were grouped in phytosociological alliances, and average representation of alien species was calculated for each alliance (expressed as the percentage of alien species per relevé) (Kowarik 1988, 1990a).

In Fig. 4, the vegetation units are grouped as follows: segetal vegetation, ruderal vegetation, grassland vegetation, riverine vegetation, water and mire vegetation, shrub and forest vegetation. For each vegetation unit, the dominant life form is shown. From the pattern of alien species which are divided in archaeophytes and neophytes, the following conclusions can be derived:

1. In contrast to the early stages of old field succession in non-urban areas, including less than 15% aliens of the total species number (examples in Rejmánek 1989), the proportion of alien species is much higher in the segetal and ruderal vegetation (up to 60% in the Aphanion or Sisymbrion).

2. The highest percentages of alien species are found in vegetation units with annuals as the dominating life form, *i.e.*, all units of segetal vegetation and the pioneer stages of ruderal vegetation (Sisymbrion/Salsolion).
3. Comparing archaeophytes and neophytes as early and later immigrants shows distinct differences: archaeophytes prevail in the vegetation of arable fields, neophytes in ruderal vegetation units. High percentages of archaeophytes are only found in vegetation units characterized by annuals.
4. As expected, there are two trends concerning the moisture gradient and the stage of succession. Linkola (1916) found less alien species in wet compared to mesic meadows. The data in Fig. 4 support the generally accepted relevance of the moisture gradient for the establishment of alien species (Rejmánek 1989). The vegetation of mires is virtually free of alien species, and there are only a few in aquatic vegetation. In grassland vegetation, the trend is clear: starting from 25% in the Armerion elongatae on dry sites, the percentage of alien species decreases to only 4% on wet sites (meadows of the Calthion and Molinion).
5. In the course of succession the percentage of alien species decreases conspicuously: In the ruderal vegetation, the annual phase starts with 59% aliens. The Dauco-Melilotion which is already dominated by monocarpic perennials, still includes 44% of alien species, but in the perennial ruderal vegetation (dominated by both grasses or herbs) the average percentage of aliens decreases to about one third. Grassland vegetation generally includes less than 30% of aliens. In the Quercion robori-petraeae as the most frequent woody vegetation in the fringe of the city (dominated by oaks and Scotch pine), even less than 10% of all species are aliens. (These 10% include the North American *Prunus serotina*, a species which was able to establish dense populations after having been planted by foresters, Starfinger 1991).
6. Nevertheless, on a much higher level, compared to areas beyond the impact of urbanization, the percentage of alien species seems to decrease during succession as expected. However, woody vegetation on urban sites may include a surprisingly high amount of aliens: 33% of total species number have been found in urban stands of *Robinia pseudoacacia.* These stands (which have not been planted) represent the oldest stages of old field succession on urban sites which developed in the centre of Berlin after the WW II. Their age varies between 25 and 35 years.

May alien species persist in the urban woody vegetation?

In most studies on alien species as dominants or co-dominants in urban woody vegetation, their association with other species has been analysed (*e.g.*, Kohler and Sukopp 1964a,b; Kowarik and Böcker 1984; Becher and Brandes 1985; Kunick 1990; Diesing and Gödde 1989; Schmitz 1991). Although most of the studied stands represent later stages of successional series, only a little information is available on the role of alien shrub and tree species through succession. The results shown in Fig. 4 confirm a decreasing proportion of alien species during succession as a general trend. However, the existence of more than 30-year-old successional stages which are both rich in and dominated by alien species (such as the urban stands of *Robinia pseudoacacia* in Berlin) does not fit this theory. Obviously, alien species may persist longer than expected in urban old field successional series.

The succession on urban wastelands in Berlin may serve as an unintended long-term experiment on the performance of alien compared to native species over more than four decades of succession. Vast sites were destroyed during and after the WW II, and due to the political situation in Berlin, many of them in the western part of the city have remained abandoned until today. Because of the total destruction, both native and alien species had to immigrate in order to colonize the open sites. After more than four decades of succession the woody vegetation in five wastelands (totally 98.7 ha) has been analysed (see Kowarik 1992a,c for details).

On average, 45% of the forests are dominated by aliens, 55% by native species. *Robinia pseudoacacia,* native to North America, is the most successful alien tree with *Betula pendula* as its native counterpart. However, there are extreme differences between single areas: alien woody species dominate between 5% and 74% of the woody vegetation, even in areas with comparable site conditions (such as derelict railway areas).

The vegetation of one railway area of about 20 ha had been mapped twice in 10 years. The analyses of the vegetation maps revealed that the area covered by woody vegetation virtually doubled in ten years from 37% to 69%. However, the proportion of alien and native trees as dominants of forest vegetation remained almost unchanged, with 37:63 in 1991 compared to 38:62 in 1981. Obviously, both *Robinia* and *Betula* may persist and enlarge the area occupied over time.

But how long may *Robinia* dominate ruderal forests? Demographic studies show urban *Robinia*-stands being far from resistent to the invasion by other woody species (Kowarik 1990b): a strikingly high number of 38 tree, 35 shrub, and 4 species of woody climbers have been found in *Robinia*-stands, and about half of these species are aliens in Berlin. Compared to native *Betula pendula*, the stands of *Robinia* harbour more individuals of, mainly shade tolerant, tree and shrub species. Focusing upon the performance of shade tolerant and tall growing trees as potential competitors of *Robinia,* indicates future succession trends. On average, shade tolerant trees (mainly *Acer platanoides, A. pseudoplatanus*) already represent 83% of all tree individuums in the herb layer (<0.9m), 57% in the shrub layer (<5m), but only 9% in the tree layer (>5m). Most of the potential competitors are still confined to height classes up to 7m, and in some stands, resulting from the distance to seed sources, competitors are completely missing in the shrub or tree layers.

These results indicate a strikingly slow change in the dominance structure of urban *Robinia*-stands in Berlin which contrasts clearly with the performance of *Robinia* in native American forests. In the Appalachians, it is replaced almost totally by other tree species within 20-30 years (Boring and Swank 1984). The immigration success of native species as potential competitors, however, indicates future changes in the dominance of tree species but this is presumably not a question of years, but of decades.

Discussion

Both historical and spatial analyses of the occurence of alien species in the urban environment have shown that non-native species are strongly promoted by urbanization in central European cities. In rural floras of Brandenburg, for example, there are less than 20% aliens, but from the outskirts to the city centre of Berlin, the percentage of alien species increases from about 30 to 50% of all species (Table 3). Comparisons with floristic data from the 19th century revealed increasing numbers of alien species (Scholz 1960; Klotz 1984; Sudnik-Wójcikowska 1987; Jackowiak 1989), and increasing frequencies of alien species. It is mainly neophytes as the later immigrants (since 1500 AD) that are promoted by the effects of urbanization (Tables 1, 2).

Even if there is no doubt on the general trend, it has to be questioned which of the diverse factors usually summarized under the term 'urbanization' may be responsible for the enhancement of alien species. I will discuss the role of three of them:
a. the availability of specific 'urban' niches,
b. the usually high level of disturbance in the urban environment, and
c. the minor isolation of seed sources.

Finally, some conclusions on the role of alien species during the course of succession will be derived from the results summarized in this paper.

Availability of specific 'urban' niches

The changes in environmental factors on urban-rural gradients may provide specific niches which could be realized by aliens, either exclusively or better than by native species. There are some examples of urban-industrial habitats which are virtually exclusively colonized by alien species: *Inula graveolens* and *Chenopodium botrys* on dumps and other industrial sites in North-Rhine-Westphalia (Gödde 1984; Dettmar and Sukopp 1971), the latter species also on burning dumps in Lille (Lampin 1969), and on heavily disturbed calcareous sites in the post-war Berlin (Sukopp 1971). *Buddleja davidii* is often the only woody pioneer colonizing crevices in abandoned housing areas in London (Burton 1983), on gravel in railway areas (Schmitz 1991), or on dumps as dominant species or as a co-dominant of *Betula pendula* (Dettmar 1991). These examples of successfully colonizing alien species refer to sites which may be extreme both in terms of stress (*e.g.*, low pH-values, limited water supply, high temperatures) or disturbance (*e.g.*, mechanical perturbation of soils). They support the view of Trepl (1993) who explained the often stated minor success of alien species in colonizing extreme habitats (*e.g.*, Rejmánek 1989) by the isolation from seed sources rather than by the limited invasibility of these sites.

Rising temperatures, often combined with a limited water supply, are characteristic factors which usually change directionally on rural-urban gradients. Considering the origin of many alien species in warmer areas (Scholz 1960; Sarisaalo-Taubert 1963), their increasing presence in the city centre(s) (usually coinciding with the centre(s) of the urban heat archipel) may be interpreted as a result of the pre-adaptation to urban conditions. In Berlin, for example, the spatial pattern of *Ailanthus altissima* coincides with the city zonation (Table 3) and spatial structure of the heat archipel (Kowarik and Böcker 1984). There is some evidence, that the establishment of urban heat islands which exacerbated the effects of generally rising temperatures since the 1850's has in some cases enabled or, at least, promoted the invasion of species with high requirements for temperature (Kowarik 1995).

The relevance of temperature on rural-urban gradients is stressed by a corresponding pattern on elevation gradients. Usually, there are less alien species with increasing elevation (Frenkel 1970; Fox and Fox 1986; Brandes 1989), and in both cases, increasing percentages of alien species coincide with rising temperatures. Considering that varying temperatures are only one part of the diversification of urban site conditions stresses the role of specific 'urban' niches in opening invasion windows for alien species.

Man-made disturbance

The increasing proportion of alien species on the rural-urban gradient (up to 50% of all species in the centre of Berlin, Table 3) may be explained by a corresponding change in the intensity of human impact which is presumably highest in city centres. But there are usually less disturbed 'islands' in the centre (*e.g.*, parks, cemeteries) and also heavily disturbed areas in the outskirts (*e.g.*, refuse pits, industrial plants). Thus, the mosaic structure of many cities causes difficulties in interpreting the results of spatial analysis in terms of varying human impact.

However, the theory of disturbance (in the sense of Grime 1979) as promoting the establishment of alien species by reducing the competition of other species and by creating open space (*e.g.*, Crawley 1987) can be tested by using the hemeroby approach (hemeroby as an expression of the man-made components of disturbance, see Kowarik 1990a). Considering the vegetation of more than 5000 sites reveals distinctive differences in the floristic inventory between the nine groups of sites (Fig. 2): there are more aliens compared to native species on sites affected by higher intensities of man-made disturbance.

On the community level, a strong correlation between the level of man-made disturbance and the percentage of alien species has been found (Fig. 3). This trend is crystal-clear when considering vegetation units on a higher aggregated level, see Fig. 3 which refers to phytosociological alliances. However, not all, but many of these alliances include communities which are subjected to a different intensity of man-made disturbance. The more detailed analysis on the level of single phytosociological alliances, however, revealed a less clear, or even lacking, significant correlation between man-made disturbance and the percentage of alien species (Table 4). In general, these results confirm the role of disturbance as a driving force in enhancing alien species. In some vegetation units however, different amounts of alien species may not be sufficiently explained by a varying level of disturbance.

Isolation from seed sources

Considering the role of cities as centres of species' immigration (Sukopp 1976; Jäger 1977; Kowarik 1990a) could explain the high numbers of alien species in central European cities. However, in many cases good immigration conditions (*i.e.*, availability of dispersal vectors and of seed sources, both from unintentionally introduced diaspores and from planted individuals) coincide with the availability of specific urban sites which are often subjected to a high level of disturbance (*e.g.*, urban railway areas). Though the effects of disturbance, availability of specific niches, and the lack of isolation from seed sources can hardly be distinguished from each other.

The role of immigration conditions compared to the relevance of individual features ('ideal' invaders) or community characteristics (invasibility) still seems to be underestimated, although profoundly discussed by some authors (*e.g.*, Schroeder 1972; Trepl 1984, 1993; Orians 1986). Linkola's landmark study from 1916 includes one possibility of testing the relevance of the immigration conditions: a comparison between the flora of isolated settlements surrounded by forests with that of settlements in rural areas which were subjected to the cultural influence of humans for longer. Linkola (1916) found distinctly less alien species in isolated villages. These results indicate the limiting role of the immigration conditions.

Supposing that urban wastelands are less isolated from seed sources of alien species than rural wastelands, comparisons between these and between different urban areas may serve as another approach to elucidate the specifity of the urban environment in terms of immigration conditions. The results shown in Fig. 4 reveal a much higher percentage of alien species in various urban vegetation units compared to non-urban old field successional series and to rural wastelands (Rejmánek 1989; Prach 1994).

Focussing the proportion of alien *versus* native tree species as dominants of several urban wastelands in Berlin showed a broad variance in the area covered by *Robinia pseudoacacia* or by *Betula pendula.* If both the native *Betula* and the alien *Robinia* may become dominant during succession on similar sites (derelict railway areas), the presence or absence of both species can be explained neither by the resistence of preceding successional stages nor by limited resources. Both species can share the same resources, both are able to establish dense stands. This example stresses the role of immigration conditions which seem to be a decisive factor for the dominance of *Robinia* or *Betula* on urban wastelands in Berlin. Considering the virtually unlimited availability of *Betula* seeds, the potential dominance of *Robinia* is limited by the effective dispersal of this species. Conversely, the minor role of *Robinia* in non-urban wasteland succession (Prach 1994) may be due to limited seed sources, less effective dispersal, and, in colder areas, by the higher temperature requirements of *Robinia*.

Impatiens parviflora is a good example of a species whose spread from urban to rural sites is enhanced by improved dispersal conditions: it needed about 50 years to arrive in forests, and today, *Impatiens* is the most common alien species in central European forests (Trepl 1984). Although the invasion of non-urban forests or other rural habitats may not be exclusively a question of resistance or invasibility, but also a question of time. This hypothesis is supported by the higher presence of alien species in forests adjacent to urbanized areas (Asmus 1981; Moran 1984) and by the switch of species, formerly confined to settlements, to rural habitats (*e.g., Amaranthus retroflexus, Sisymbrium loeselii* in the district of Halle, Grosse 1987), and even to natural habitats (*e.g., Reynoutria* species, *Heracleum mantegazzianum*, and *Impatiens glandulifera,* Pyšek and Prach 1993 in the Czech Republic).

The role of alien species in succession

The data on the presence of alien species in the vegetation of Berlin shown in Fig. 4 confirm a decrease of alien species from early successional stages, usually dominated by annuals and monocarpic perennials with up to 60% aliens of all species, to perennial stages with 30-40% aliens. During the course of old field succession, Rejmánek (1989) found an exponential decline of aliens. It started from about 14% at the beginning and fell to less than 4% after about ten years (see also Osbornová *et al.* 1989). As opposed to these successional seres, the decrease in urban vegetation is far from exponential. Also, older successional stages may include a percentage of alien species which is more than twice that of the initial stages of non-urban successional series: a third of all species in the perennial ruderal vegetation of Berlin are aliens despite the dominance of grasses, herbs or even *Robinia pseudoacacia* which is the most frequent alien tree in the woody vegetation of urban wastelands. The oldest stems in urban *Robinia* stands are about 35 years old, and although the demographic results indicate changes in dominance, the switch to other prevailing tree species (presumably the shade tolerant *Acer platanoides* and *A. pseudoplatanus*) will again be a question of decades. Thus, considering both dominance and species richness (in % of all species), alien species may persist much longer than expected in urban successional series.

Acknowledgments

I thank Petr Pyšek for the comments on the manuscript, Lois Child for improving my English, and Ursula Jonczyk for preparing the figures.

References

Aey, W. 1990. Historical approaches to urban ecology. Methods and first results from a case study (Lübeck, West-Germany). In: H. Sukopp, S. Hejný and I. Kowarik (eds.), Urban Ecology, pp. 113-129. SPB Academic Publ., The Hague.

Ascherson, P. 1864. Flora der Provinz Brandenburg, der Altmark und des Herzogthums Magdeburg. Abt. 2. Specialflora von Berlin. 1034 pp. Berlin.

Asmus, U. 1981. Der Einfluß von Nutzungsänderung und Ziergärten auf die Florenzusammensetzung stadtnaher Forste in Erlangen. Ber. Bayer. Bot. Ges. 52: 117-121.

Baker, H.G. 1965. Characteristics and modes of origins of weeds. In: H.G. Baker and G.L. Stebbins (eds.), The Genetics of Colonizing Species, pp. 141-172. Academic Press, London.

Becher, R.D. and Brandes, D. 1985. Vergleichende Untersuchungen an städtischen und stadtnahen Gehölzbeständen am Beispiel von Braunschweig. Braunschw. Naturk. Schr. 2: 309-339.

Böcker, R., Auhagen, A., Brockmann, H., Kowarik, I., Scholz, H., Sukopp, H. and Zimmermann, F. 1991. Liste der wildwachsenden Farn- und Blütenpflanzen von Berlin (West) mit Angaben zur Gefährdung der Sippen, zum Zeitpunkt ihres ersten spontanen Auftretens und zu ihrer Etablierung im Gebiet sowie zur Bewertung der Gefährdung. Landschaftsentwickl. Umweltforsch. S 6: 57-88.

Boring, L.R. and Swank, W.T. 1984. The role of black locust (*Robinia pseudoacacia*) in forest succession. J. Ecol. 72: 749-766.

Brandes, D. 1983. Flora und Vegetation der Bahnhöfe Mitteleuropas. Phytocoenologia 11: 31-115.

Brandes, D. 1989. Geographischer Vergleich der Stadtvegetation in Mitteleuropa. Braun-Blanquetia 3: 61-67.

Bullock, P. and Gregory, P. (eds.) 1991. Soils in the Urban Environment. Blackwell Scientific Publ., Oxford.

Burton, R.M. 1983. Flora of the London Area. London Natural History Society, London.

Connell, J.H. 1979. Tropical rain forests and coral reefs as open non-equilibrium systems. In: R.M. Anderson, B.D. Turner and L.R. Taylor (eds.), Population Dynamics, pp. 141-163. Blackwell Scientific Publ., Oxford.

Crawley, M.J. 1987. What makes a community invasible? In: M.J. Crawley, P.J. Edwards and A.J. Gray (eds.), Colonization, Succession and Stability, pp. 629-654. Blackwell Scientific Publ., Oxford.

Dettmar, J. 1991. Industrietypische Flora und Vegetation im Ruhrgebiet. Diss. Bot. 191: 1-397.

Dettmar, J. and Sukopp, H. 1991. Vorkommen und Gesellschaftsanschluß von *Chenopodium botrys* L. und *Inula graveolens* (L.) Desf. im Ruhrgebiet (Westdeutschland) sowie im regionalen Vergleich. Tuexenia 11: 49-65.

Diesing, D. and Gödde, M. 1989. Ruderale Gebüsch- und Vorwaldgesellschaften nordrheinwestfälischer Städte. Tuexenia 9: 225-251.

Egler, F.E. 1961. The nature of naturalization. In: Recent Advances in Botany, pp. 1341-1345. University of Toronto Press, Toronto.

Falinski, J.B. (ed.) 1971. Synanthropization of Plant Cover. II. Synanthropic Flora and Vegetation of Towns Connected with their Natural Conditions, History and Function. Mater. Zakl. Fitosocjol. Stosowanej UW 27: 1-317.

Fox, M.D. and Fox, B.J. 1986. The susceptibility of natural communities to invasions. In: R.H. Groves and J.J. Burdon (eds.), Ecology of Biological Invasions: An Australian Perspective, pp. 57-66. Australian Academy of Science, Canberra.

Frenkel, R.E. 1970. Ruderal Vegetation along Californian Roadsites. University of California Press, Berkeley.

Gödde, M. 1984. Zur Ökologie und pflanzensoziologischen Bindung von *Inula graveolens* (L.) Desf. in Essen. Natur und Heimat 44(4): 101-108.

Grime, J.P. 1979. Plant Strategies and Vegetation Processes. 222 pp. John Wiley and Sons, Chichester.

Grosse, E. 1987. Anthropogene Florenveränderungen in der Agrarlandschaft nördlich von Halle (Saale). Hercynia N.F. 24: 179-209.

Gutte, P., Klotz, S., Lahr, C., and Trefflich, A. 1987. *Ailanthus altissima* (Mill. Swingle) - eine vergleichend pflanzengeographische Studie. Folia Geobot. Phytotax. 22: 241-262.

Hobbs, R.J. 1989. The nature and effects of disturbance relative to invasions. In: J.A. Drake, H.A. Mooney, F. di Castri, R.H. Groves, F.J. Kruger, M. Rejmánek and M. Williamson (eds.), Biological Invasions: A Global Perspective, pp 389-405. John Wiley and Sons, Chichester.

Jackowiak, B. 1989. Dynamik der Gefäßpflanzenflora einer Großstadt am Beispiel von Poznan/Polen. Braun-Blanquetia 3: 89-98.

Jäger, E. 1977. Veränderungen des Artenbestandes von Floren unter dem Einfluß des Menschen. Biol. Rundschau 15: 287-300.

Jalas, J. 1955. Hemerobe und hemerochore Pflanzenarten. Ein terminologischer Reformversuch. Acta Soc. Fauna Flora Fenn. 72(11): 1-15.

Klotz, S. 1984. Phytoökologische Beiträge zur Charakterisierung und Gliederung urbaner Ökosysteme, dargestellt am Beispiel der Städte Halle und Halle-Neustadt. Thesis Martin-Luther-Universität, Halle-Wittenberg.

Klotz, S. 1990. Species/area and species/inhabitants relations in European cities. In: H. Sukopp, S. Hejný and I. Kowarik (eds.), Urban Ecology, pp. 99-103. SPB Academic Publ., The Hague.

Kohler, A. and Sukopp, H. 1964a: Über die Gehölzentwicklung auf Berliner Trümmerstandorten. Ber. Deutsch. Bot. Ges. 76: 389-406.

Kohler, A. and Sukopp, H. 1964b. Über die soziologische Struktur einiger Robinienbestände im Stadtgebiet von Berlin. SBer. Ges. Naturforsch. Freunde (N.F.) 4(2): 74-88.

Kowarik, I. 1986. Vegetationsentwicklung auf innerstädtischen Brachflächen - Beispiele aus Berlin (West). Tuexenia 6: 75-98.

Kowarik, I. 1988. Zum menschlichen Einfluß auf Flora und Vegetation. Theoretische Konzepte und ein Quantifizierungsansatz am Beispiel von Berlin (West). Landschaftsentwickl. Umweltforsch. 56: 1-280.

Kowarik, I. 1990a. Some responses of flora and vegetation to urbanization in central Europe. In: H. Sukopp, S. Hejný and I. Kowarik (eds.), Urban Ecology, pp. 45-74. SPB Academic Publ., The Hague.

Kowarik, I. 1990b. Zur Einführung und Ausbreitung der Robinie (*Robinia pseudoacacia* L.) in Brandenburg und zur Gehölzsukzession ruderaler Robinienbestände in Berlin. Verh. Berliner Bot. Ver. 8: 33-67.

Kowarik, I. 1991. The adaptation of urban flora to man-made perturbations. In: O. Ravera (ed.), Terrestrial and Aquatic Ecosystems: Perturbation and Recovery, pp. 176-184. Ellis Horwood, London.

Kowarik, I. 1992a. Einführung und Ausbreitung nichteinheimischer Gehölzarten in Berlin und Brandenburg und ihre Folgen für Flora und Vegetation. Ein Modell für die Freisetzung gentechnisch veränderter Organismen. Verh. Bot. Ver. Berlin Brandenburg, Beiheft 3: 1-188.

Kowarik, I. 1992b. Das Besondere der städtischen Flora und Vegetation. Schriftenr. Deutsch. Rat Landespflege 61: 33-47.

Kowarik, I. 1992c. Zur Rolle nichteinheimischer Arten bei der Waldbildung auf innerstädtischen Standorten in Berlin. Verh. Ges. Ökol. 21: 207-213.

Kowarik, I. 1995. Time lags in biological invasions with regard to the success and failure of alien species. In: P. Pyšek, K. Prach, M. Rejmánek and M. Wade (eds.), Plant Invasions - General Aspects and Special Problems, pp. 15-38. SPB Academic Publ., Amsterdam.

Kowarik, I. and Böcker, R. 1984. Zur Verbreitung, Vergesellschaftung und Einbürgerung des Götterbaumes (*Ailanthus altissima* (Mill.) Swingle) in Mitteleuropa. Tuexenia 4: 9-29.

Kunick, W. 1974. Veränderungen von Flora und Vegetation einer Großstadt, dargestellt am Beispiel von Berlin (West). Thesis Technische Universität Berlin.

Kunick, W. 1982a. Zonierung des Stadtgebietes von Berlin (West). Ergebnisse floristischer Untersuchungen. Landschaftsentwickl. Umweltforsch. 14: 1-164.

Kunick, W. 1982b. Comparison of the flora of some cities of the central European lowlands. In: R. Bornkamm, J.A. Lee and M.R.D. Seeward (eds.), Urban Ecology. 2nd European Ecological Symposium, pp. 13-22. Blackwell Scientific Publ., Oxford.

Kunick, W. 1990. Spontaneous woody vegetation in cities. In: H. Sukopp, S. Hejný and I. Kowarik (eds.), Urban Ecology, pp. 167-174. SPB Academic Publ., The Hague.

Kutschkau, H. 1982. Rückgang und Ausbreitung in der Gefäßpflanzenflora von Berlin (West) seit 1860. Thesis FU Berlin.

Kuttler, W. 1988. Spatial and temporal structures of the urban climate - a survey. In: K. Grefen and H. Löbel (eds.), Environmental Meteorology, pp. 305-333. Kluwer Academic Publ., Dordrecht.

Lampin, P. 1969. La Végétation Pionniere d'un Terril en Combustion. 67 pp. Université de Lille, Fac. Sci., Lille.

Landolt, E. 1991. Distribution patterns of flowering plants in the city of Zurich. In: G. Esser and D. Overdieck (eds.), Modern Ecology: Basic and Applied Aspects, pp. 807-822. Elsevier Science Publ., Amsterdam.

Linkola, K. 1916. Studien über den Einfluß der Kultur auf die Flora in den Gegenden nördlich vom Ladogasee. Acta Soc. Fauna Flora Fenn. 45(1).

McDonnel, M.J. and Pickett, S.T.A. 1990. Ecosystem structure and function along urban-rural gradients: an unexploited opportunity for ecology. Ecology 71: 1232-1237.

Moran, M.A. 1984. Influence of adjacent land use on understory vegetation of New York forests. Urban Ecol. 8: 329-340.

Müller, N. 1987. *Ailanthus altissima* (Miller) Swingle und *Buddleja davidii* Franchet. Zwei adventive Gehölze in Augsburg. Bayer. Bot. Ges. 8: 105-107.

Orians, G.H. 1986. Site characteristics favoring invasions. In: H.A. Mooney and J.A. Drake (eds.), Ecology of Biological Invasions of North America and Hawai, pp. 133-148. Springer-Verlag, New York.

Osbornová J., Kovářová M, Lepš, J. and Prach, K. (eds.) 1989. Succession in Abandoned Fields. Studies in Central Bohemia, Czechoslovakia. Kluwer Academic Publ., The Hague.

Pickett, S.T.A. and White, P.S. (eds.) 1985. The Ecology of Natural Disturbance and Patch Dynamics. Academic Press, Orlando.

Prach, K. 1994. Succession of woody species in derelict sites in Central Europe. Ecol. Engin. 3: 49-56.

Punz, W., 1992. Stadtökologie. Forschungsansätze und Perspektiven. Schr. Ver. Verbreitung Naturwiss. Kenntnisse Wien 132: 89-120.

Pyšek, A. and Pyšek, P. 1988. Standörtliche Differenzierung der Flora der westböhmischen Dörfer. Folia Mus. Rer. Natur. Bohem. Occid., Plzeň. Botanica 28: 1-52.

Pyšek, P. 1989. Archaeophytes and neophytes in the ruderal flora of some Czech settlements. Preslia 61: 209-226 (in Czech).

Pyšek, P. 1992. Settlement outskirts - may they be considered as ecotones? Ekológia (ČSFR) 11: 273-286.

Pyšek, P. 1993. Factors affecting the diversity of flora and vegetation in central European settlements. Vegetatio 106: 89-100.

Pyšek, P. 1995. On the terminology used in plant invasion studies. In: P. Pyšek, K. Prach, M. Rejmánek and M. Wade (eds.), Plant Invasions - General Aspects and Special Problems, pp. 71-81. SPB Academic Publ., Amsterdam.

Pyšek, P. and Prach, K. 1993. Plant invasions and the role of riparian habitats: a comparison of four species alien to central Europe. J. Biogeogr. 20: 413-420.

Pyšek, P. and Pyšek, A. 1990. Comparison of the vegetation and flora of the West Bohemian villages and towns. In: H. Sukopp, S. Hejný and I. Kowarik (eds.), Urban Ecology, pp. 105-112. SPB Academic Publ., The Hague.

Rebele, F. 1986. Die Ruderalvegetation der Industriegebiete von Berlin (West) und deren Immissionsbelastung. Landschaftsentwickl. Umweltforsch. 43: 1-224.

Reidl, K. 1993. Zur Gefäßpflanzenflora der Industrie- und Gewerbegebiete des Ruhrgebietes. Ergebnisse aus Essen. Decheniana 146: 39-55.

Reidl, K. and Dettmar, J. 1993. Flora und Vegetation der Städte des Ruhrgebiets, insbesondere der Stadt Essen und der Industrieflächen. Ber. Deutsch. Landeskd. 67: 299-326.

Rejmánek, M. 1989. Invasibility of plant communities. In: J.A. Drake, H.A. Mooney, F. di Castri, R.H. Groves, F.J. Kruger, M. Rejmánek and M. Williamson (eds.), Biological Invasions: A Global Perspective, pp 369-388. John Wiley and Sons, Chichester.

Roy, J. 1990. In search of the characteristics of plant invaders. In: F. di Castri, A.J. Hansen and M. Debussche. K (eds.), Biological Invasions in Europe and the Mediterranean Basin, pp. 335-352. Kluwer Academic Publ., Dordrecht.

Sarisaalo-Taubert, A. 1963. Die Flora in ihrer Beziehung zur Siedlung und Siedlungsgeschichte in den südfinnischen Städten Provoo, Loviisa und Hamina. Ann. Bot. Soc. Zool. Bot. Fenn. Vanamo 35: 1-190.

Schmitz, J. 1991. Vorkommen und Soziologie neophytischer Sträucher im Raum Aachen. Decheniana 144: 22-38.

Scholz, H. 1960. Die Veränderungen in der Berliner Ruderalflora. Ein Beitrag zur jüngsten Florengeschichte. Willdenowia 2: 379-397.

Schroeder, F.-G. 1969. Zur Klassifizierung der Anthropochoren. Vegetatio 16: 225-238.

Starfinger, U. 1991. Population biology of an invading tree species - *Prunus serotina*. In: A. Seitz and V. Loeschke (eds.), Species Conservation: A Population-Biological Approach, pp. 171-183. Birkhäuser, Basel.

Sudnik-Wojcikowska, B. 1987. Dynamik der Warschauer Flora in den letzten 150 Jahren. Gleditschia 15: 7-23.
Sukopp, H. 1971. Beiträge zur Ökologie von *Chenopodium botrys* L. 1. Verbreitung und Vergesellschaftung. Verh. Bot. Ver. Prov. Brandenburg 108: 3-25.
Sukopp, H. 1972. Wandel von Flora und Vegetation in Mitteleuropa unter dem Einfluß des Menschen. Ber. Landwirtsch. 50: 112-130.
Sukopp, H. 1976. Dynamik und Konstanz in der Flora der Bundesrepublik Deutschland. Schr. R. Vegetationskde. 10: 9-27.
Sukopp, H. 1979. Vorläufige systematische Übersicht von Pflanzengesellschaften Berlins aus Farn- und Blütenpflanzen. 16 pp. (unpublished manuscript).
Sukopp, H. 1981. Ökologische Charakteristika der Großstadt. In: Tagungsbericht 1. Leipziger Symposium Urbane Ökologie, pp. 5-12, Leipzig.
Sukopp, H. (ed.) 1990: Stadtökologie. Das Beispiel Berlin. 455 pp. Reimer, Berlin.
Sukopp, H., Auhagen, A., Bennert, W., Böcker, R., Hennig, U., Kunick, W., Kutschkau, H., Schneider, C., Scholz, H., and Zimmermann, F. 1982. Liste der wildwachsenden Farn- und Blütenpflanzen von Berlin (West) mit Angaben zur Gefährdung der Sippen und Angaben über den Zeitpunkt der Einwanderung in das Gebiet von Berlin (West). Landschaftsentwickl. Umweltforsch. 11: 19-58.
Sukopp, H., Hejný, S. and I. Kowarik (eds.) 1990. Urban Ecology. 282 pp. SPB Academic Publ., The Hague.
Sukopp, H. and Trepl, L. 1987. Extinction and naturalization of plant species as related to ecosystem structure and function. Ecol. Studies 61: 245-276.
Sukopp, H. and P. Werner 1982. Nature in Cities. A Report and Review of Studies and Experiments Concerning Ecology, Wildlife and Nature Conservation in Urban and Suburban Areas. 94 pp. Council of Europe Nature and Environment Series 28. Strasbourg.
Thellung, A. 1912. La flore adventice de Montpellier. Mém. Soc. Sci. Nat. Cherbourg 38: 622-647.
Thellung, A. 1918/19. Zur Terminologie der Adventiv- und Ruderalfloristik. Allg. Bot. Zschr. 24/25(9-12): 36-42.
Trepl, L. 1983. Zum Gebrauch von Pflanzenarten als Indikatoren der Umweltdynamik. SBer. Ges. Naturforsch. Freunde Berlin N.F. 23: 151-171.
Trepl, L. 1990a. Research on anthropogenic migration of plants and naturalization. Its history and current state of development. In: H. Sukopp, S. Hejný and I. Kowarik (eds.), Urban Ecology, pp. 75-97. SPB Academic Publ., The Hague.
Trepl, L. 1990b. Zum Problem der Resistenz von Pflanzengesellschaften gegen biologische Invasionen. Verh. Berliner Bot. Ver. 8: 195-230.
Trepl, L. 1993. Zur Rolle interspezifischer Konkurrenz bei der Einbürgerung von Pflanzenarten. Arch. Nat. Lands. 33: 61-84.
Webb, D.A. 1985. What are the criteria for presuming native status? Watsonia 15: 231-236.
Wittig, R. 1989. Methodische Probleme der Bestandsaufnahme der spontanen Flora und Vegetation von Städten. Braun-Blanquetia 3: 21-28.
Wittig, R., Diesing, D. and Gödde, M. 1985. Urbanophob - Urbanoneutral - Urbanophil. Das Verhalten der Arten gegenüber dem Lebensraum Stadt. Flora 177: 265-282.
Wittig, R., König, H. and Rückert, E. 1989. Nutzungs- und baustrukturspezifische Analyse der ruderalen Stadtflora. Braun-Blanquetia 3: 69-79.
Wittig, R., Sukopp, H. and Klausnitzer, B. 1993. Die ökologische Gliederung der Stadt. In: H. Sukopp and R. Wittig (eds.), Stadtökologie, pp. 271-318. Fischer Verlag, Stuttgart.

POST-FIRE RECOVERY AND INVASION BY ALIEN PLANT SPECIES IN A SOUTH AMERICAN WOODLAND-STEPPE ECOTONE

Miriam Gobbi, Javier Puntieri and Susana Calvelo
Centro Regional Universitario Bariloche, cc 1336, 8400 Bariloche, Argentina

Abstract

The role of alien species in early plant succession was studied in two sites of the Subantarctic Forests Region of Argentina which had been affected by an extensive fire in 1987. One of these sites was dominated by the shrub morph of *Nothofagus antarctica* and the other one by the tree morph of that species. Vegetation cover was assessed one to three years and five years after the fire in 25 to 35 1 m^2 plots in each site. The degree and vigour of resprouting of *Nothofagus antarctica* were evaluated. There was high variability in the species present in the plots. The ordination of species along two axes employing Canonical Correspondence Analysis with year and site as covariables accounted for 8% of the variability in species cover. Species richness was initially higher on the shrub-morph site but after five years it was higher on the tree-morph site due mostly to the co-occurrence of native and alien species in the latter. Diversity was similar in both sites after five years. Plant cover only differed between sites after five years of secondary succession, when it was higher on the tree-morph site (50% on average) than on the shrub-morph site (20%). One year after the fire the most important species in both sites was *Muehlenbeckia hastulata* (a native climbing semi-shrub), but this species almost disappeared from both sites four years later. By the fifth year the most common species were *Phacelia secunda* (a native perennial herb) and *Diostea juncea* (a native shrub) in the shrub-morph site and *Rumex acetosella* (an alien perennial herb) and *Verbascum thapsus* (alien biennial herb) in the tree-morph site. In both cases, these species made up more than 50% of the total plant cover. Alien species represented 3% of the total cover on the shrub-morph site and 72% on the tree-morph site. The degree of resprouting of *Nothofagus antarctica* from root crown was similar in both sites, but its vigour was higher in the tree-morph site. Differences in early post-fire succession between formerly tree-dominated and shrub-dominated areas could be associated with the presence of cattle 25 years ago (before the whole area was fenced). Cattle may have assisted the dispersal of alien species in the tree-dominated area. The possible role of alien species in the future evolution of these communities is discussed.

Introduction

In South America the Subantarctic Forests Region (SFR) occupies a strip of land on both slopes of the Andes from about 36°S (Northern Patagonia) down to 55°S in Tierra del Fuego. One of the characteristic plant genera of this phytogeographical region is *Nothofagus*, which makes up most of its forest mass. The oriental border of this region contacts the Patagonian steppe. A remarkable precipitation gradient takes place from the areas of maximum forest development (>2500 mm) to the steppe (<500 mm).

Pollen analyses indicate that the limit between forest and steppe has moved considerably since the last full glaciation (Markgraf *et al.* 1986; Markgraf 1987; Veblen and Markgraf 1988). The most recent fluctuations were, apparently, affected by human activities (Veblen and Lorentz 1988), which is also believed to be the cause of the sharp increase in the last centuries of the alien species *Rumex acetosella* and *Plantago lanceolata* (Markgraf 1987). In Argentina, however, the SFR has been detached from the main stream of alien species invasions, which has historically been

Plant Invasions - General Aspects and Special Problems, pp. 105-115
edited by P. Pyšek, K. Prach, M. Rejmánek and M. Wade

associated with agricultural activities centered around the more temperate and fertile pampas (Mazocca *et al.* 1976; Mack 1989). At present, frequencies of intentional fires and clear-cutting and grazing on eastern areas of the SFR are at a high level, in particular due to the pressure of the increasing human population in areas surrounding National Parks (which include an important part of the SFR). This, together with the lack of both restrictions to the introduction of alien species (*e.g.*, ornamental plants) and control of species already invading in and around National Parks may be setting the scene for a new wave of alien plant invasions in this region.

One of the plant communities most affected by human activities is that of *Nothofagus antarctica*. This species is considered an early successional species (Veblen *et al.* 1977; Markgraf 1987) and the most tolerant of all South American *Nothofagus* species to poor growth conditions (Alberdi *et al.* 1985). It is also associated with post-fire stands, especially those where sexual reproduction is precluded, due to which occupies those areas where the SFR meets the Patagonian steppe (Auer 1958). This species has the capacity to sprout vigorously from its root crown (Veblen and Markgraf 1988). *Nothofagus antarctica* has a high phenotypic plasticity: it may de-

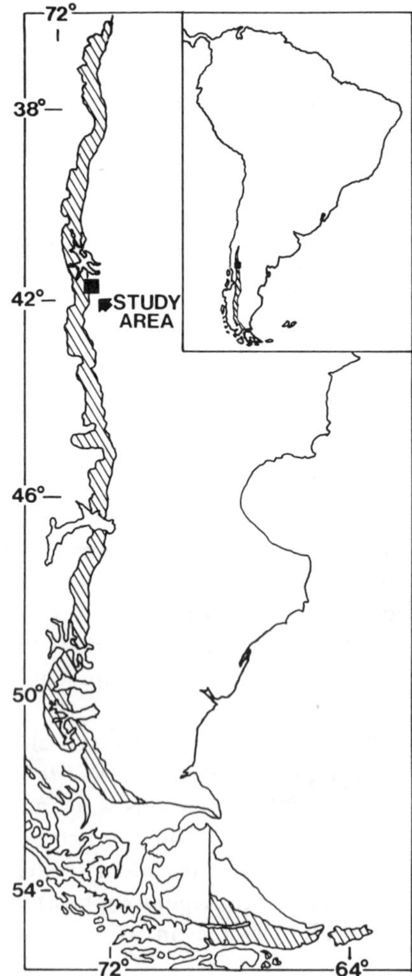

Fig. 1. Location of the study area. The hatched area represents the Subantarctic Forest Region.

velop as a tree up to 15 m high, as a tall shrub or as a chamaephyte (Ramírez *et al.* 1985).

This paper refers to the early secondary succession of the plant community in two neighbouring sites where two different morphs of *Nothofagus antarctica,* shrub and tree, were dominant until 1987. In February 1987, a fire started 2 km away from San Carlos de Bariloche (Argentina), the largest human settlement within the area of National Parks of Northern Patagonia, covering 3000 ha of land including, among other communities, shrublands and forests of *Nothofagus antarctica* close to the border between the SFR and the steppe. Part of the area affected is out of the National Park. Its owner had kept that area free of cattle for about 25 years before the fire and did not cut the dead wood after the fire (Fig. 1). A long term study of the process of succession in that area of the SFR was set up in order to gain insight into the past evolution and future development of these forests.

Methods

In 1988, 1989, 1990 and 1992, after the seasonal peak of vegetative development of most plant species (March-April), 30 1 m^2 plots were set up along a randomly oriented transect in each of the selected sites and the percent projected cover of each species was estimated in each plot. Species and plots were ordered using Canonical Correspondence Analysis (program CANOCO, Ter Braak 1987). Two covariables, year and area (= growth form of *Nothofagus antarctica*), were included in the analysis. The significance of the first axis of the ordination was assessed by means of a Monte Carlo simulation based on 100 permutations (a procedure of CANOCO).

Species diversity was measured using Shannon-Wiener index (Peet 1974).

The degree of resprouting of *Nothofagus antarctica* from the stem base was estimated as a proportion of a random sample of 50 individuals (or ramets) that resprouted in 1988 and 1989, the persistence of the burned shoots allowing their identification. The vigour of resprouting was assessed by measuring the height of the new shoots in 1988, 1989 and 1992.

Soil samples were taken from both areas in winter, spring and summer at three levels: 0-5 cm, 6-15 cm and 16-30 cm. Water content of soil samples at 1/3 bar, 15 bar and atmospheric pressure (natural drainage) was measured. Hydric constants and soil pH were not expected to vary between such closely located sites, given the uniformity of soils within the whole region (derived from volcanic ashes) (A. Marcolín, Soil Laboratory, Instituto Nacional de Tecnología Agropecuaria), and all soil samples obtained for each depth were pooled before analyzing. Nomenclature follows Correa (1969-1988) and Dimitri (1974).

Results

The soils of the sites correspond to an andosol, characterized by an unstructured upper stratum. Water content values appear in Table 1. The availability of water did not change notably from winter to summer below 5 cm depth, and decreases in the surface layer were such that levels reached were not low enough for plants to be under water stress.

Table 1. Soil characteristics in the study area in the year after the fire.

	Depth (cm)		
	0 - 5	6 - 15	16 - 30
Water content (%)			
1/3 bar	38.0	25.3	25.8
15 bar	12.5	13.9	13.9
1 atm winter	46.5	18.2	22.3
spring	33.1	21.5	25.0
summer	21.0	18.4	23.1

A total of 39 species of vascular plants appeared in the study area including those present in the transects (33 species). Sixty percent of all species were registered in the shrub-morph site and 85% in the tree-morph site.

In the shrub-morph site early regeneration was dominated by the climbing semi-shrub *Muehlenbeckia hastulata,* the perennial herb *Phacelia secunda* and the native shrubs *Nothofagus antarctica, Schinus patagonicus* and *Diostea juncea* (Table 2), all of which recovered from rootstock. By the end of the period of study, *Muehlenbeckia hastulata* died out in the whole area. The semi-shrub *Senecio filaginoides* became important in terms of cover in 1992.

Table 2. Mean cover (%) of the most important higher plants found in the study sites dominated by the shrub-morph or by the tree-morph of *Nothofagus antarctica* within five years of a fire.

Species	Life form	Shrub-morph site				Tree-morph site			
		1988	1989	1990	1992	1988	1989	1990	1992
Nothofagus antarctica	shrub/tree	1.9	<1.0	1.7	1.9	2.0	2.7	5.3	<1.0
Diostea juncea	shrub/tree	2.4	1.9	6.3	4.7	-	2.3	-<1.0	
Schinus patagonicus	shrub/tree	3.1	7.1	3.5	-	<1.0	<1.0	<1.0	2.7
Berberis buxifolia	shrub	<1.0	<1.0	4.0	<1.0	-	<1.0	<1.0	<1.0
Senecio filaginoides	semi-shrub	-	-	-	4.2	-	-	4.4	3.3
Muehlenbeckia hastulata	semi-shrub	16.4	13.9	25.5	<1.0	34.1	16.8	23.6	-
Phacelia secunda	perennial herb	8.5	2.0	2.7	6.1	-	<1.0	<1.0	<1.0
Solidago chilensis	perennial herb	-	-	5.4	<1.0	-	<1.0	<1.0	<1.0
Rumex acetosella	perennial herb	-	-	-	<1.0	-	-20.6	19.3	
Verbascum thapsus	biennial	-	-	-	<1.0	-	<1.0	4.3	16.2

At the beginning of the study in the tree-morph site *Muehlenbeckia hastulata* was overall the dominant species with *Nothofagus antarctica* being the dominant shrub species (Table 2). Five years after the fire the dominant species at this site were the alien herbs *Rumex acetosella* and *Verbascum thapsus.* At the end of the study period, the resprouting shrubs, *Schinus patagonicus* and *Ribes cucullatum*, and the semi-shrub *Senecio filaginoides* were, together with *Nothofagus antarctica*, the most important native species. As in the shrub-morph site, all *Muehlenbeckia hastulata* plants had died out after five years of re-growth.

Species richness was initially higher in the shrub-morph site than in the tree-morph site (Mann-Whitney's U=4.83, $P<0.001$) but this situation reversed four years later (U=2.66, $P<0.001$; Fig. 2a). Diversity was higher in the shrub-morph site in 1988

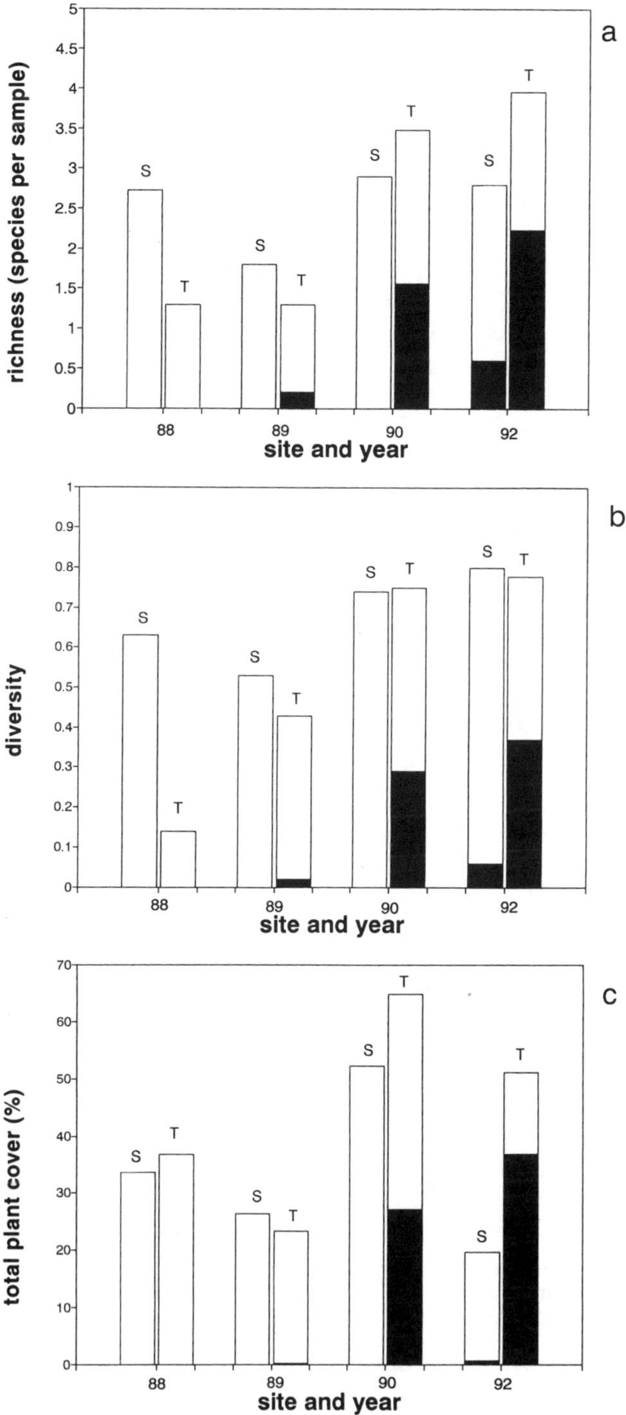

Fig. 2. Plant species *(a)* richness, *(b)* diversity (expressed as Shannon-Wiener index), and *(c)* cover of native (empty bars) and alien (solid bars) plant species within five years following a fire in sites previously dominated by the shrub-morph (S) and the tree-morph (T) of *Nothofagus antarctica.*

(U=5.44, *P*<0.001) but did not differ significantly (P>0.05) between sites from then on (U=1.67, 0.17 and 1.90 for 1989, 1990 and 1992, respectively; Fig. 2b). Total plant cover was similar between sites in 1988, 1989 and 1990 (U=0.38, 0.27 and 0.50, respectively, P>0.1), and was higher for the tree-morph site in 1992 (U=4.38, *P*<0.001; Fig. 2c). At the shrub-morph site total plant cover decreased from about 60% in 1990 to about 25% in 1992, mostly due to the significant decrease in the cover of *Muehlenbeckia hastulata*. Alien species made a higher contribution in the tree-morph site to richness, diversity and plant cover than in the shrub-morph site.

The variability in species cover between sample units was very high and only a low percent could be explained by the first two axes of CANOCO (Table 3). However, the first axis of that ordination was significant (Monte Carlo permutations, F=12.3, *P*<0.01). The ordination of species placed most native shrubs (*Schinus patagonicus, Diostea juncea, Ribes cucullatum, Berberis buxifolia, Senecio subumbellatus*) in a group associated with the shrub-morph community. *Nothofagus antarctica*, however, was not in this group due to its low cover (<5%), relative to that of other native shrubs, in the last two years of sampling (Fig. 3). Most alien species appeared in a group together with some native species associated with the later stages sampled in this study and with the tree-morph community. All native species in this

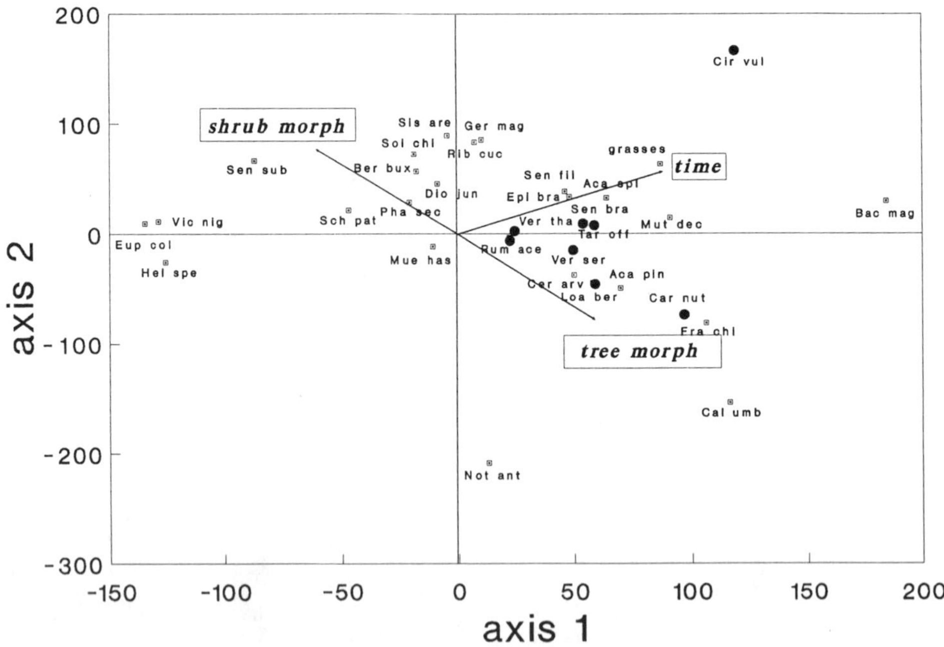

Fig. 3. Biplot ordination of species and covariables (time and growth form, shrub or tree) of *Nothofagus antarctica*, obtained by means of CANOCO. Dots: native species, black circles: species alien for the Subantarctic Forests Region. Species codes: Aca pin: *Acaena pinnatifida*, Aca spl: *Acaena splendens*, Bac mag: *Baccharis magellanica*, Ber bux: *Berberis buxifolia*, Cal umb: *Calandrinia umbellata*, Car nut: *Carduus nutans*, Cer arv: *Cerastium arvense*, Cir vul: *Cirsium vulgare*, Dio jun: *Diostea juncea*, Epi bra: *Epilobium brachycarpum*, Eup col: *Euphorbia collina*, Fra chi: *Fragaria chiloense*, Ger mag: *Geranium magellanicum*, Hel spe: *Heliotropium* sp., Loa ber: *Loasa bergii*, Mue has: *Muehlenbeckia hastulata*, Mut dec: *Mutisia decurrens*, Not ant: *Nothofagus antarctica*, Pha sec: *Phacelia secunda*, Rib cuc: *Ribes cucullatum*, Rum ace: *Rumex acetosella*, Sch pat: *Schinus patagonicus*, Sen bra: *Senecio bracteolatus*, Sen fil: *Senecio filaginoides*, Sen sub: *Senecio subumbellatus*, Sis are: *Sisyrinchium arenarium*, Sol chi: *Solidago chilensis*, Tar off: *Taraxacum officinale*, Ver ser: *Veronica serpyllifolia*, Ver tha: *Verbascum thapsus*, Vic nig: *Vicia nigricans*.

Table 3. Summary for the first two axes of CANOCO analysis with time as ordinal variable and plant community (dominated by shrub- or tree-morph of *Nothofagus antarctica*).

	Axes	
	1	2
Eigenvalues	0.482	0.247
Species-environment correlations	0.789	0.626
Cumulative variance explained (%)		
of species data	5.4	8.1
of species-environment relationship	66.1	-

group are perennial herbs (*Acaena* spp.) or semi-shrubs (*Senecio* spp.) characteristic of degraded and dry areas in the region.

The percentage recovery of shrub-morph *Nothofagus antarctica* was 53% in 1988 and 74% in 1989 and that of tree-morph *Nothofagus antarctica* was 58% in 1988 and 67% in 1989. Differences between sites were not significant in either year (χ^2=0.74 for 1988 and χ^2=0.72 for 1989; P>0.1). The resprouting shoots of *Nothofagus antarctica* were higher in the tree-morph site than in the shrub- morph site: 50 cm *vs.* 35 cm in 1988 (U=3.71, *P*<0.001) and 72 cm *vs.* 50 cm in 1989 (U=3.04, *P*<0.001).

Discussion

Early succession after disturbance

This study shows the importance of resprouting shrub and semi-shrub species in early stages of post-fire development in the Subantarctic Forest Region. Both shrub and tree morphs of *Nothofagus antarctica* are able to regenerate vegetatively, and diferences in vigour (judged by the height of shoots) are evident soon after resprouting, thus suggesting a genetic difference between the two morphs. Other re-sprouting woody species (like the shrubs *Diostea juncea* and *Schinus patagonicus*) soon outgrow *Nothofagus antarctica*. Among the most important species found in the study area, the native semi-shrub *Muehlenbeckia hastulata* and the perennial herb *Phacelia secunda*, both native of the Subantarctic Forest Region, also recovered from rootstock. Seedling establishment of native species, on the other hand, is considered unimportant in burned places previously dominated by species capable of vegetative re-growth after fire. A clear difference between study sites in the kind of recovery is caused by the important increase in cover of alien species and in the number of native herbs and semi-shrubs characteristic of degraded areas (like *Acaena splendens* and *Acaena pinnatifida*) in the site previously dominated by the tree morph of *Nothofagus antarctica.* It appears that, in that site, opportunistic species took advantage of the demise of *Muehlenbeckia hastulata,* which was, until 1990, the dominant species. Although plant cover was notably higher in the tree-morph site, species richness and, in particular, diversity, remained similar between both sites, showing that only a few species contributed to the higher cover in that site.

Two hypotheses could be put forward to explain the difference between sites in the degree of success of alien species:

1. The site once dominated by *Nothofagus antarctica* trees, was, because of its open understorey, frequented by cattle before the whole area including all of this site

was fenced, enabling a seed bank of alien species like *Rumex acetosella* and *Verbascum thapsus* to build up. Conversely, the vegetation in the area of *Nothofagus antarctica* shrubs may have been too dense for cattle to roam in.
2. Differences in species composition between sites determined that the number of species able to sprout from rootstock was higher in the previous shrubland and this restricted the probability of alien species becoming established in that site.

The first hypothesis is supported by other studies which have found that in Argentina *Nothofagus antarctica* forests were the most affected by livestock when compared with other forests dominated by other *Nothofagus* spp., whereas shrublands dominated by *Nothofagus antarctica* were the least frequented by livestock when compared with other shrublands (Veblen *et al.* 1992). The similarity between sites with respect to the degree of plant cover in the first two years after the fire would make the second hypothesis less likely. This also suggests that differences between sites were not due to the level of radiation at ground level, which is known to affect both seed germination and seedling establishment (Fenner 1978; Grime 1979).

The community that resulted in the area previously dominated by the shrub morph of *Nothofagus antarctica* agrees with the observations made by other authors, who associated the high, broom-like shrub *Diostea juncea*, one of the dominant species found here, with shrublands affected by fires (Willis 1914; Dimitri 1962). Those communities where *Diostea juncea* is dominant occupy areas considered to be degraded by the activity of cattle (Dimitri 1962; Correa Luna 1969; Gallopín 1978). This view is not supported by the data obtained here, since *Diostea juncea* was less represented in the area apparently most affected by livestock. This apparent contradiction may be only a matter of time scale: this and other native shrubs are present in the area now dominated by alien herbs and may, as they grow, become dominant in a few years. The species of native shrubs already present in the area are closer to those found in more humid woodlands than to steppe species. The soil analyses indicated no stressful conditions for plants in the first growth season after the fire (particularly in the soil deeper than 5 cm). Good water retention in the soil might be a key factor in the development of a plant community with affinities to woodland vegetation despite the ecotonal character of the study area and the presence of dry grassland areas in its proximity.

Alien species in northern Patagonia: former and current invaders

Rumex acetosella and *Verbascum thapsus,* the dominant species in the site previously dominated by the tree morph of *Nothofagus antarctica* at the end of study period, are alien species of Eurasian origin and now have a worldwide distribution. *Rumex acetosella* is able to propagate vegetatively (Grime *et al.* 1988) and could be considered a naturalized species throughout Argentina and, indeed, in the Subantarctic Forest Region. It has a high frequency even in apparently undisturbed areas (Gobbi 1994). This was probably one of the earliest successfully invading species in nothern Patagonia from outside South America since it appears massively in pollen deposits of that region in the stratum corresponding to the last centuries (Markgraf 1987). *Rumex acetosella* is known to develop a persistent seed bank in grazed areas of the ecotone between the Subantarctic Forest Region and the Patagonian steppe (Ghermandi 1992). Because of its relatively low competitive ability (Grime 1979; Grime *et al.* 1988), its presence may not represent a severe restriction to the future development of the community. *Verbascum thapsus*, on the other hand, is a more localized alien species in Argentina: it is important only in Patagonia (Mazocca *et al.* 1976) and appears only along open roadsides and heavily degraded areas. Since it forms very dense populations due to the poor dispersal of seeds from the parent plant, its presence may hinder the establishment of other species and thus the future development of the community (Salisbury 1961). The biennials, *Cirsium vulgare* and *Carduus nutans*, and the annual *Epilobium brachycarpum,* were relatively unimportant in the study site until 1992 (and were still so in 1994, as suggested by a recent visit by the authors to the site) and are the most important species in a neighbouring burned *Nothofagus pumilio* stand (M. Gobbi, J. Puntieri and S. Calvelo, unpublished data). Both thistles are among the first weeds colonize agricultural lands in Argentina (Mack 1989) and are, since 1963, considered agricultural pests by law. They are also important weeds worlwide (Holm *et al.* 1977). *Epilobium brachycarpum,* on the contrary, is a North American species which, so far as we know, has a limited distribution both worldwide and in Argentina (Solomon 1988). In northern Patagonia, however, this species is rapidly spreading along roadsides and in disturbed areas (J. Puntieri, personal observation).

The long history of natural disturbance in the form of landslides, fires, vulcanism and glaciers in northeastern Patagonia should have been major forces on the evolution of plant species in this region. In fact, vegetation recovery after natural disturbance is well known for this region (*e.g.,* Veblen 1985). Nevertheless, alien plant species are becoming significant contributors to the vegetation cover in the Subantarctic Forest Region, particularly after human-related disturbance (see Brion *et al.* 1988; Puntieri 1991). Human activities may have an effect on the response of the vegetation to a disturbance to which native species are adapted. In this sense, the presence in that region of biennial, rosette forming, fast growing alien species like *Verbascum thapsus, Cirsium vulgare* and *Carduus nutans*, a life form adapted to cattle mowing and trampling (Salisbury 1961; Frenkel 1977), may reflect a selective pressure that operated 25 years ago. It is interesting to note the lack of native species in that category of the life-form spectrum within the study area. The combination of relatively low weight (both in mass and density) of native herbivores, dry summers and high frequency of fires in the Subantarctic Forest Region may have restricted the possibilities of evolution of biennial competitive and ruderal species (*sensu* Grime 1979). Moreover, since the Subantarctic Forest Region of Argentina is mostly surrounded by the predominantly xeromorphic vegetation of the Patagonian steppe (to the east) and the mountain-top vegetation of the Andes (to the west), the invasion of naturally-dispersed mesomorphic, fast growing species from other temperate regions would be relatively low. Humans may therefore be not only introducing a new form of disturbance to a region but also aiding the dispersal of plant's adaptive answer to that disturbance.

Acknowledgments

We thank Bernardo Benroth for enabling us to use his land for this study and Arrigo Marcolín and the Instituto Nacional de Tecnología Agropecuaria for carrying out the soil analyses. We are indebted to Petr Pyšek for his help in data processing and Edy

Rapoport, Petr Pyšek and an anonymous reviewer for useful comments on the manuscript. The Institute of Applied Ecology, Kostelec nad Černými lesy, Czech Republic, facilitated the CANOCO program and computing facilities.

References

Alberdi, M., Romero, M., Ríos, D. and Wenzel, H. 1985. Altitudinal gradients of seasonal frost resistance in *Nothofagus* communities of Southern Chile. Oecol. Plantarum 6: 21-30.

Auer, V. 1958. The Pleistocene of Fuego-Patagonia. Part II. The history of the flora and vegetation. Ann. Acad. Sci. Fenn. 50: 1-239.

Brion, C., Grigera, D., Puntieri, J. and Rapoport, E. 1988. Plantas exóticas en bosques de *Nothofagus*, comparaciones preliminares entre el norte de la Patagonia y Tierra del Fuego. Monogr. Acad. Nac. Ciencias Exactas, Físicas y Naturales 4: 37-47.

Correa, M.N. 1969-1988. Flora Patagónica. Parts II-V and VII. Colección Científica, Instituto Nacional de Tecnología Agropecuaria, Volume VIII, Buenos Aires.

Correa Luna, H. 1969. Cuenca del Río Manso Superior, Parque Nacional Nahuel Huapi. Servicio Nacional de Parque Nacionales. Unpublished report (deposited at the Intendencia de Parques Nacionales, San Carlos de Bariloche, 8400 Argentina).

Dimitri, M.J. 1962. La flora Andino-Patagónica. An. Parques Nac. 9: 4-115.

Dimitri, M.J. 1974. Pequeña flora ilustrada de los Parques Nacionales Andino-Patagónicos. An. Parques Nac. 13: 1-122.

Fenner, M.J. 1978. A comparison of the abilities of colonizers and closed-turf species to establish from seed in artificial swards. J. Ecol. 66: 953-963.

Frenkel, R.E. 1977. Ruderal Vegetation along some California Roadsides. Ed. 2. 163 pp. University of California Press, Berkeley.

Gallopín, G. 1978. Estudio ecológico integrado de la cuenca del Río Manso Superior (Río Negro, Argentina). I Descripción general de la cuenca. An. Parques Nac. 14: 161-230.

Ghermandi, L. 1992. Caracterización del banco de semillas de una estepa del noroeste de Patagonia. Ecol. Austral. 2: 39-46.

Gobbi, M. 1994. Regeneración de la vegetación en incendios recientes de bosques de 'ciprés de la cordillera' (*Austrocedrus chilensis*) en el área del Parque Nacional Nahuel Huapi. Medio Ambiente (in press).

Grime, J.P. 1979. Plant Strategies and Vegetation Processes. 222 pp. John Wiley and Sons, Chichester.

Grime, J.P., Hodgson, J.G. and Hunt, R. 1988. Comparative Plant Ecology. 742 pp. Unwin Hyman, London.

Holm, L.G., Plucknett, D.L., Pancho, J.V. and Herberger, J.P. 1977. The World's Worst Weeds: Distribution and Biology. 609 pp. Hawai University Press, Honolulu.

Mack, R.N. 1989. Temperate grasslands vulnerable to plant invasions: characteristics and consequences. In: J.A. Drake, H.A. Mooney, F. di Castri, R.H. Groves, F.J. Kruger, M. Rejmánek and M. Williamson (eds.), Biological Invasions: A Global Perspective, pp. 155-179. John Wiley and Sons, Chichester.

Markgraf, V., Bradbury, J.P. and Fernández, J. 1986. Bajada de Rahue, Province of Neuquen, Argentina: an interstadial deposit in northern Patagonia. Palaeogeogr. Palaeoclimatol. Palaeoecol. 56: 251-258.

Markgraf, V. 1987. Paleoenvironmental changes at the northern limit of the Subantarctic *Nothofagus* forest, lat. 37 S, Argentina. Quat. Res. 28: 119-129.

Mazocca, A., Marsico, O.J. and Del Puerto, O. 1976. Manual de Malezas. Ed. 3. Hemisferio Sur. Buenos Aires.

Peet, R.K. 1974. The measurement of species diversity. Ann. Rev. Ecol. Syst. 5: 285-307.

Puntieri, J.G. 1991. Vegetation response on a forest slope cleared for a ski-run with special reference to the herb *Alstroemeria aurea* Graham *(Alstroemeriaceae)*, Argent. Biol. Conserv. 56: 207-221.

Ramírez, C., Correa, M., Figueroa, H. and San Martín, J. 1985. Variación del hábito y habitat de *Nothofagus antarctica* en el centro sur de Chile. Bosque 6: 55-73.

Salisbury, E. 1961. Weeds and Aliens. The New Naturalist. Collins, London.

Solomon, J.C. 1988. *Epilobium*. In: M.N. Correa (ed.), Flora Patagónica. Vol. 8. Part V. pp. 274-284. INTA, Buenos Aires.

Ter Braak, C.J.F. 1987. CANOCO - a FORTRAN program for canonical community ordination by (partial) (detrended) (canonical) correspondence analysis, principal components analysis and redundancy analysis. 90 pp. TNO, Wageningen.

Veblen, T.T., Ashton, D.H., Schlegel, F.M. and Veblen, A.T. 1977. Plant succession in a timberline depressed by vulcanism in south-central Chile. J. Biogeogr. 4: 275-294.

Veblen, T.T. and Lorentz, D.C. 1988. Recent vegetation changes along the forest/steppe ecotone of Northern Patagonia. Ann. Ass. Am. Geographers 78: 93-111.
Veblen, T.T. and Markgraf, V. 1988. Steppe expansion in Patagonia? Quat. Res. 30: 331-338.
Veblen, T.T., Kitzberger, T. and Lara, A. 1992. Disturbance and forest dynamics along a transect from Andean rain forest to Patagonian shrubland. J. Veget. Sci. 3: 507-520.
Willis, B. 1914. Northern Patagonia. Seribner Press, New York.

THE INVASIVE COMPONENT OF A RIVER FLORA UNDER THE INFLUENCE OF MEDITERRANEAN AGRICULTURAL SYSTEMS

M. Teresa Ferreira[1] and I.S. Moreira[2]
[1]*Departamento de Engenharia Florestal and* [2]*Departamento de Botânica e Engenharia Biológica, Instituto Superior de Agronomia, Tapada da Ajuda, 1399 Lisboa Codex, Portugal*

Abstract

The basin of river Sorraia spreads throughout Central Portugal and most of its area is occupied by cork-oak forest and by rice and other water-consuming Mediterranean crops. The vegetation of the river Sorraia and its direct tributaries were surveyed during the growth seasons of 1986-89. All species within the banks were recorded and separated into two groups, lotic (*i.e.*, aquatic and wetland) and terrestrial, both including an alien component. The invasiveness of South American alien *Myriophyllum aquaticum* was studied in detail in a river segment 23 km in length. The frequency of occurrence of alien plants found in the river increased downstream and can be related to habitat modifications and disturbance caused by the intensification of surrounding agricultural systems and human activities. Some aliens show a high degree of spatial invasiveness in river environments that are highly modified by agriculture, notably those affected by straightening and heavy siltation. The two alien species showing such invasiveness are *Paspalum paspalodes* and *Myriophyllum aquaticum*, which are currently displacing indigenous species in many Portuguese river systems.

Introduction

Botanists have designated the plants introduced from other parts of the world as aliens, adventive, introduced, exotic, invaders or weeds (Elton 1958; Forcella and Harvey 1983; Groves and Burdon 1986; Roy 1990; see Pyšek 1995 for a review). The terms adventive and weed are also used by herbologists to designate unwanted, potentially harmfull plants (Harlan and De Wet 1965; Godinho 1984; Salva and Bermejo 1988). In fact, alien plants frequently become weeds (Ashton and Mitchell 1989; Holzner 1982; Guilherm 1991), and their spreading potential is thought to be related to the anthropogenic disturbance of habitats (Salisbury 1964; Mack 1985; Fox and Fox 1986; Quézel *et al.* 1990; Lépart and Debussche 1991).

Relations between invasion and disturbance are complex and they depend on the type and frequency of the disturbance, the environmental constraints and the biology of the particular species concerned (Lépart and Debussche 1991). Agriculture, farming and associated activities are ubiquitous factors of disturbance and invasive plants can even be related to the type of cultivation and the duration of the rotation (Guilherm 1991). On the other hand, river habitats and riverine ecosystems have intrinsic features of intermediate to high disturbance (Wissmar and Swanson 1990) and are considered easily invasible (Di Castri 1990). Often, a large number of alien plants can be found in European inland waterways and rivers (Quézel *et al.* 1990). This paper attemps to evaluate possible relations between riparian alien plants and the intensity of agricultural activities in southern Portuguese areas.

Plant Invasions - General Aspects and Special Problems, pp. 117-127
edited by P. Pyšek, K. Prach, M. Rejmánek and M. Wade

Study area

The river Sorraia is the largest Portuguese tributary in the Tagus basin, draining a catchment area of 7652 km^2 and occupying a large part of the central region of the country (Fig. 1). Its direct tributaries are the rivers Erra and Sor draining from the northeast and Divor and Raia from the southwest. The basin of river Sorraia possesses Mediterranean features, with 80% of the annual rainfall occurring in a period of three winter months, with mild winters, an annual average summer temperature of 16-18°C, and natural summer drought during an average of four months per year. The basin spreads throughout a series of plains of low altitude (up to 300 m), mostly occupied by coarse and fine Cenozoic deposits and Quaternary alluvia. However, ancient geological materials like granites, quartzites and schists are found in some parts of the heads and middle courses.

About 60% of the area is used for forestry activities related to *Quercus suber, Pinus pinaster* and *Eucalyptus globulus.* However, the planaltic heads where miocenic materials dominate, a large part of the middle-river areas and all the lower valleys are used for agriculture, especially rice, maize, orchards and tomato production. Rice fields are the dominant landscape type in the floodplains of rivers Sorraia and Raia, and also in the heads of rivers Erra and Divor. Although the heads of Erra and Divor represent large crop areas of rice and maize, the number of sites surrounded by this type of land use increases downstream.

All these rivers are regulated (Fig. 1). A large part of the outflow from the reservoirs (Montargil, Maranhao, Furadouro and Gameiro) is distributed to the system of irrigation canals that cross the Sorraia Valley. The outflow of Divor reservoir irrigates the rice and maize fields in the heads of the Divor river. These and other smaller reservoirs also used for irrigation, control the river's flow, determine the number of weeks of river dryness, or flood the river bed for short periods during summer. Agricultural activities also affect the river system by direct manipulation of the river channels including straightening, dredging, embankment, weed cutting and cattle

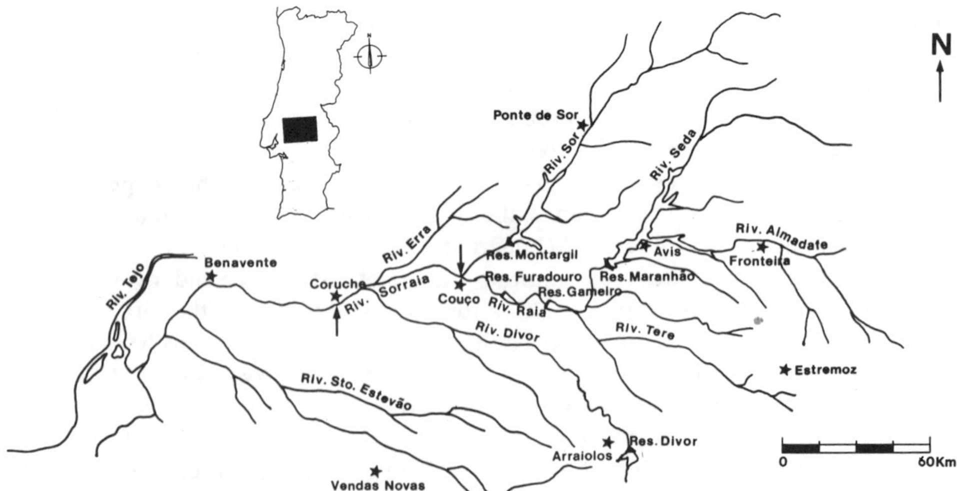

Fig. 1. Hydrographic basin of Sorraia showing the largest reservoirs. Human settlements are indicated by asterisks (only those whose population exceeds 2000 are shown). The river reach between Couco and Coruche, in which the distribution of *Myriophyllum aquaticum* was studied, is indicated by arrows.

grazing along many stretches. River water and sediments have high values of cations and nutrients and heavy siltation occurs in many sites under agricultural influence (Ferreira 1992).

Human settlements are small and scattered; Fig. 1 shows those with more than 2000 inhabitants. The industries are few and also small, most of them associated with the washing and processing of crops and farm animals. The human pressure upon the river corridors is thus essentially of agricultural origin.

Methods

In the period 1986-89, 116 sampling sites each of 250 m in length were established sequentially throughout the basin in rivers Erra (15 sites), Sor (36 sites), Divor (30 sites), Raia and Sorraia (35 sites). About 9% of the total river length was surveyed, the longest distance between sites being 7.3 km and average distance was 4.3 km. River width varied between 10-100 m.

All plant species present in the channel (inner bank, bed and water) were identified following Vasconcellos (1969), Tutin *et al.* (1964-80), Franco (1971, 1984), and Valdés *et al.* (1987). Based on Franco (1971, 1984), a separation was made between the group of aquatic and wetland plants, *i.e.*, lotic species, and a group of terrestrial plants also occurring within the river corridor. Lotic plants refer to those which are consistently associated with the water environment or areas of natural moisture. Alien species were distinguished in both groups. Relative frequency was expressed for each species as a percentage of the total number of sites in which the species was recorded. Density was estimated using a 5-degree scale, ranging from a few shoots (value 1) to dense stands (value 5). Cover was estimated visually in each site using a 5-degree scale, ranging from 1 (0-20%) to 5 (80-100%).

For each site, the distance to the source (DSOU) was determined and its order number ascribed according to Strahler (1957). A nominal variable was considered, expressing the type of land use (HUSE) surrounding each river segment: *(1)* forest and game; *(2)* forest, cattle and small orchards; *(3)* cattle and small areas of irrigated crops; *(4)* large areas of irrigated crops and rice fields.

The technique of Canonical Correspondence Analysis (CANOCO, Ter Braak 1987) and TWINSPAN (Hill 1979) were used to analyse the data. The reciprocal information analysis, proposed by Godron (1966, 1968) and described in Daget and Godron (1982), was used to relate alien species to their preferencial longitudinal position and the type of land use prevailing in the drainage area. The reciprocal information between a species and a variable indicates the strength of their association. When the species is equitatively distributed in all the classes considered for the variable, their reciprocal information value is low. The centroid of a species indicates its ecological preference and represents its maximum weighted occurrence in the environmental gradient contained in the classes considered for the variable. The reciprocal information between a species and a variable is given by an equation

$$RI = \sum_{1}^{nk} \frac{U(K)}{nr} \log 2 \left(\frac{U(K)}{R(K)} \frac{nr}{U(E)} \right) + \sum_{1}^{nk} \frac{V(K)}{nr} \log 2 \left(\frac{V(K)}{R(K)} \frac{nr}{V(E)} \right)$$

where nr is the number of sites sampled, R(K) the number of sites sampled in class k, U(K) and V(K) the number of presences and absences of a species in the class k of classes nk considered for the variable, and U(E) and V(E) the probability of presence or absence of the species. The centroid of a species is the mean coordinate weighted by the frequency of its records in each class considered for a variable, expressed as

$$C = \Sigma (N_i C_i) / \Sigma N_i$$

where C_i is the coordinate of the class i and N_i is the frequency of records in the class i. The DSOU classes considered were: *(1)* up to 5 km; *(2)* 6 to 20 km; *(3)* 21 to 50 km; *(4)* 51 to 80 km; *(5)* 81 to 120 km; *(6)* more than 120 km. The programme INFMUT was used for the reciprocal information analysis (Bacou and Lépart 1984).

Furthermore, the invasiveness of the alien species *Myriophyllum aquaticum* was studied during August 1990 on a 23 km long reach of the river Sorraia, between Couco and Coruche (Fig. 1) by complete mapping and measurement of its cover. This particular reach shows the typical lowland features of the Sorraia river. Total cover of indigenous hydrophytes and surface of the submerged bed were also determined. Cover was recorded along sections 0.5 km in length.

Table 1. Lotic (*i.e.*, aquatic and wetland) and terrestrial alien species of the Sorraia basin. Area of origin, life form, frequency of occurrence and invasiveness are given for each species. Relative frequency is shown in parentheses. Estimation of invasiveness was based on average density in sites where the species occurred.

Species	Origin	Life form	Occurrence	Invasiveness
Lotic				
Ammannia coccinea	Tropical America	annual	unfrequent (4%)	medium
Azolla caroliniana	Tropical America	perennial	frequent (18%)	low
Bidens frondosa	North America	annual	very frequent (35%)	medium
Cyperus eragrostis	Tropical America	perennial	very frequent (20%)	medium
Echinochloa oryzicola	SE and Central Asia	annual	rare (2%)	very low
Eclipta prostrata	Tropical America	annual	rare (2%)	low
Eichhornia crassipes	South America	perennial	unfrequent (3%)	high
Lindernia dubia	North America	annual	frequent (7%)	medium
Lindernia procumbens	Tropical Asia	annual	rare (<0.1%)	very low
Myriophyllum aquaticum	South America	perennial	very frequent (39%)	very high
Paspalum dilatatum	South America	perennial	rare (0.5%)	very low
Paspalum paspalodes	Tropical America	perennial	very frequent (48%)	very high
Salix fragilis	North America	perennial	rare (<0.1%)	very low
Terrestrial				
Acacia dealbata	SE Australia	perennial	rare (1.6%)	very low
Ailanthus altissima	East China	perennial	rare (<0.1%)	very low
Amaranthus albus	North America	annual	unfrequent (2.5%)	low
Amaranthus retroflexus	North America	annual	frequent (17%)	medium
Arundo donax	Asia	perennial	very frequent (16%)	medium
Conyza albida	Brasil	annual	rare (<0.1%)	low
Conyza bonariensis	South America	annual	frequent (11.4%)	medium
Conyza canadensis	North America	annual	unfrequent (1%)	Low
Cydonia oblonga	SW and Central Asia	perennial	rare (<0.1%)	very low
Datura stramonium	North America	annual	unfrequent (2.8%)	low
Eucalyptus camalduensis	Australia	perennial	rare (0.5%)	very low
Eucalyptus globulus	Tasmania	perennial	rare (<0.1%)	very low
Oxalis pes-caprae	South Africa	annual	rare (0.9%)	low
Phytolacca americana	North America	annual	rare (<0.1%)	very low

Results

A total of 437 species were found in the river corridors of the Sorraia basin, but only 213 were considered lotic species. Only 27 of these species (6.2%) were alien (Table 1). Of the 14 terrestrial aliens, five are phanerophytes occasionally established in the upper bank (*e.g., Acacia dealbata* and *Eucalyptus camalduensis*). The others are annuals, and include two species highly invasive on sand bars of lower courses, *Amaranthus retroflexus* and *Conyza bonariensis. Arundo donax* is the only species frequent throughout the basin and it can occupy considerable areas on banks.

Some of the lotic alien plants found have had little success as invaders of riverine habitats, like *Paspalum dilatatum* and *Lindernia procumbens,* which are only occasionally recorded. *Eichhornia crassipes* and *Azolla caroliniana* are flotants with optimum conditions in lentic habitats and can show large cover in situations of low water velocity, though the former needs greater depths to establish and occurs only in large rivers. *Lindernia dubia, Cyperus eragrostis* and *Ammannia coccinea* can achieve high densities on wet sand habitats. *Myriophyllum aquaticum* and *Paspalum paspalodes* show the highest frequency of occurrence and cover.

All aliens were preferentially associated with the mid and lower courses (Fig. 2) and with sites disturbed by agricultural activities (Fig. 3). Moreover, the aliens with greater invasiveness tend to have the high reciprocal information values towards HUSE.

Large deposits of fine materials (silt and organic mud) exist only below reservoirs, on the upper course of rivers Divor and Erra, associated with areas of rice and maize crops. The river is used as a drainage channel and is subject to straightening, nutrient enrichment, siltation and periodic dredging. On the other hand, beds dominated by

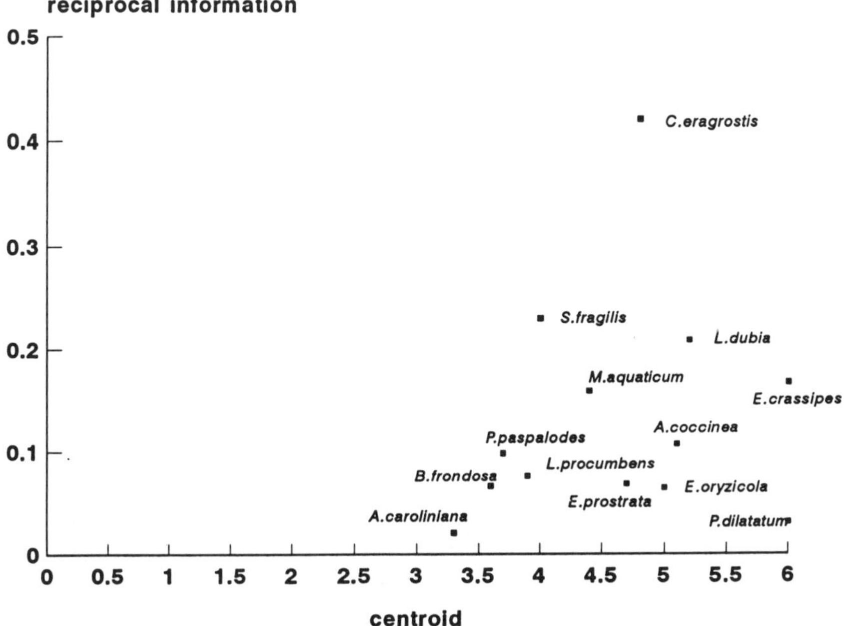

Fig. 2. Centroid position and reciprocal information between the alien lotic species and the variable distance to the source (DSOU). Full species names are given in Table 1.

sand materials characterise wide downstream sites which can have the same type of surrounding land use but also have important riparian zones and less direct interference. Five species show large cover in both environments. These are the indigenous

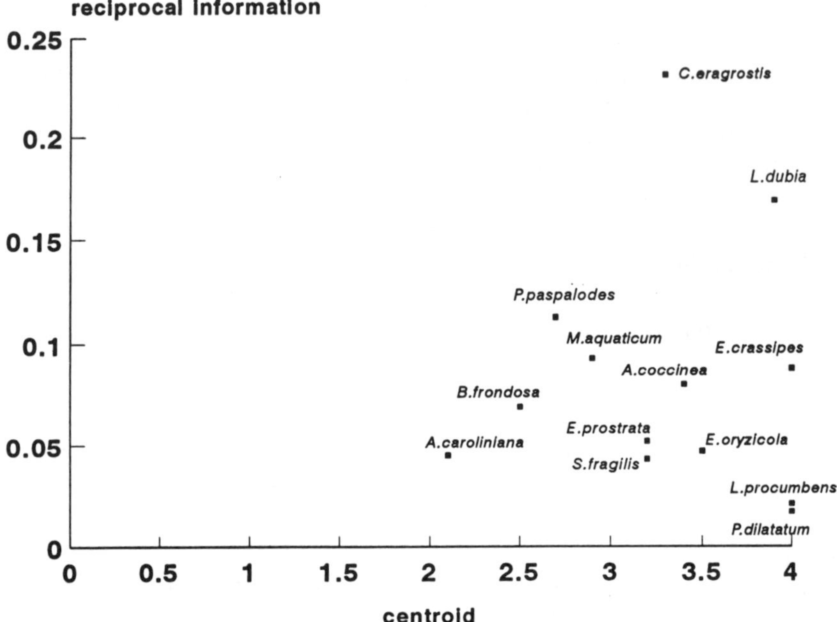

Fig. 3. Centroid position and reciprocal information between the alien lotic species and the variable land use (HUSE). Full species names are given in Table 1.

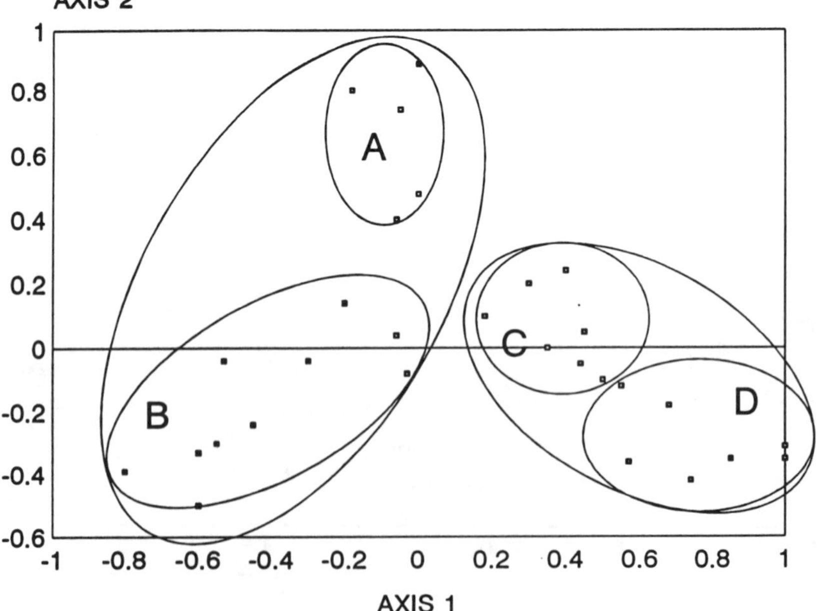

Fig. 4. Spatial distribution of sites studied on the river Divor, as yielded by CANOCO. Groups A, B, C and D represent TWINSPAN grouping at the second level of division. See text for explanation.

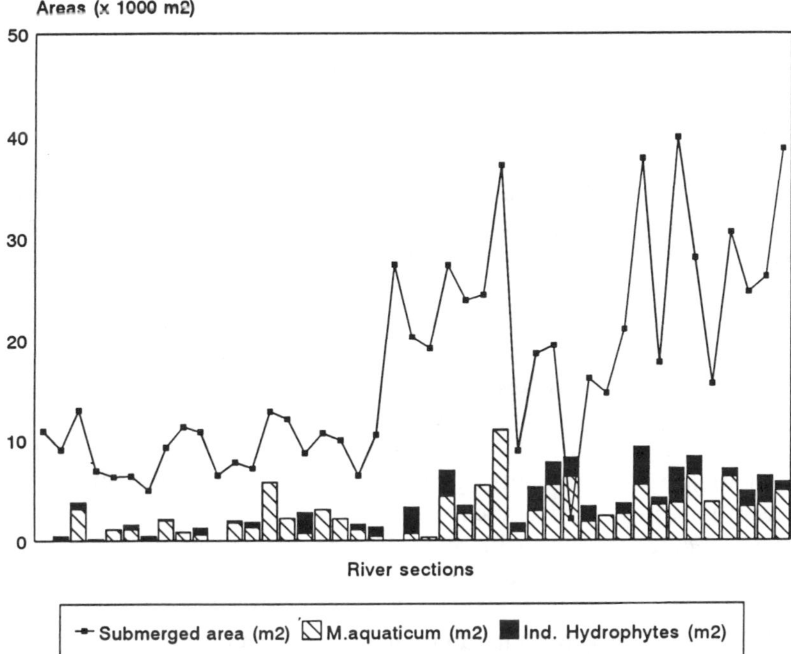

Fig. 5. Cover of the alien *Myriophyllum aquaticum* and the total cover of indigenous hydrophytes in 0.5 km sections of the reach between Couco and Coruche. Total submerged area is also shown. area.

Typha latifolia, *Sparganium erectum* and *Cyperus longus* but especially the aliens *Paspalum paspalodes* and *Myriophyllum aquaticum*. They cover completely the available surface near the margin and in water.

The abundance of these species and of a few others such as *Polygonum lapathifolium, Callitriche stagnalis* and the alien *Bidens frondosa* resulted in the floristic proximity of sites from the upper and lower course when multivariate treatments were performed. In the river Divor, five sites in the upper course have been highly modified for agricultural purposes. Fig. 4 shows the spatial distribution of all sites from the river Divor in the plain created by the first and second axis of the correspondence analysis, together explaining 33.9% of variability of the system. Also shown are the groups obtained by the hierarchic division through TWINSPAN.

On the second level of division, TWINSPAN separated four groups of sites: two groups disturbed by agriculture on the upper course (group A) and lower course (group B) and two groups from forested areas and low intensity agricultural use of drainage basins (groups C and D, respectively). However, the grouping of A and B was statistically stronger and occured in the first division. TWINSPAN indicator species for groups A and B were the alien species *Myriophyllum aquaticum* and *Bidens frondosa*.

Fig. 5 shows the cover of *Myriophyllum aquaticum* and the total cover of indigenous hydrophytes in the selected river reach. The total submerged river area is also shown. *M. aquaticum* dominated the indigenous hydrophytic flora in all but two sections; in several sections, it is the only hydrophyte present.

Discussion

Alien species are consistently associated with downstream sites in the Sorraia basin and with the intensification of agricultural activities surrounding the river corridor. The down-stream increase in number of alien species was also reported from the river Adour, Southern France (Tabacchi and Planty-Tabacchi 1994), and related to the human interference upon river corridors and fragmentation of the riparian forest (Chauvet and Décamps 1989), creating new habitats for herbaceous colonization. Exotic species tend to dominate the pioneer patches in the riparian zones after floods (Décamps 1993).

Tabacchi *et al.* (1990) found that about 13% of the species in the river Adour, Southern France, were aliens. In the present study, however, the total number of alien species recorded was relatively low. This paucity is probably because only the channel was considered, and the higher parts of the valley bottomland, such as floodplains and terraces, were excluded. Had the enlarged river corridor been examined, the proportion of aliens would probably increase due to the areas representing transition to the adjacent terrestrial environments. These are strongly affected by human activities, intensely cultivated and with many crop habitats, where S- and R-strategists (Grime 1977) are numerous. However, river ecosystems require more specialized life forms.

In central Europe, river valleys harbour an important number of alien species. For example, according to Sýkora (1990), two thirds of the Polish exotic flora can be found in riverine habitats. Some authors relate this peculiarity to the favourable microclimatic conditions of these systems (Quézel *et al.* 1990), others to the disturbance of natural riverbank communities (Sýkora 1990). However, Mediterranean rivers have highly variable flow conditions throughout the year and frequently no flow at all during summer. As Di Castri (1990) points out, Mediterranean ecosystems are highly resistant to invasions because of their geological and evolutionary history of frequent natural disturbances and early man-related history of exogenous disturbance.

Though 14 alien terrestrial species were found, most of them are rare or infrequent, with the exception of *Amaranthus retroflexus* and *Conyza bonariensis,* which occur frequently in lowland river habitats. When dry habitats occur within the river corridor, in lateral sand bars, islands or upper banks, they become occupied by non-riverine species. Most of the species appearing in these instream habitats are annual S- and R-strategists (Grime 1977), surviving on periodically flooded and waterlogged soils and germinating when the river alluvia and sand deposits dry out. This is the case for most terrestrial aliens found, while alien phanaerophytes are planted or associated with plantations and orchards.

Riverine alien plants like *Cyperus eragrostis* and *Bidens frondosa* are becoming progressively more common in the river margins. *Bidens frondosa* already shows a much larger distribution area than *Bidens tripartita,* its indigenous congener. Some of these alien species are common as crop-accompanying species in rice fields and orchards near-by the river (Rosa and Espírito-Santo 1984; Carvalho *et al.* 1993) namely *Ammannia coccinea, Echinochloa orizycola, Eclipta prostrata, Lindernia dubia* and *Paspalum dilatatum*, suggesting an on-going process of colonization. When wet and waterlogged areas occur inside the river corridor, hygrophytes and eventually some of the emergents occupy these habitats. They are usually C- and R-strategists, with optimum development during limosal conditions and subsequent subsistence through long-term terrestrial conditions.

Many of the alien species found were introduced into Mediterranean areas between the 16th and 18th centuries, namely *Ailanthus altissima, Paspalum paspalodes, Acacia dealbata, Eucalyptus globulus, Amaranthus retroflexus, Conyza canadensis* and *Oxalis pes-caprae* (LeFloch' 1991). Others are common invaders throughout the world, such as *Eichhornia crassipes* and *Myriophyllum aquaticum* (Ashton and Mitchell 1989).

Aquatic alien species were few, being however more important in terms of cover than species richness. Two perennial species, *Myriophyllum aquaticum* and *Paspalum paspalodes* show the greater frequency of occurrence and the highest invasive potential. These species were the most ubiquitous and important invaders found in the Sorraia basin and in Portuguese rivers and channels in general, at the expense of indigenous hydrophytes and near-shore graminoid assemblages. They are C-strategists, strong competitors for space with the ability to grow rapidly, produce large amounts of biomass, and, under favourable conditions, exploite optimally habitat resources .

Relatively little is known about the date of introduction and the rates of plant spread in the Mediterranean basin (LeFloch' 1991) and pratically no consistent records exist from Portugal. Many plants show a rather extended latent phase from the time of their introduction to the time when they become invasive. *Myriophyllum aquaticum* was first reported in mid-thirties, but it was as late as the seventies when it became invasive in all Portuguese river lowlands (Duarte *et al.* 1984). No earlier records are available from the Sorraia region for the invasive aquatic aliens. However, natural plant communities were probably dominated by indigenous species which are still present, such as *Potamogeton fluitans, P. crispus* and *Ceratophyllum demersum*. Nowadays their total cover rarely exceeds that observed for *Myriophyllum aquaticum*.

The extension of the geographical range of a species has been studied by some authors, while others have searched for the characteristics of a successful weed (Elton 1958; Baker 1965; Fox and Fox 1986; Ashton and Mitchell 1989; Roy 1990). Comparatively fewer authors have studied the role of environmental characteristics and type of habitats conditioning the success of invasion (Forcella and Harvey 1983; Bergelson *et al.* 1993; Pyšek and Prach 1993). For example, affinity to riparian habitats differed among four alien species studied during particular phases of invasion in central Europe (Pyšek and Prach 1993). Most authors agree that spreading potential is related to human disturbance, habitat instability and the initial stages of plant succession (Fox and Fox 1986; Lépart and Debussche 1991). Aquatic systems disturbed by human activities appear to be particularly vulnerable to plant invasions (Ashton and Mitchell 1989) and lowland rivers and river corridors are even more disturbed habitats, often exposed to enormous human pressure. The role of river ecotones and river habitats in the invasive behaviour of alien species should receive more attention, namely in order to evaluate the relative importance of the stochasticity of invasions, interactions between alien and indigenous species and the type and importance of disturbance regimes.

References

Ashton, P.J. and Mitchell, D.S. 1989. Aquatic plants: patterns and modes of invasion, attributes of invading species and assessment of control programmes. In: J.A. Drake, H.A. Mooney, F. di Castri, R.H. Groves, F.J. Kruger, M. Rejmánek and M. Williamson (eds.), Biological Invasions: A Global Perspective, pp. 111-154. John Wiley and Sons, Chichester.

Bacou, A.M. and Lépart, J. 1984. INFMUT, Dispositif et Mode de Utilisation en Libre Service de la Bibliothéque Infeco Implantée sur Micro-ordinateur Corail en Vue de la Gestion et du Traitement de Fichiers de Rélevés Phyto- et Zoo-écologiques. 280 pp. C.N.R.S., Montpellier.

Baker, H.G. 1965. Characteristics and modes of origin of weeds. In: H.G. Baker and G.L. Stebbins (eds.), The Genetics of Colonizing Species, pp. 147-169. Academic Press, New York.

Bergelson, J., Newman, J.A. and Floresroux, E.M. 1993. Rates of weed spread in spatially heterogeneous habitats. Ecology 74: 999-1011.

Carvalho, C.J., Espírito-Santo, M.D. and Moreira, I. 1993. Estudo preliminar das infestantes vernais de pomares de citrinos de Santiago do Cacém. Actas do I Congresso de Citricultura, pp. 451-461. Lisboa.

Chauvet, E. and Décamps, H. 1989. Lateral interactions in a fluvial landscape: the river Garonne, France. J. North Am. Benthol. Soc. 8: 9-17.

Daget, P. and Godron, M. 1982. Analyse Frequencielle de l'Écologie des Éspèces dans les Communautés. 282 pp. Masson, Paris.

Décamps, H. 1993. River margins and environmental changes. Ecol. Applic. 3: 441-445.

Di Castri, F. 1990. On invading species and invaded ecosystems: the interplay of historical chance and biological necessity. In: F. di Castri, A.J. Hansen and M. Debussche (eds.), Biological Invasions in Europe and the Mediterranean Basin, pp. 3-16. Kluwer Academic Publ., Dordrecht.

Duarte, C., Agusti, S. and Moreira, I. 1984. Water hyacinth (*E. crassipes* Marts Solms.) and watermilfoil (*M. aquaticum* (Vell.) Verdc.) in Portugal. In: Proceedings of the European Weed Research Society, 3rd Symposium on Weed Problems in the Mediterranean Area, pp. 667-674. Lisboa.

Elton, C.S. 1958. The Ecology of Invasions by Animals and Plants. 181 pp. Methuen, London.

Ferreira, M.T. 1992. Estrutura e dinâmica das comunidades de macrófitos aquáticos da bacia hidrográfica do Sorraia. 402 pp. Thesis. Instituto Superior de Agronomia, Lisboa.

Forcella, F. and Harvey, S.J. 1983. Eurasian weed infestation in Western Montana in relation to vegetation and disturbance. Madrono 30: 102-109.

Fox, M.D. and Fox, B.J. 1986. The susceptibility of natural communities to invasion. In: R.H. Groves and J.J. Burdon (eds.), Ecology of Biological Invasions: An Australian Perspective, pp. 57-66. Cambridge University Press, Sydney.

Franco, J.A. 1971. Nova Flora de Portugal (Continente and Acores), Vol. I. 648 pp. Author's edition. Lisbon.

Franco, J.A. 1984. Nova Flora de Portugal (Continente and Acores), Vol. II. 660 pp. Author's edition. Lisbon.

Godinho, I. 1984. Les définitions d'adventice et de mauvaise herbe. Weed Res. 24: 121-125.

Godron, M. 1966. Application de la théorie de l'information a l'étude de l'homogenéité et de la structure de la végétation. Oecol. Plant. 2: 187-197.

Godron, M. 1968. Quelques applications de la notion de fréquence en écologie végetale (recouvrement, information mutuelle entre espèces et facteurs écologiques, échantillonage). Oecol. Plant. 3: 185-212.

Grime, J. 1977. Evidence for the existence of three primary strategies in plants and its relevance to ecological and evolutionary theory. Am. Natur. 111: 1169-1194.

Groves, R.H. and J.J. Burdon (eds.), 1986. Ecology of Biological Invasions: An Australian Perspective. Cambridge University Press, Cambridge.

Guilherm, J.L. 1991. Weed invasions in agricultural areas. In: R.H. Groves and F. di Castri (eds.), Biogeography of Mediterranean Invasions, pp. 379-392. Cambridge University Press, Cambridge.

Harlan, J.R. and De Wet, J.M.J. 1965. Some thoughts about weeds. Econ. Botany 19: 16-24.

Hill, M.O. 1979. TWINSPAN, a Fortran Program for Arranging Multivariate Data in a Ordered Two-Way Table by Classification of the Individuals and Attributes. 47 pp. Cornell University, Ithaca, NY.

Holzner, W. 1982. Concepts, categories and characteristics of weeds. In: W. Holzner and M. Numata (eds.), Biology and Ecology of Weeds, pp. 3-20. Dr. W. Junk Publ., The Hague.

LeFloch' E. 1991. Invasive plants of the Mediterranean Basin. In: R.H. Groves and F. di Castri (eds.), Biogeography of Mediterranean Invasions, pp. 67-80. Cambridge University Press, Cambridge.

Lépart, J. and Debussche, M. 1991. Invasion processes as related to succession and disturbance. In: R.H. Groves and F. di Castri (eds.), Biogeography of Mediterranean Invasions, pp. 159-178. Cambridge University Press, Cambridge.

Mack, R.N. 1985. Invading plants: their potential contribution to population biology. In: J. White (ed.), Studies on Plant Demography, pp. 127-142. Academic Press, London.

Pyšek, P. 1995. On the terminology used in plant invasion studies. In: P. Pyšek, K. Prach, M. Rejmánek and M. Wade (eds.), Plant Invasions - General Aspects and Special Problems, pp. 71-81. SPB Academic Publ., Amsterdam.

Pyšek, P. and Prach, K. 1993. Plant invasions and the role of riparian habitats: a comparison of four species alien to Central Europe. J. Biogeogr. 20: 413-420.

Quézel, P., Barbero, M., Bonin, G. and Loisel, R. 1990. Recent plant invasions in the circum-mediterranean region. In: F. di Castri, A.J. Hansen and M. Debussche (eds.), Biological Invasions in Europe and the Mediterranean Basin, pp. 51-60. Kluwer Academic Publ., Dordrecht.

Rosa, M.L. and Espírito-Santo, M.D. 1984. Evolution des mauvaises herbes des rizières portugaises. In: Proceedings of the European Weed Research Society, 3rd Symposium on Weed Problems in the Mediterranean Area, pp. 411-417. Lisboa.

Roy, J. 1990. In search of the characteristics of plant invaders. In: F. di Castri, A.J. Hansen and M. Debussche (eds.), Biological Invasions in Europe and the Mediterranean Basin, pp. 335-332. Kluwer Academic Publ., Dordrecht.

Salisbury, E. 1964. Weeds and Aliens. 388 pp. Collins, London.

Salva, A.P. and Bermejo, J.E. 1988. Concepto de mala hierba. ITEA 75: 47-69.

Strahler, A.N. 1957. Quantitative analysis of watershed geomorphology. Am. Geophys. Union Trans. 38: 913-920.

Sýkora, K.V. 1990. History of the impact of man on the distribution of plant species. In: F. di Castri, A.J. Hansen and M. Debussche (eds.), Biological Invasions in Europe and the Mediterranean Basin, pp. 37-50. Kluwer Academic Publ., Dordrecht.

Tabacchi, E. and Planty-Tabacchi, A. 1994. Évolution longitudinale de la végetation du corridor de l'Adour. Botánica Pirenaico-Cantábrica 5: 455-468.

Tabacchi, E., Planty-Tabacchi, A. and Décamps, O. 1990. Continuity and discontinuity of the riparian vegetation along a fluvial corridor. Landscape Ecol. 5: 9-20.

Ter Braak, C.J.F. 1987. CANOCO, a Fortran program for canonical community ordination for (parcial) (detrended) (canonical) correspondence analysis, principal component analysis and redundancy analysis. Version 2.1. 90 pp. IWIS-TNO, Wageningen.

Tutin, T.G., Heywood, V.H., Burges, N.A., Moore, D.M., Walentine, D.H., Walters, S.M. and Webb, D.A. 1964-1980. Flora Europaea. Vol. I-IV. Cambridge University Press, London.

Valdés, B., Talarera, S. and Fernandez-Galiano, E. 1987. Flora Vascular da Andaluccia Occidental. 1680 pp. Ketres Editora, Barcelona.

Vasconcellos, J. 1969. Plantas Aquáticas, Anfíbias e Ribeirinhas. 164 pp. Ministério da Agricultura, Lisboa.

Wissmar, R.C. and Swanson, F.J. 1990. Landscape disturbances and lotic ecotones. In: R. Naiman and H. Décamps (eds.), The Ecology and Management of Aquatic-Terrestrial Ecotones, pp. 65-89. Man and the Biosphere Series, Vol IV. The Parthenon Publ. Group, Paris.

III
BIOLOGY AND ECOLOGY OF INVASIVE PLANTS

THE INVASIVE NATURE OF *FALLOPIA JAPONICA* IS ENHANCED BY VEGETATIVE REGENERATION FROM STEM TISSUES

John H. Brock[1], Lois E. Child[2], Louise C. de Waal[3] and Max Wade[2]
[1]*School of Planning and Landscape Architecture, Arizona State University, Tempe, AZ 85287-2005, USA;* [2]*International Centre of Landscape Ecology, Department of Geography, Loughborough University, Loughborough, Leicesteshire LE11 3TU, United Kingdom;* [3]*School of Applied Sciences, Division of Environmental Science, University of Wolverhampton, Wolverhampton WW1 1SB, United Kingdom*

Abstract

Three greenhouse experiments were conducted with *Fallopia japonica* (Japanese knotweed) during the months of May, June, and September 1992, on plant material collected from Loughborough, Leicestershire in the United Kingdom. The experiments tested the hypothesis that *F. japonica,* known to regenerate from rhizomes, could produce new plants from stem tissues. Three portions of stems were harvested with each having two nodes. One segment was cut just above the soil surface, the middle segment at about 0.5 m height, and the upper segment from greater or equal to 1.0 m height. Treatments common to all experiments were stems in water, stems buried with a slight covering of potting media, and stems oriented vertically with one node buried and one node aerial. Stems in water, and those buried in vermiculite/peat potting media provided about a 60% success rate in generating new shoot growth. Shoots placed vertically produced on average 11% shoot regeneration. Stems in shade had slightly greater shoot regeneration rate compared with those subjected to sunlight. Adventitious roots were produced near the base of some new shoots about 21 days following initial growth. Stem segments harvested in the autumn produced both vegetative and floral shoots. Floral shoots were observed on about 20% of the nodes. Based on these results and given good microclimate conditions, stems of *F. japonica* can produce new shoots, roots and regenerate new plants. As stem tissue can be readily transported by water and human activities, the potential to form new colonies of *F. japonica* from stem tissues needs to be recognized by land managers.

Introduction

Fallopia japonica (Houtt.) Ronse Decraene (Japanese knotweed) native to Japan, Taiwan, and northern China is a member of the *Polygonaceae* family. It is also known by the synonyms *Reynoutria japonica* Houtt. and *Polygonum cuspidatum* Sieb. and Zucc. (Bailey 1989). This alien species is widely distributed in the British Isles (Conolly 1977), and much of central and northern Europe (Tutin *et al.* 1964). In North America, it is known from northwestern California (Hickman 1993), into the Pacific Northwest (Figueroa 1988), in several provinces of Canada and in many of the eastern and northcentral states of the United States that are dominated by temperate deciduous forest ecosystems (Fernald 1950). The plant was brought into the United States in 1890s from England for its ornamental value (Pridham and Bing 1975).

F. japonica is highly invasive and recognized as a noxious weed in most of its distribution. King (1966) states that it is "one of the most persistent and aggressive of all perennial weeds". It is most common on sites disturbed by human activity such as industrial areas, transportation corridors, and sites where soil movement during construction is common. It is a special problem along water courses where native,

Plant Invasions - General Aspects and Special Problems, pp. 131-139
edited by P. Pyšek, K. Prach, M. Rejmánek and M. Wade

riparian vegetation can be replaced by dense stands of *F. japonica* (Child *et al.* 1992). While recognized as very invasive, in the British Isles, it does not produce viable seeds (Bailey 1989) as only female plants have been recorded to date. In the United States, Meade and Locandro (1979) noted no seedling regeneration, indicating a lack of sexual reproduction.

The primary propagules of *F. japonica* are parts of its extensive rhizome system, of which even small pieces can establish new plants (Palmer 1990, Beerling 1990). Brock and Wade (1992) found 40% regeneration of new shoots from rhizome segments 1 cm in length, with as little as 0.7 g fresh weight. Given the large amount of below ground level biomass has been calculated to be 14,000 kg/ha to a depth of 25 cm (Brock 1994), the potential to spread *F. japonica* by movement of soil material is substantial.

Up to 1992, when this research was initiated at the International Centre of Landscape Ecology, Loughborough University of Technology, United Kingdom, information about stem vegetative regeneration of *F. japonica* was conjectural or anecdotal. Figueroa (1988) stated that shoots from parts of *F. japonica* stems can develop into plants, which was apparently based on field observation of stems floating in water. Locandro (1978) observed that even internode stem tissue on occasion produced adventitious roots, but did not mention regeneration of shoot material.

The purpose of this research was to determine whether or not *F. japonica* could produce new shoots from stem tissue. If regeneration from stem tissue were proved it would help to explain the presence of apparently juvenile stands of *F. japonica* found in established landscapes without recent importation, or movement of soil material.

Material and methods

Stems of *F. japonica* were collected 7 May 1992 on the banks of Wood Brook (near Forest Road) (National grid reference SK 523 183) in the town of Loughborough, Leicestershire, United Kingdom, and on 8 June and 2 September 1992 from a roadside stand on Snell's Nook Lane west of Loughborough (National grid reference SK 504 174). Experiments were conducted for about 30 days to test regeneration capability of *F. japonica* from harvests of stem tissues made in late spring, early summer, and autumn. Greenhouse planting was completed either on the day of, or the day following stem harvest.

Stem material was divided into three segments, each with two nodes, to represent lower (L), middle (M), and upper (U) portions of the stems. The lower segments were cut below the first node above ground level to include the next node, the middle segments were cut at approximately 0.5-1.0 m height, and the upper segments were taken from >1.5 m height characterised by that part of the plant with active branching. Segments and internode length and stem diameter (cm) were measured and fresh weight (g) of the stem segments determined. Segments of each stem portion (L, M, U) were planted in flats (approximately 30 x 40 x 10 cm) filled with vermiculite and potting soil; segments were grouped by portion (L, M, U) in units of 5 and randomly assigned to 3 replicate positions within the flats, giving $N = 15$ observations per treatment. Air temperatures were kept at 22 to 25°C. Ambient light was used for the 'sun' treatments, while shade treatments consisted of flats placed below the greenhouse benches. Specific treatments in the three experiments (May, June, September) are presented in Table 2. The different treatments included segments laid horizontally on the surface of the greenhouse soil mix, segments placed in a vertical position with one node buried and one node exposed to the air (burial/aerial), horizontal segments partially buried, and those designated as buried were covered with a few milimetres of soil mix. The water treatment consisted of floating labelled stem segments in a flat. The water was changed every 2 or 3 days to keep it fresh and discourage algal growth. During the process of the experiments, some treatments were eliminated and others added (see Table 2) as observations were made and new questions posed. All watering was undertaken using rain water at room temperature.

Following evidence of new shoots, data collected on alternate days included date of shoot emergence, height of stems, and number of leaves per shoot. Observations during the 30-35 day growth periods also included presence of adventitious roots forming on the new shoots and presence of floral structures. The experiments were harvested 30-35 days following planting.

Data were entered to Statview or Systat spreadsheets and analyzed using descriptive statistics. Analysis of variance was used where appropriate. Where significant differences were indicated at P=0.05, t-tests and least significant difference (lsd) statistics were calculated to aid mean separation (Steel and Torrie 1960).

Results and discussion

Pattern of vegetative growth

The average length of *F. japonica* stem segments cut from the upper portion of the parent shoot was approximately 4 cm less in length and the diameter 0.9 cm less compared to dimensions of segments from the lower shoot. Average fresh weight of lower stem segments was 4.2 times greater compared with upper stem segments (Table 1). Internode length averaged 15 cm for all segments, with the middle segments having the greatest average length of 19 cm.

Table 1. Average length, diameter, and fresh weight of segments of *Fallopia japonica* stems harvested in May and June 1991 from stands in the vicinity of Loughborough, Leicestershire, United Kingdom.

	Stem segment		
Characteristic	Lower	Middle	Upper
Length (cm)	23.0	22.8	18.7
Diameter (cm)	1.9	1.6	1.0
Fresh weight (g)	53.7	35.6	12.8

While there are conjectural reports that *F. japonica* might produce new shoots from internode tissues (Locandro 1978), this was not confirmed with these greenhouse trials. None of the internode segments produced vegetative growth except for some signs of adventitious root growth which was limited to internode material located near the soil surface. These growth points initiated no further growth over the 30-day observation period.

New vegetative growth was only observed in the presence of axillary buds. New shoots were found to begin emerging from axillary buds within 4 days following planting (Fig. 1a). The modal value for new shoot emergence from the *F. japonica* axillary buds was 10 days.

Buds on stem segments that produced new shoots in general had one new shoot per node, however in a minority of cases, several new shoots emerged from the same node (Fig. 2). Since each stem segment contained two nodes, shoot regeneration data were recorded with relation to node position. Across all experiments, 85% of the upper nodes produced new shoots, while 73% of the lower buds produced new shoots. These data were not significantly different in a pooled *t*-test (Steel and Torrie 1960). During the course of the experiments many of these new shoots died, either from dessication of the parent tissues, or in some cases the new shoot showed disease symptoms similar to 'damping off' conditions. The average survival of shoots formed from axillary buds 30 days' post-planting across all experiments was 33%.

Data were also summarized as to regeneration success for lower, middle, and upper stems segments across the three experiments. Mean shoot regeneration was very similar, with values of 33, 33, and 29% calculated for lower, middle and upper stem segments. Since there were no significant differences, data presented in later sections of this paper have been pooled by node and stem segment position of the parent stem.

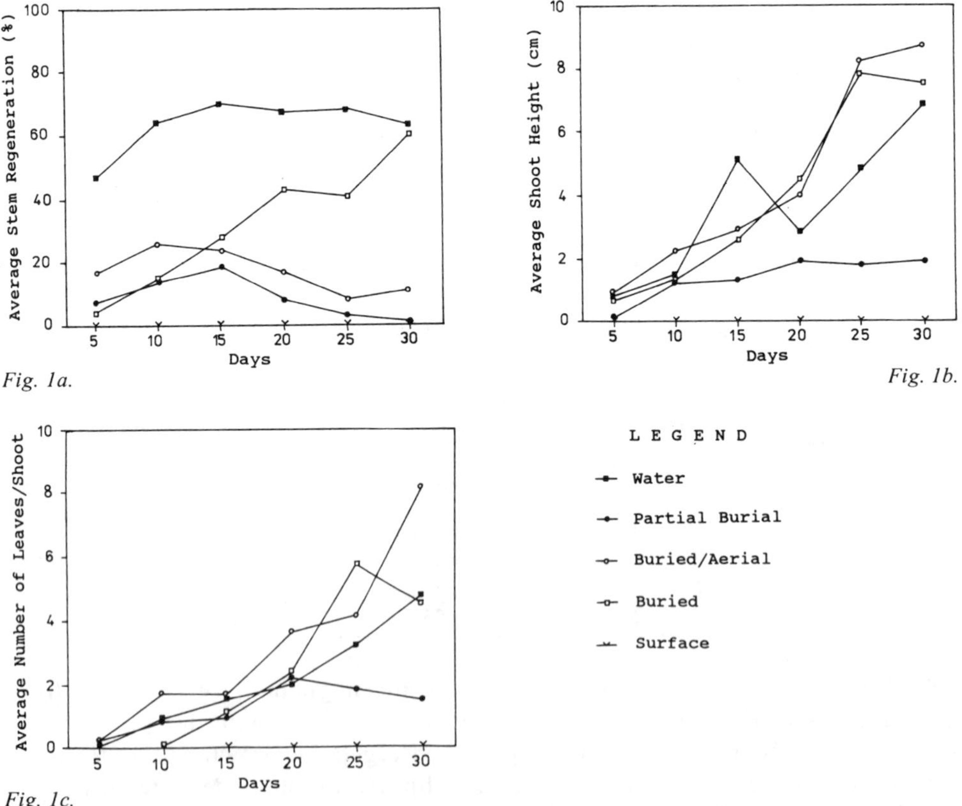

Fig. 1. Effects of treatments on regeneration and shoot development of *Fallopia japonica* stem segments. *(a)* Regeneration success expressed as a percentage of stem segments producing new shoots. *(b)* Average shoot height (cm). *(c)* Average number of leaves per shoot. Days following planting in a greenhouse are shown. Data are averages of late spring, early summer and autumn experiments.

Fig. 2. Shoot regeneration of *Fallopia japonica* from a segment of stem tissue. Note the presence of adventitious roots forming near the base of the new shoots.

Treatment effects

Shoot regeneration

Regeneration of shoots of *F. japonica* was greatest for stem segments in the water and buried treatments (Fig. 1a), in either sun or shade (Table 2). The pooled average regeneration success for the water treatment was 63% (range = 50-80%). Shoot regeneration success of stem segments that were buried continued to increase over the 30 days. The pooled average regeneration for the buried treatment was at 60% (range = 39-73%). Stem segments assigned to the buried/aerial and particularly buried treatments had decreasing levels of regeneration success with time (Fig. 1a).

Shoot regeneration was never realized for stem segments placed on the surface of the soil mix. While some buds developed shoot growth, these new shoots wilted and died usually within a few days. Stem segments placed on the surface were brown and desiccated within about 14 days. Partially buried segments produced a low number of shoots. No shoots survived the late spring experiment and only 3% survived in the early summer planting (Table 2). Stem segments with any aerial exposure had significantly lower regeneration which was also true for the stems assigned to the bur-

Table 2. Treatments applied to segments of *Fallopia japonica* stems in three greenhouse experiments conducted in late spring (May), early summer (June), and autumn (September) 1992. Data are mean values (n=15) for percentage shoot regeneration (Shoots), height (cm), and number of leaves per shoot (Leaves) at approximately 30 days following planting. When comparing particular parameters among treatments and across experiments, mean values followed by the same letter are not significantly different at P=0.05 using the lsd statistical comparison.

Experiment:	Late spring	Early summer	Autumn
Treatment:	Water	Water	Water-sun
Shoots	50.0^{a}	58.0^{a}	80.0^{a}
Height	10.4^{b}	9.0^{b}	3.6^{b}
Leaves	7.5^{ab}	5.6^{b}	2.3^{b}
Treatment:	Partially buried	Partially buried	Water-shade
Shoots	0.0^{c}	3.0^{b}	64.0^{a}
Height	0.0^{c}	3.9^{b}	1.5^{bc}
Leaves	0.0^{c}	1.7^{b}	1.3^{bc}
Treatment:	Buried/aerial	Buried/aerial	Buried/aerial
Shoots	14.0^{b}	5.0^{b}	13.0^{b}
Height	27.2^{a}	4.8^{b}	2.5^{b}
Leaves	13.3^{a}	4.7^{b}	1.1^{bc}
Treatment:	Surface	Buried	Buried-sun
Shoots	0.0^{c}	39.0^{ab}	73.0^{a}
Height	0.0^{c}	7.4^{b}	7.5^{b}
Leaves	0.0^{c}	5.3^{b}	3.7^{b}
Treatment:		Surface	Buried-shade
Shoots		0.0^{c}	67.0^{a}
Height		0.0^{c}	3.4^{b}
Leaves		0.0^{c}	3.7^{b}

ied/aerial treatment. The pooled average regeneration of new shoots for buried/aerial segments was 10.6% (Table 2). While regeneration was minimal, some of the most vigorous growth was observed from new shoots in the buried/aerial treatment. This was primarily from the buried lower node where the new shoot developing with adventitious roots made direct contact with the soil material.

Shade treatments did not benefit shoot regeneration. Average regeneration of new shoots was 22% in sun treatments compared to 24% in shade treatments (Table 2).

Dispersal by water of *F. japonica* stems in riparian habitats is likely to be a common event and could be aiding the expansion of the range of this invasive plant. This dispersal strategy should be of concern to any personnel managing river corridors. Likewise, any vegetation/land management activity during the growing season that would result in a shallow covering of stems of *F. japonica* may also contribute to its spread. The additive effect of water transported stems covered by sediments or debris, during a flood event, could further ensure the development of new colonies of *F. japonica*. However, based on this greenhouse research, stems floating in water and then deposited on a surface, such as a sand bar, may be less successful in generating new colonies because stem segments placed on a surface without covering desiccated.

Shoot height

Average shoot height increased rapidly for those *F. japonica* plants regenerating in the buried/aerial and buried treatments (Fig. 1b). The most rapid shoot increase was observed to be made on shoots that had successfully generated adventitious roots. The maximum shoot of about 80 cm was measured in the late spring buried/aerial treatment. That treatment contained several rapidly growing shoots, which heavily influenced the average height values, especially considering its low percentage regeneration rate. The maximum shoot height observed in the autumn experiment was 52.5 cm from a shoot in the buried shade treatment. The overall mean height of shoots in the three experiments was approximately 12 cm.

Because of the large variance associated with the shoot height data, only large statistical differences among treatments could be found (Table 2). The greatest average shoot height (27.2 cm) was found in the late spring buried/aerial treatment. This growth was primarily the result of successful rooting and vigorous shoot development of a few new shoots emerging from the buried node. Average shoot heights for stem segments in water were comparable for the late spring and early summer (10.4 and 9.0 cm, respectively), and, as for all treatments, shoot height was lowest in the autumn experiment (Table 2).

A growth rate of 2.9 mm/day was calculated for new shoots developing from segments of *F. japonica* under these greenhouse conditions. A report from Iowa State University (Anonymous 1980) reported an infield growth rate of 50 to 100 mm in 24 hours. Wolf (1971) reported a mean shoot increase of *F. japonica* of 46.5 mm/day. In both of those cases, the shoots were emerging from an intact rhizome system and plant colony. Brock and Wade (1992) reported that shoots regenerating from *F. japonica* rhizome segments had a daily height increase of 4.8 mm.

Number of leaves

Leaf formation lagged behind the rate of stem elongation. The average number of new leaves per shoot was less than two at 15 days following planting, and in the successful treatments rapidly increased over the duration of the study (Fig. 1c). The buried/aerial treatment had the greatest number of leaves at 30 days following planting, because of the successful development of a few robust shoots. A more usual response was for new shoots to have 4 to 5 leaves by the end of the experimental period, as shown by data in the water and buried treatments (Fig. 1c).

The number of leaves was greatest in the late spring, and lowest in the autumn experiments. This could be due to the production of floral shoots as an immediate response rather than vegetative growth of the stem segments harvested in the autumn. The shift in growth response was most likely triggered by a decreasing photoperiod.

Root regeneration

Adventitious roots developed at the base of new shoots and became evident at about 20 days into the experiments. On some stem segments, adventititious roots developed from root primordia evident on the harvested stem. Root development in the spring experiment occurred most commonly in the water and buried/aerial treatments. The presence of adventitious roots greatly aided shoot development and growth, especially shoots in the buried/aerial treatment. The highest percentage of adventitious root generation was observed in the autumn experiment for stem segments buried in the soil mix exposed to sun (Table 3).

Summarizing the three experiments, overall adventitious root development, regardless of vigour, was 9.6% in the late spring experiment, 12.8% in the early summer study, and 27% in the autumn experiment.

A value of 4.8% for shoots of about 30 days of age with vigorous root development was reported by Brock and Wade (1992) for new shoots developing from *F. japonica* rhizome segments. The increase in adventitious root development in the autumn experiment could be from a seasonal change in growth regulators in the stems of *F. japonica*.

Table 3. Percentage adventitious root development on new shoots of *Fallopia japonica* from stem segments placed in a greenhouse as influenced by time and treatments. (- = treatment not included in experiment.)

	Experiment		
Treatment	Late spring	Early summer	Autumn
Water-sun	20	27	10
Water-shade	-	-	14
Buried-sun	-	27	59
Buried-shade	-	-	43
Buried/aerial	22	9	9
Partially buried	0	0	-
Surface	0	0	0

Floral stems

Many of the *F. japonica* stem segments in the autumn experiment gave rise to flowering shoots. Initially, the water-sun and water-shade treatments produced 25% and 21% flowering shoots, respectively. These treatments, however, had only 10% and 4% floral shoots at harvest time. A similar trend was observed for the other treatments, except in the buried-sun treatment where 7% floral shoots were observed emerging initially and had 7% floral shoots when the experiment ended.

The position of the bud (lower or upper) on the stem segment did not appear to influence the presence or absence of floral shoots. Floral shoot development was predominantly from the upper portion of the *F. japonica* parent stem. The overall values for development of floral shoots were 3% for lower, 17% for middle, and 33% for upper stem segments.

Conclusions

This research documented that *F. japonica* can produce new shoots from stem tissues. The most successful media for shoot regeneration was water. At the end of the 30-day observation periods, the height of new shoots was approximately 6 cm and each shoot supported 4 to 5 leaves. Stem segments harvested in any part of the growing season produced some new shoots but the autumn samples gave the greatest number. No difference in shoot regeneration success was found among the different locations of stem fragments from the parent stem (lower, middle, upper), and between buds located on the lower or upper node position of the stem segment.

Even without viable seeds (Bailey *et al.* 1995), *F. japonica* has a potentially successful strategy for reproduction from cut stems. In the present study, the overall average shoot regeneration was 33%, with 16% of those new shoots having developed adventitious roots. This provides a stem segment of *F. japonica* containing at least two intact buds with a potential of 5.3% viability to produce a new shoot with adventitious roots. Brock (unpublished data) found the average fresh weight of *F.*

japonica in the autumn to be 1511 g/m^2 (or 937 g/m^2 oven-dried). The average fresh weight of *F. japonica* stem segments used in these experiments was 34 g. The conservative potential for a stand of *F. japonica* is to produce (1511 g.m^{-2}/34 g^{-1}=) 44 stem segments/m^2. Given a viability factor of 5.3%, a yield of 2.3 new shoots/m^2 could be expected.

The stem regeneration potential for new clones of *F. japonica* clearly exists, and is enhanced when moist weather conditions and cutting treatments are juxtaposed during the growing period. This is not uncommon in the United Kingdom, and throughout the known European and North American distribution of *F. japonica*. There is a need to develop a comprehensive and integrated vegetation management strategy for this invasive species, with an associated precaution to prevent its spread by both stem and rhizome tissues.

References

Anonymous. 1980. Japanese polygonum (*Polygonum cuspidatum* Sieb. and Zucc.). Weed Control. Co-operative Extension Service E5. Iowa State University, Ames, IA.

Bailey, J.P. 1989. Cytology and Breeding Behaviour of Giant Alien *Polygonum* Species in Britain. PhD Thesis, University of Leicester.

Bailey, J.P., Child, L.E. and Wade, M. 1995. Assessment of the genetic variation and spread of British populations of *Fallopia japonica* and its hybrid *Fallopia* x *bohemica*. In: P. Pyšek, K. Prach, M. Rejmánek and M. Wade (eds.), Plant Invasions: General Aspects and Special Problems, pp. 141-150. SPB Academic Publ., Amsterdam.

Beerling, D.J. 1990. The Ecology and Control of Japanese Knotweed (*Reynoutria japonica* Houtt.) and Himalayan Balsam (*Impatiens glandulifera* Royle) on River Banks in South Wales. 138 pp. Thesis, University of Wales, Cardiff.

Brock, J.H. 1994. Technical note: standing crop of *Fallopia japonica* in the autumn of 1991 in the United Kingdom. Preslia 66(4).

Brock, J.H. and Wade, P.M. 1992. Regeneration of Japanese knotweed (*Fallopia japonica*) from rhizomes and stems: observations from greenhouse trials. In: Proceedings of the 9th International Symposium on the Biology of Weeds, pp. 85-94. Dijon.

Child, L.E., De Waal, L., Wade, P.M. and Palmer, J.P. 1992. Control and management of *Reynoutria species* (Knotweed). Aspects Appl. Biol. 29: 295-307.

Conolly, A.P. 1977. The distribution and history in the British Isles of some alien species of *Polygonum* and *Reynoutria*. Watsonia 11: 291-311.

Fernald, M.L. (ed.) 1950. Gray's Manual of Botany. Edn. 8. American Book Company, New York.

Figueroa, P.F. 1988. Japanese Knotweed (*Polygonum cuspidatum*), a Potential Noxious Weed. Weyerhaeuser Research Report. Technical Report No. 050-3910/9, 9 pp.

Hickman, J.C. (ed.) 1993. Higher Plants of California. University of California Press, Berkeley.

King, L.J. 1966. Weeds of the World: Biology and Control. Interscience Publ., New York.

Locandro, R.R. 1978. Weed watch, Japanese bamboo - 1978. Weeds Today 9: 21-22.

Meade, J.A. and Locandro, R.R. 1979. Japanese bamboo. The Journal. Cooperative Extension Service, pp. 1-4. Rutgers University, New Brunswick.

Palmer, J.P. 1990. Japanese knotweed (*Reynoutria japonica*) in Wales. In: The Biology and Control of Invasive Plants, pp. 80-85. Proceedings of the Conference of the Industrial Ecology Group of the British Ecological Society. Cardiff.

Pridham, A.M.S. and Bing, A. 1975. Japanese bamboo. Plants and Gardens 31: 56-57.

Steel, R.G.D. and J.H. Torrie. 1960. Principles and Procedures of Statistics. McGraw-Hill, New York.

Tutin, T.G., Heywood, V.H., Burges N.A., Vallentine, D.H., Walters, S.M. and D.A. Webb (eds.) 1964. Flora Europaea 1. Cambridge University Press, Cambridge.

Wolf, F.T. 1971. The growth rate of *Polygonum cuspidatum*. J. Tennessee Acad. Sci. 46: 80.

ASSESSMENT OF THE GENETIC VARIATION AND SPREAD OF BRITISH POPULATIONS OF *FALLOPIA JAPONICA* AND ITS HYBRID *FALLOPIA* X *BOHEMICA*

John P. Bailey[1], Lois E. Child[2] and Max Wade[2]
[1]*Department of Botany, Leicester University, Leicester LE1 7RH, United Kingdom;*
[2]*International Centre of Landscape Ecology, Department of Geography, Loughborough University, Loughborough, Leicestershire LE11 3TU, United Kingdom*

Abstract

The management of *Fallopia japonica* (Japanese knotweed) and the hybrid *Fallopia* x *bohemica* is posing serious problems to environmental managers throughout Europe. The ability to differentiate between these plants and the knowledge of the genetic constitution of these taxa are important in the understanding of the autecology of these plants and would be very useful in the context of any proposed biological control programme. The identification, on morphological grounds, of the taxa is discussed and useful differential features are described for British material. The results of a survey of the distribution of the hybrid *F.* x *bohemica* in British Isles are also presented. Results are reported on the first attempts of the application of Polymerase Chain Reaction Random Amplified Polymorphic DNA (PCR RAPD) techniques to *F.* x *bohemica* and *F. sachalinensis* material from the British Isles, and *F. japonica* from the British Isles, Japan, and the United States of America. In the British Isles, *Fallopia japonica* var. *japonica* has a very limited genetic base.

Introduction

Since the introduction of *Fallopia japonica* var. *japonica* (Japanese knotweed) from Leiden in the 1840s (Beerling *et al.* 1994), the spread of this vigorous invasive rhizomatous perennial plant has caused growing concern to environmental managers throughout Europe (De Waal *et al.* 1994). Another taxon *F.* x *bohemica (F. japonica* var. *japonica* x *F. sachalinesis*) is gradually emerging as a small but significant component of the Japanese knotweed population. Many problems are caused by these taxa, from restricting access to stream and riverbanks and damaging flood defence structures to outcompeting and excluding native vegetation and blighting potential development land. These plants have extended their range in many European countries to become problems in urban amenity areas, on wasteground, railway embankments and road verges (Child *et al.* 1992).

In this paper *Fallopia* section *Reynoutria* covers *F. japonica* and its varieties, *F.* x *bohemica* and *F. sachalinensis*. The term Japanese knotweeds covers only *F. japonica* var. *japonica* and *F.* x *bohemica*.

Although the spread of these species is largely by vegetative means, with the potenial of 0.7 g of *F. japonica* rhizome to give rise to a new plant (Brock and Wade 1992); the role of seed dispersal should not be overlooked. Hybrids have been recorded in the wild at sites where both parents, *F. japonica* and male-fertile *F. sachalinensis* were present. This posed several questions. Had the hybrid originated spontaneously from seed in the wild or had it been spread vegetatively from discarded rhizome material? If only hybrid seed is available in the British Isles, what role does

Plant Invasions - General Aspects and Special Problems, pp. 141-150
edited by P. Pyšek, K. Prach, M. Rejmánek and M. Wade

such seed play in the increasing distribution of the plants? How widespread is the hybrid throughout the British Isles and should its distribution also be a cause for concern for environmental managers?

A joint project between the universities of Leicester and Loughborough was initiated to discover the extent of genetic variation within the populations of *Fallopia* section *Reynoutria* taxa and to assess the distribution and invasion potential of the hybrid *F.* x *bohemica* in the British Isles.

In Britain all *Fallopia japonica* var. *japonica* plants examined so far bear only male-sterile flowers (Bailey 1994) and are octoploid with 88 chromosomes. There is also a very limited range of morphological variation found when considered against that found in native Japanese plants as determined from herbarium specimens. These facts lead to the conclusion that the British population has a very limited genetic base. Two important considerations stem from this observation, firstly biological control agents would be more effective if the populations were clonal. Secondly, the presence of enormous stands of male-sterile plants has led to hybridization with related taxa on a massive scale, mainly with *Fallopia baldschuanica* (Bailey 1988).

Bailey (1989) describes the breeding behaviour of many large *Fallopia* species in the British Isles. Due to the apparent lack of male fertile plants in the British Isles any seed set on *F. japonica* plants is a product of hybridization.

Hybridization and the role of seed production in the spread of these taxa

Due to the limited genetic variation found in British populations of *Fallopia* section *Reynoutria* taxa it is fairly easy to understand the hybrids that are produced between the Japanese and giant knotweeds. Only the male-sterile form of *F. japonica* var. *japonica* is known, whilst both *F. japonica* var. *compacta* and *F. sachalinensis* exist as hermaphrodite and male-sterile (female) plants. Some years ago a series of artificial hybrids involving *F. japonica* var. *japonica* (male-sterile), *F. japonica* var *compacta* (male-sterile and hermaphrodite) and *F. sachalinensis* (male-sterile and hermaphrodite) were made at Leicester and have reached maturity. These have been invaluable in determining the constitution of the various hybrid clones found in the British Isles.

Although *F. japonica* var. *japonica* is known to occur at both the tetraploid and octoploid chromosome levels (2n = 44, 2n = 88) it is only the octoploid that is known

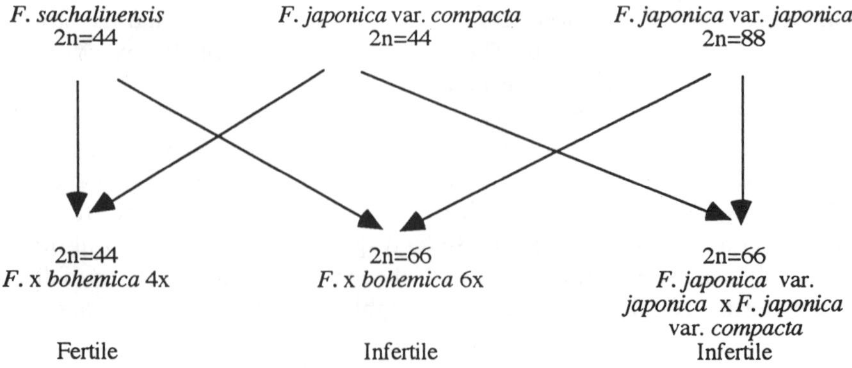

Fig. 1. Hybridization between *F. japonica* and *F. sachalinensis*.

in Britain. *F. japonica* var. *compacta* and *F. sachalinensis* in contrast are both tetraploid in the British Isles (Bailey and Stace 1992). It can then be seen (Fig. 1), that hybrids between *F. japonica* var. *japonica* and *F. japonica* var. *compacta* and between *F. japonica* var. *japonica* and *F. sachalinensis* can be readily recognised on the basis of their intermediate chromosome number of 2n = 66. The great majority of British hybrid plants can, on cytological and morphological grounds, be attributed to the hybrid constitution *F. japonica* var. *japonica* x *F. sachalinensis* = *F.* x *bohemica* 6x.

In addition to the hexaploid hybrids, it is also possible to have tetraploid forms of *F.* x *bohemica* (Fig. 1). If the result of a hybridization event in Britain the hybrid would be expected to be of the constitution *F. japonica* var. *compacta* x *F. sachalinensis* (since *F. japonica* var. *compacta* is the only tetraploid *F. japonica* known from this country). The hybrid *F. japonica* var. *japonica* x *F. japonica* var. *compacta* is known only from one site in Cornwall.

The role of seed in the spread of hybrid taxa

It is well established that in Britain at least, *F. japonica* var. *japonica* is spread exclusively by vegetative means by transport of stem and rhizome fragments, and that any seed produced is hybrid. Large amounts of viable seed are indeed produced by *F. japonica* var. *japonica* in years without early frosts, but this seed is almost exclusively the result of pollination by *Fallopia baldschuanica* (Bailey 1988). So although *F. japonica* var. *japonica* cannot reproduce itself by seed in Britain, the seed that it produces may be important in the spread of hybrid *Fallopia* taxa. At the very limited number of sites where *F. japonica* var. *japonica* grows adjacent to a hermaphrodite *F. sachalinensis,* viable seed of *F.* x *bohemica* is produced. At one such site, Caerynwch Hall in Wales, *F.* x *bohemica* plants at various stages of maturity were recorded and young seedlings were found growing through the considerable accumulation of leaf litter. As these plants were growing next to a river bank there is a good chance of hybrid seed being dispersed downstream.

Seed collected from gardens is often a source of hybrid plants. Since *F. japonica* and *F. sachalinensis* are self-incompatible (Bailey 1994), seed collected from open-pollinated plants is invariably hybrid. Seed said to be from *F. japonica* on botanic garden seed exchange lists is often *F.* x *bohemica*, and in the early stages of this work was a serious source of confusion. Again any seed saved by gardeners in this country would be expected to be hybrid and this must be an important source of hybrid plants, particularly if they are raised in a glasshouse and planted out later.

Hybrid seed may also be produced by further rounds of hybridization via backcrossing of existing hybrids with either *F. japonica* var. *japonica* or *F. sachalinensis,* or indeed between hybrids (Fig. 2). Examples of seed production from such sources are known, though there are as yet no reports of germination in the wild. Bailey and Stace (1991) reported that whilst all the hexaploid hybrids had a very low fertility the tetraploid hybrids had a fertility comparable with the tetraploid species. It has been found that, in spite of the very low fertility of these hexaploid hybrids, small amounts of fertile pollen are produced that are in turn capable of producing viable seed. Plants produced by backcrossing with the parental taxa could be expected to exhibit an almost continuous range of morphological variation between the original

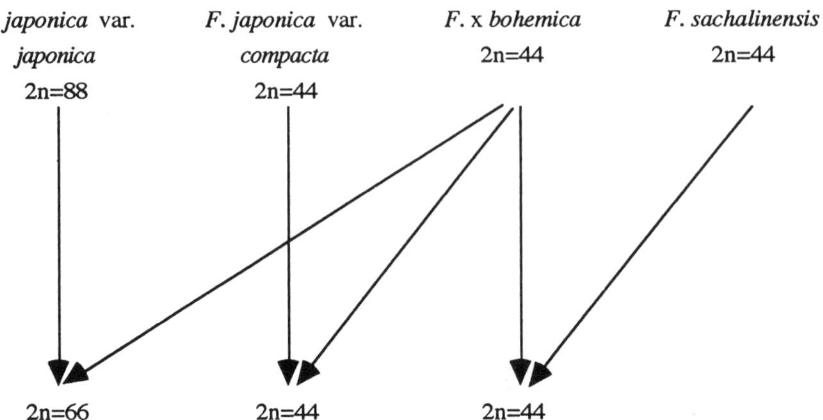

Fig. 2. Possible backcross derivatives of tetraploid *F.* x *bohemica.*

parents, making identification of taxa very difficult. It is thought that such a series of secondary hybridizations account for the unusual range of plants found in and around Cirencester, Gloucestershire.

As mentioned earlier, the main type of seed produced in Britain is hybrid with *F. baldschuanica* and it is possible that large amounts of viable hybrid seed are formed, but what is the fate of such seed? Certainly some of it is eaten by birds, and plants in Leicester have been seen completely stripped of their seed by *Passer domesticus* (house sparrows). Very few reports of germinating seed or seedlings are known from Britain, and work at Leicester is currently looking into the conditions governing both the germination and survival of these seeds in the wild. It must be stressed that it is the critical first winter that is important, since the mature plants are quite hardy. Only one hybrid is known with *F. baldschuanica* and that is in London where it has established itself naturally (Bailey 1988).

Factors affecting the spread of *F. japonica* hybrids in Britain are firstly lack of suitable pollen, very few male-fertile clones are found, and secondly the edaphic and climatic factors which seem to determine whether the seed germinates in the first place and whether it survives the winter. It is possible that in countries with a different climate and greater abundance of male-fertile clones, seed production is an important factor in the spread of these hybrids. The only continental clones that have been examined cytologically and morphologically, at Aachen in Germany, did not fit in with the pattern of variation that characterizes British populations (J. Bailey, unpublished data).

Hybrid survey

A survey to assess the distribution of hybrids in the British Isles was carried out over a single growing season with the assistance of Botanical Society of the British Isles (BSBI) members along with subscribers to the Arboricultural Advisory and Information Service (AAIS). A postal survey was conducted which consisted of a recording sheet giving the characteristics of the two parent species (*F. japonica* var. *japonica* and *F. sachalinensis*) and the hybrid *F.* x *bohemica.*

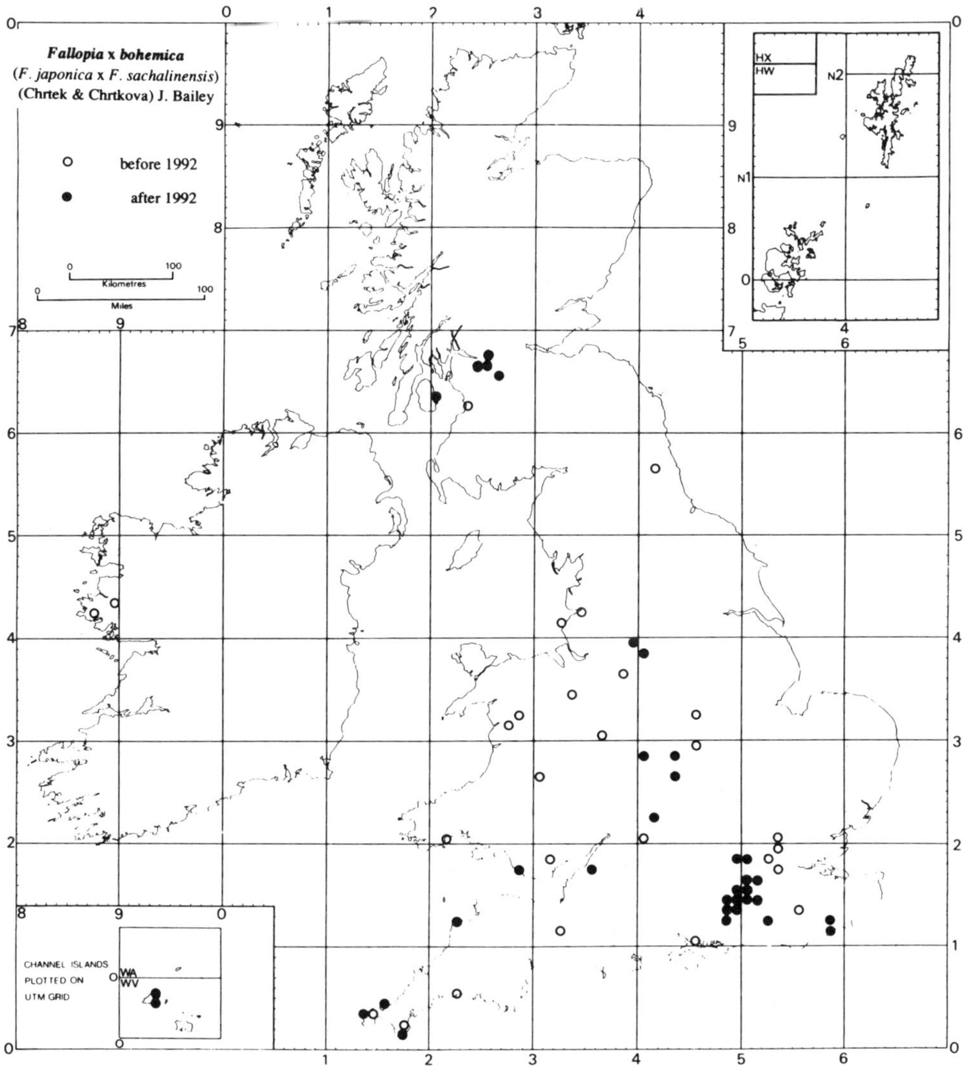

Fig. 3. Distribution map of *Fallopia* x *bohemica* in the British Isles.

A total of 4,700 questionnaires were circulated and from the responses, thirty three new *F.* x *bohemica* locations were identified throughout the British Isles. The distribution of the hybrid was plotted on a standard Biological Records Centre, Institute of Terrestrial Ecology, base map, with presence recorded on a 10 km square basis. The results can be seen in Fig. 3. A number of 10 km squares contain more than one record and the survey highlighted the presence of hybrid clusters in certain areas such as Surrey; Cirencester, Gloucestershire; Caerynwch Hall and Amroth, Wales.

Identification of *Fallopia* x *bohemica* plants

The main features for identification of the hybrid *F.* x *bohemica* and its parents, *F. sachalinensis* and *F. japonica* var *japonica* are shown in Table 1.

Table 1. Identification of *Fallopia* section *Reynoutria* taxa.

	Species		
	Fallopia sachalinesis (F. Schmidt ex Maxim.) Ronse Decraene 2n = 44	*Fallopia japonica* var. *japonica* (Houtt.) Ronse Decraene Syn. *Reynoutria japonica* 2n = 88	*Fallopia* x *bohemica* (Chrtek and Chrtková) J. Bailey (*F. sachalinensis* x *F. japonica*) 2n = 66 or 44
Height			
	Striking, gigantic plant to 4 m tall	Large plant, 2-3 m tall	Habit intermediate, 2.5-4m tall
Leaf characters			
	Basal leaves ovate to oblong, base cordate	Leaves ovate, acuminate, base truncate	Leaves intermediate in size and shape, tip acuminate weakly to moderately cordate at base
	Up to 40 cm long and 22 cm wide	10-15 cm long	Up to 23 cm long and 19 cm wide
	Length: width ratio *c.* 1.5	Length: width ratio 1-1.5	Length: width ratio 1.1-1.8
	Undersides of leaves with scattered, long, flexuous hairs (trichomes)	Undersides of leaves entirely glabrous	Undersides of larger leaves with numerous short, stout hairs (trichomes), easily visible with a hand lens
Sex expression			
	Male-fertile flowers (with exserted anthers) and male-sterile flowers (with small, empty included anthers and well developed stigmas) borne on separate plants	Flowers usually male-sterile	Male-fertile and male-sterile flowers borne on separate plants

Apart from the difference in leaf shape noted by Chrtek and Chrtková (1983), there are a number of more cryptic features that may be used (Bailey and Conolly 1991). These are all based on examination of the morphology of the lower epidermis of the leaf. Firstly, the leaf of *F. sachalinensis* is much larger and thinner than that of *F. japonica*; probably as a consequence of this the *F. sachalinensis* leaf has a very much thicker cuticle, and this is visible in both light and SEM pictures as a distinctive reticulate pattern. *F. japonica* in comparison has a smooth cuticle, and the few ridges or reticulations found are restricted to the vicinity of the stomata. *F.* x *bohemica* has epidermal striations that are clearly intermediate. The second morphological difference is in the trichomes of the lower epidermis. *F. sachalinensis* has scattered long hairs, one cell wide and up to 14 cells long, which are clearly visible to the naked eye. They are lightly cutinized and tend to collapse in dried specimens. The microscopic appearance of *F. japonica* trichomes is of single heavily cutinized cells slightly protruding from the leaf veins, they are not visible to the naked eye. The *F.* x *bohemica* (4x and 6x) hybrids have trichomes that are completely different from those of either parent. The combination of heavy striae from the *F. japonica* parent with the additional length of the *F. sachalinensis* hairs create a unique range of trichomes, often with a distinct swollen basal cell and projecting at an acute angle. These hairs or prickles are usually distributed evenly over the veins of the lower epidermis and can be seen by eye, especially if the leaf is bent double and a vein viewed against the light. It is not currently possible to distinguish the 4x and 6x cytotypes of *F.* x *bohemica* except by chromosome number.

With all these morphological characters it is important that comparable regions of leaf are used and only fully mature large stem leaves are used as the differences are not so clearcut in the upper leaves, especially those subtending the inflorescence. These characters have been confirmed by comparison with the type specimen of *F. sachalinensis* and the Houttyn type specimen of *F. japonica*.

It is important to emphasize that whilst these differences work well on British populations it is by no means certain that they would be equally applicable to mainland Europe. The cytological and morphological characters do not give a fine enough degree of discrimination to allow the identification of say, a particular clone of *F.* x *bohemica* (6x). In order to be able to chart the spread of such clones and to assess the likely success of future biological control, alternative techniques such as those of molecular ecology need to be employed.

Molecular markers

Whilst gross morphology, sex expression and chromosome number give some clues to the identity and potential origin of the different clones, it is necessary to develop a method with more discriminatory power for establishing the precise genetic make-up of a particular clone, whether it be *F. japonica* var. *japonica, F. sachalinensis* or *F.* x *bohemica.* Various techniques exist with the potential for answering such questions. Two of these, isoenzymes and PCR RAPDs have been employed in the current research programme. These two techniques operate at different levels in the molecular hierarchy, isoenzymes at the enzyme level, RAPDs at the more fundamental DNA sequence level. Both techniques aim to pinpoint polymorphisms which are useful in marking a particular genotype.

DNA techniques

A number of researchers have used RAPDs for examining the extent of genetic variation in populations including some which are vegetatively propagated (Chalmers *et al.* 1992).

In this study, serious problems were encountered in extracting clean DNA from *Fallopia* taxa, and after a series of experiments a modification of the CTAB method of Doyle and Doyle (1987) was employed.

For the RAPD reaction 2.5 μl of sample DNA with a concentration between 1 and 20 ng/μl was used, with 0.5 unit Boehringer Taq polymerase per reaction. Operon primers F2 and F4 gave good results:

Cycle 1	94°C for 5 min
Cycles 2-46	94°C for 30 sec
	35°C for 1 min
	72°C for 2 min

Reactions were then kept at 72°C for 5 min and held at 25°C, before running out on a 2% agarose gel and visualization with ethidium bromide.

All accessions of octoploid *F. japonica* var. *japonica,* gave a remarkably similar banding pattern (3 main bands with the slowest being the most intense) (Fig. 4, lanes 4, 6, 7, 8, 9, 10 and 12). Most notable were the American accessions P169 and P170.

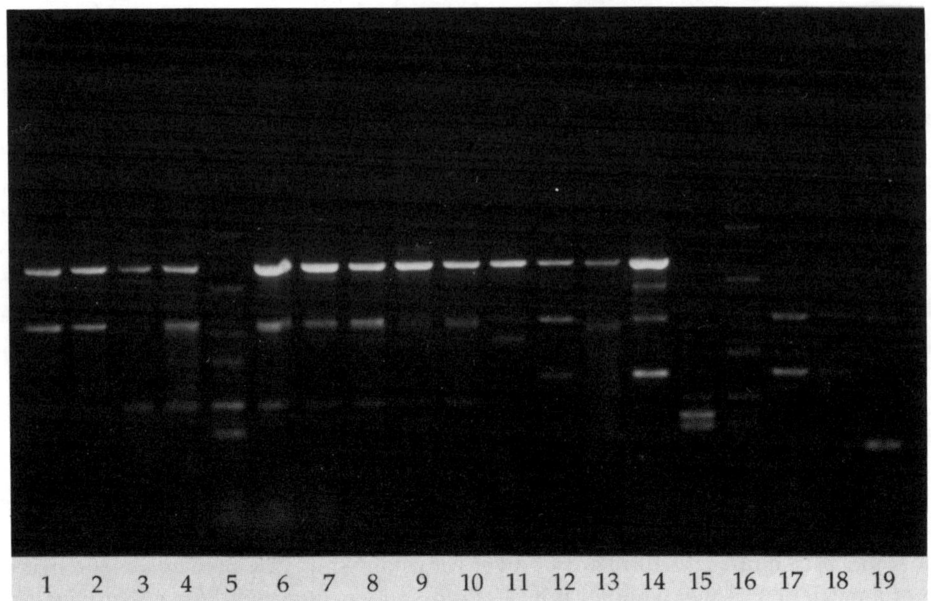

Fig. 4. RAPD gel of *Fallopia* section *Reynoutria* taxa.

Lane	Code	Taxon	Chr. No.	Locality
1	BR7	*F.* x *bohemica*		Brithdir, Wales
2	BR6	*F.* x *bohemica*		Brithdir, Wales
3	BR5	*F.* x *bohemica*		Brithdir, Wales
4	BR3	*F. japonica* var. *japonica*		Brithdir, Wales
5	BR1	*F. sachalinensis*		Brithdir, Wales
6	P199	*F. japonica* var. *japonica*	2n = 88	Chienhall, Cornwall
7	P192	*F. japonica* var. *japonica*	2n = 88	Swansea, Wales
8	P170	*F. japonica* var. *japonica* male-sterile	2n = 88	New York
9	P169	*F. japonica* var. *japonica* male-fertile	2n = 88	New York
10	P12	*F. japonica* var. *japonica*	2n = 88	Leicester
11	P164	*F. japonica* x *F. baldschuanica*	2n = 54	Haringay, London
12	P105d	*F. japonica* var. '*uzenesis*'	2n = 88	Japanese seed
13	P78a	*F. jap.* var. *compacta* x *F. sachalinensis*	2n = 44	Artificial hybrid
14	P134	*F. japonica* var. '*terminalis*'	2n = 44	Hachijo Is. Japan
15	P57	*F. sachalinensis* male-sterile	2n = 44	Howey, Wales
16	P55	*F. sachalinensis* male-fertile	2n = 44	Nant-Y-Frith, Wales
17	P193	*F. japonica* var. *compacta*		Heads of Valley, Wales
18	P174	*F. japonica* var. *compacta*	2n = 44	Brecon, Wales
19	P2	*F. japonica* var. *compacta*	2n = 44	Bracken Hill, Kent

The male-sterile plant gave an identical banding pattern to the British accessions, but the unique male-fertile clone had an additional slow-moving band. P164, a hybrid between *F. japonica* var. *japonica* and *F. baldschuanica* was also distinguished by a single additional band to the standard *F. japonica* var. *japonica* pattern. The *F.* x *bohemica* hybrids (lanes 1, 2, 3 and 13) are not readily distinguishable with this primer and generally resemble the *F. japonica* var. *japonica* pattern more than the *F. sachalinensis* pattern.

The *F. sachalinensis* pattern with this primer gives as many as 8 major bands (lanes 5, 15 and 16). Lanes 15 and 16 represent male-sterile and hermaphrodite plants respectively, there is a greater difference in banding pattern between these two accessions than there is between those in lanes 5 and 16.

Lanes 15-17 show 3 different British *F. japonica* var. *compacta* accessions, as in the case of *F. sachalinensis* there are significant differences in banding between these plants. Lanes 12 and 14 are from wild collected seed from Japan, and are listed uncritically under the varietal names which they were sent under. They share two bands with the octoploid *F. japonica* var. *japonica* plants, one of which is also shared with two of the *F. japonica* var. *compacta* accessions.

Future prospects

Behind the dramatic spread of *F. japonica* var. *japonica* (Conolly 1977) lies the lesser known and frequently overlooked hybrid *F.* x *bohemica*. A survey form sent to British botanists has added more than 30 new sites of this hybrid, some of them occupying considerable areas. There is no reason to suspect that the hybrid is any less invasive than *F. japonica* var. *japonica*, in addition it is frequently male-fertile and thus a source of pollen for back-crossing with either parent. There is no doubt that the distribution map (Fig. 3) is a gross underestimate of the occurrence of this taxon in the British Isles. Clusters of records in London, Surrey and Cornwall represent the efforts of groups of keen local botanists capable of identifying the hybrid with confidence. If such groups were active throughout the whole country a great many more hybrid clones would be discovered. Studies are also urgently needed into the relative vigour, resistance to herbicides and invasibility of *F.* x *bohemica* compared to *F. japonica* var. *japonica*.

It can be seen from the very preliminary RAPD results that male sterile octoploid *F. japonica* var. *japonica* examined, from a wide geographical range has the same banding pattern. This is in stark contrast to *F. japonica* var. *compacta* and *F. sachalinensis* where much smaller samples gave a considerable variation in banding pattern. This is strong, but incomplete evidence towards the limited nature of the genetic base of *F. japonica* var. *japonica* in the British Isles. RAPDs are clearly a powerful tool in the elucidation of genetic base of Japanese knotweeds, and with suitable primers it should be possible to detect not only *F.* x *bohemica* but also look at introgression of the *F. sachalinensis* genome into *F. japonica* var. *japonica,* a task which is probably impossible to achieve by non-molecular means.

It is also important that the genetic base of Japanese knotweeds is studied on a European basis to see what lessons are to be learned in terms of the role of introgression and seed set in their spread over a region with wider climatic range.

Acknowledgments

The authors acknowledge the support of the Joint Leicester and Loughborough Universities Research Fund for this project. We would also like to thank Pauline Bablak who carried out much of the DNA work, and John Newbury and Parminda Singh for their invaluable assistance in optimizing the RAPD reactions. We are extremely grateful to all those AAIS and BSBI members who generously contributed to the hybrid survey.

References

Bailey, J.P. 1988. Putative *Reynoutria japonica* Houtt. x *Fallopia baldschuanica* (Regel) Holub hybrids discovered in Britain. Watsonia 17: 163-164.

Bailey, J.P. 1989. Cytology and Breeding Behaviour of Giant Alien *Polygonum* Species in Britain. Thesis, University of Leicester.

Bailey, J.P. 1994. Reproductive biology and fertility of *Fallopia japonica* (Japanese knotweed) and its hybrids in the British Isles. In: L.C. de Waal, L.E. Child, P.M. Wade and J.H. Brock (eds.), Ecology and Management of Invasive Riverside Plants, pp. 141-158. John Wiley and Sons, Chichester.

Bailey, J.P. and Conolly, A.P. 1985. Chromosome numbers of some alien *Reynoutria* species in the British Isles. Watsonia 15: 270-271.

Bailey, J.P. and Conolly, A.P. 1991. Alien species of *Polygonum* and *Reynoutria* in Cornwall 1989-1990. Bot. Cornwall Newsl. 5: 33-46.

Bailey, J.P. and Stace, C.A. 1991. Chromosome number, morphology, pairing, and DNA values of species and hybrids in the genus *Fallopia* (*Polygonaceae*). Plant Syst. Evolut. 180: 29-59.

Beerling D.J., Bailey, J.P. and Conolly A.P. 1994. Biological flora of the British Isles: *Fallopia japonica.* J. Ecol. (in press).

Brock, J.H. and Wade, P.M. 1992. Regeneration of Japanese knotweed (*Fallopia japonica)* from rhizomes and stems: observations from greenhouse trials. IXème Colloque Internationale sur la Biologie des Mauvaises Herbes, 1992, Dijon.

Chalmers, K.J., Waugh, R., Sprent, J.I., Simons, A.J. and Powell, W. 1992. Detection of genetic variation between and within populations of *Gliricidia sepium* and *G. maculata* using RAPD markers. Heredity 69: 465-472.

Child, L.E., De Waal, L.C., Wade, P.M. and Palmer, J.P. 1992. Control and management of *Reynoutria* species (knotweed). Aspects Appl. Biol. 29: 295-307.

Chrtek, J. and Chrtková, A. 1983. *Reynoutria* x *bohemica,* a new hybrid from the dock family. Čas. Nár. Muz. Praha, Ser. Nat. 152: 120 (in Czech).

Conolly, A.P. 1977. The distribution and history in the British Isles of some alien species of *Polygonum* and *Reynoutria*. Watsonia 11: 291-311.

De Waal, L.C., Child, E.L., Wade, P.M. and Brock, J.H. (eds.) 1994. Ecology and Management of Invasive Riverside Plants. 217 pp. John Wiley and Sons, Chichester.

Doyle, J.J. and Doyle, J.L. 1987. A rapid DNA isolation procedure for small quantities of fresh plant tissue. Phytochem. Bull. 19: 11-15.

REYNOUTRIA SACHALINENSIS IN EUROPE AND IN THE FAR EAST: A COMPARISON OF THE SPECIES ECOLOGY IN ITS NATIVE AND ADVENTIVE DISTRIBUTION RANGE

Herbert Sukopp and Uwe Starfinger
Institut für Ökologie, Technische Universität Berlin, Schmidt-Ott Straße 1, D 12165 Berlin, Germany

Abstract

Reynoutria sachalinensis is native to the Far East from where it was introduced to Europe and Northern America. The ecology of the species is reviewed and its behaviour in its adventive distribution range is compared with that in the areas to which it is native. Information on the history of introduction, spreading dynamics, and the role of the species in vegetation is given.

Introduction

Reynoutria sachalinensis, native to the Far East, was introduced to Europe as well as to America and is spreading like its better known congener *R. japonica*. While much has been written about the spread of *R. japonica* and its impact on near-natural vegetation, relatively little is known about *R. sachalinensis*. This is especially true for studies comparing the species performance in its natural and synanthropic ranges.

The comparison of a species' ecology in its native and synanthropic ranges is a valuable tool in explaining the success of the species as an invader of new geographic regions. In the vast literature on biological invasions there are relatively few examples of this approach, *e.g.*, Sukopp and Sukopp (1988) for *Reynoutria japonica*, Starfinger (1990, 1991) for *Prunus serotina*, Sachse (1992) for *Acer negundo*. The present paper is based on observations during a visit to northern Honshu and Hokkaido made by the first author and on the little that can be found in the literature about the species. This evaluation is still in progress and our aim is to provide a collection of observations and preliminary results.

The 'type' and 'isotype' specimens of *R. sachalinensis* were collected by Weyrich in western Sakhalin near the village Nothosama and are kept in the herbarium of the Botanical Institute of St. Petersburg (LE) (Shcherbakov, personal communication). The species was first described by F.K. Schmidt as *Polygonum sachalinense* in 1859 (in Maximowicz, Primit. Fl. Amur. 233/M: Mém. Sav. Etr. Saint-Pétersbou). Today, the name *Reynoutria sachalinensis* (Schmidt Petrop.) Nakai is correct according to Buttler and Schippmann (1993). Synonyms are *Fallopia sachalinensis* (Schmidt Petrop.) Ronse Decraene, *Pleuropterus sachalinensis* (Schmidt Petrop.) Moldenke and *Tiniaria sachalinensis* (Schmidt Petrop.) Janchen.

Plant Invasions - General Aspects and Special Problems, pp. 151-159
edited by P. Pyšek, K. Prach, M. Rejmánek and M. Wade

***Reynoutria sachalinensis* in Europe**

Introduction, planting and use

Reynoutria sachalinensis was first brought to Europe in 1863 and used as a forage crop and as an ornamental in gardens and parks especially in lawns and along watercourses. Together with *R. japonica* it was recommended and planted as a fodder plant in the second half of the 19th century (Hegi 1912). It was popular with huntsmen as it was supposed to be more palatable to game than *R. japonica* and also offered good cover (Herberg 1937). This is assumed to be a major reason for its planting (*e.g.*, in Franconia, Walter 1989). According to Krausch (1987) it was planted especially in parks.

Today the two *Reynoutria* species are regarded as unsuitable for most gardens on account of their vigour and invasive potential and consequently rarely planted (Walters *et al.* 1989). Walter (1989) recommends that they should not be encouraged by further planting. Most knotweed control measures in Europe, however, are directed against *R. japonica.*

In Europe, a new use of the plant was found in its efficacy against fungal plant diseases. Leaf extracts are effective against powdery mildew of apple, *Begonia*, cucumber and wheat as well as against grey mould (*Botrytis cinerea*) of sweet peppers (Herger *et al.* 1988). Recently, it has been proposed for the decontamination of soils polluted with heavy metals as it can accumulate these in its leaves and stems.

Little is known about the insect fauna on *R. sachalinensis* in Europe. Zimmermann and Topp (1991) report that the polyphagous moth *Spilarctia lutea* and an aphid live on the plant. In this context it is interesting that *R. sachalinensis* attracts ants with extrafloral nectaries which help to protect it against herbivorous insects (Sukopp and Schick 1991).

R. sachalinensis requires two different photoperiods following each other in order to develop flowers. So, under natural conditions it flowers in autumn only, *i.e.*, in short days following long days. This type of plant is predominantly distributed on the east facing slope in the northern continents (Jäger in Schubert 1991).

It does not form mycorrhizas (Frydman 1957).

Spread and present geographic distribution

Starting from planted individuals, *R. sachalinensis* is mainly spread in two ways, the direct - even if unintentional - spread by humans, *e.g.*, dumping of garden refuse, and the transport of rhizome or stem parts by water. The role of seeds in dispersal needs further study.

The species was first recorded wild in Germany in 1869 (Hegi 1912), in the Czech Republic in 1869 (Pyšek and Prach 1993), and in Great Britain in 1896 (Stace 1991). Today, it is found in most central, western and eastern European countries though absent from the Mediterranean region (Tutin *et al.* 1993). Distribution maps exist for Germany (former FRG only, Haeupler and Schönfelder 1988), the Czech Republic (Slavík 1986), Britain (Perring and Walters 1962). The maps do not show a distinct pattern apart from some concentration along the large rivers, and in Germany it is missing from calcareous mountains. Its synanthropic distribution covers a broader range of climatic continentality than its natural distribution (Klotz 1984).

In all these countries *R. sachalinensis* is scattered to fairly common, being locally common in some areas, but always less frequent than *R. japonica.* In the Czech Republic, for example, Pyšek and Prach (1993) reported on 152 localities of *R. sachalinensis* as opposed to 515 of *R. japonica*. In the Ruhr area of western Germany it is much rarer than its congener (Dettmar 1991). In the Oberlausitz and Elbhügelland regions of central Germany 43 localities of *R. sachalinensis* are contrasted with 100 of *R. japonica* (Hardtke *et al.* 1982). In Wedding, a densely built-up district of Berlin, 53 out of 329 sample plots contained *R. japonica*, but only 9 *R. sachalinensis* (Böcker 1991). Walter (1989) lists 62 localities for *R. japonica* and 32 for *R. sachalinensis* in Upper Franconia (including cultivated stands).

Vegetation and biotopes

Spontaneously growing *R. sachalinensis* occurs in Europe mostly in two types of biotopes: in sites more or less influenced by human activities such as gardens, parks or ruderal sites in towns and villages, these are often close to planted individuals; and along watercourses under near-natural conditions. The statements about in which biotope type it is more common vary with regions:

In the Beskids it can be found mostly along watercourses, but also on roadsides (Kosmale, personal communication). In contrast, it is not known from near-natural riparian sites in Franconia but only from man made sites close to original plantings (Walter 1989). In the Czech Republic it has been long established in semi-natural sites but in riparian sites it was found for the first time more than 60 years after its first appearance elsewhere (Pyšek and Prach 1993). In the Black Forest it occurs in riparian areas (Schwabe 1989).

The factors under which the species grows vary considerably from dry and warm for ruderal sites in cities to wet and cool for riparian sites in the mountains. In the dry summers of 1987 and 1988 the plants lost many leaves on a ruderal site in Bottrop (Dettmar 1991).

Stands of *R. sachalinensis* are always very dense and tall, containing thus only few other species. In the literature 25 phytosociological relevés from western, central and southern Germany were found containing the species (Brandes 1981: 1 rel., Gödde 1986: 3 rel., Reidl 1989: 1 rel., Hetzel 1991: 5 rel., Dettmar 1991: 5 rel., Klotz and Gutte 1991: 9 rel., Schulte: 1 rel., unpublished). In these, *R. sachalinensis* always had a cover value of ‘5’ (in Braun-Blanquet scale), with the total cover being 95-100%. All the other species occurred with low cover values, mostly ‘r’ and ‘+’, never exceeding ‘2’. A total of 119 species were found co-occurring with *R. sachalinensis,* the species number per relevé was between 7 and 19, with the mean 10.9 per relevé. These numbers may result from a sampling technique that includes the edges of stands in the sample plot, so that the stands were probably poorer in species. The species most frequently accompanying *R. sachalinensis* were *Urtica dioica* (68% of the relevés), *Artemisia vulgaris* (32%), *Poa palustris* (32%), *P. trivialis* (28%), and *Aegopodium podagraria* (28%).

The ruderal stands can roughly be classified as belonging to the following phytosociological units (alliances) of the Braun-Blanquet classification system (Oberdorfer 1983): Arction or Dauco-Melilotion, the more natural stands to the Aegopodion or Petasition.

Although it shares the biotope and vegetation types with *R. japonica*, mixed stands of the two species are rare (Klotz and Gutte 1991). One such example was recorded in Darmstadt (Herger *et al.* 1988), also on river banks in the upper Rhine/Black Forest area the two species can sometimes be found together (Alberternst 1993).

Reynoutria sachalinensis in the Far East

The name of *R. sachalinensis* in Japanese is 'O Itadori', 'O' meaning big and 'Itadori' strong plant.

In the Far East, the rhizome of *R. sachalinensis* is used for treatment of several diseases, the active substances being anthraquinone derivatives (Chi *et al.* 1983). Some people in Japan eat the young stems while they are still soft. Caterpillars living in the internodia are so numerous that they are regularly used as fishing bait.

Natural geographical distribution

The native range of *R. sachalinensis* lies in eastern Asia and comprises the southern part of Sakhalin Island, the southern Kurile Islands (Kunashir and Shikotan), and the Japanese islands Hokkaido and Honshu. On Honshu it is only found on the north-western side north of the Chubu district, *i.e.,* the part close to the Japanese Sea. Distribution maps are given by Ohwi (1965), Horikawa (1976) and Morozow and Belaya (in Walter and Breckle 1991) and for Japan by Okuyama (1983).

The climate is monsoon-like though cold, it is milder than in the inner parts of the continent on equal latitudes due to warm currents in the ocean. The mean annual temperatures are between 4°C in the north and 8°C in the south, rainfall varies from 500 mm/year to over 1000 mm/year at low altitudes and is higher in the montaneous parts of the range. Rainfall maxima are in late summer and autumn (Walter and Lieth 1960).

R. sachalinensis is found from sea-level to an altitude of ca. 1050 m (Miyawaki 1987, 1988).

In contrast to *R. japonica, R. sachalinensis* on Hokkaido looks similar in size to the plants in Europe.

Vegetation and biotopes

According to Miyawaki (1987, 1988) *R. sachalinensis* forms two distinct plant communities in Japan: Angelico-Polygonetum sachalinensis Suz.-Tok. *et al.* 1956 on north Honshu and Cirsio kamtschatici-Polygonetum sachalinensis Ohba 1973 on Hokkaido. The first community is found along forest edges, roads through forests, in avalanche areas in the mountains and on coastal cliffs. It consists of *R. sachalinensis, Filipendula kamtschatica, Angelica ursina, A. edulis, Aster ageratoides*, and *Angelica pubescens* var. *matsumurae* as character species. The stands are usually 1.4 to 2.5 m tall, mostly dominated by *R. sachalinensis, Filipendula kamtschatica* or *Artemisia montana*. Further frequent species are *Petasites japonicus* var. *giganteus, Impatiens noli-tangere,* and *Equisetum arvense*. The species number varies between 12 and 30, with an average of 20. The community occurs in two subassociations: one in natural vegetation of coastal cliffs and mountain rivers which are subject to natural earth movements, and the other on road banks and other places with man-induced earth movements, representing an intermediate stage of succession. Also there are almost pure stands of *R. sachalinensis* as a pioneer vegetation on bare soils in human settlements. Fig. 1 shows an example of a natural stand of *R. sachalinensis* in north Honshu.

The Cirsio kamtschatici-Polygonetum sachalinensis is found only on Hokkaido. It grows on similar sites as the Angelico-Polygonetum sachalinensis on Honshu, *i.e.*, in half-shade of forest edges, river banks and coastal cliffs as well as on fallow fields and road banks. The most frequent species are *R. sachalinensis, Petasites japonicus* var. *giganteus, Artemisia montana, Filipendula kamtschatica* and *Angelica ursina*. Several species of this plant community reach the height of 2-3 m, the total cover is close to 100%. This plant community can be divided into three subassociations which differ in species richness (mean species numbers between 17 and 27) and site types. In total, the Cirsio kamtschatici-Polygonetum sachalinensis grows from the sea coast to the region of coniferous forests of the Vaccinio-Piceetea in altitudes of ca. 1050 m. Along rivers, the most frequent natural site type, it is not usually found directly on the gravel along the water's edge, where *Petasites japonicus* var. *giganteus* and *Salix sachalinensis* dominate, but higher on the banks in gaps between trees.

On newly formed barren sites, *R. sachalinensis* can be among the first species to colonize. The pumice flows of the 1929 eruption of the volcano Mt. Komagatake, Hokkaido, were covered with the species after 2-4 years. In spite of its density, tree species can invade the stands and eventually shade out the knotweed, which began to decrease after 35 years, giving way to a light forest of mainly *Populus maximowiczii* and *Betula platyphylla* var. *japonica* (Yoshioka 1966).

In the revegetation process on Mt. Usu, Hokkaido, after the eruptions of 1977 and 1978, *R. sachalinensis* together with *Petasites japonicus* var. *giganteus* played even more important role. Due to their ability to sprout from parts buried in as much as

Fig. 1. Reynoutria sachalinensis on a natural site. Near Hachimata in north Honshu, ca. 900 m above sea level, June 1986. Photo: A. Miyawaki.

0.5 to 1.0 m of volcanic deposits, they quickly form dense stands in large areas (Tsuyuzaki 1987).

The 'giant herb' community

Stands of *R. sachalinensis* belong to an extraordinary vegetation that has been termed 'giant herb communities' (Höchststaudengesellschaften) by Walter and Breckle (1991), who cited the comprehensive work of Morozov and Belaya (after a manuscript in Russian). This giant herb vegetation grows in northern Japan, Sakhalin, the Kuriles and the southern half of Kamtschatka between 35° and 58° N. In lower latitudes it is strictly montane, in the north it occurs in lower altitudes. The climate is moist and cool.

This vegetation is remarkable for its height, leaf area and production values. According to Numata and Yoshizawa (1975) the plants reach a height of 2-4.5 m. The biomass can reach more than 200 $t \cdot ha^{-1}$ with an energy content of close to 12,000 $MJ \cdot ha^{-1}$, more than any other herbaceous vegetation in the world. The stands are usually species poor and are dominated by a single species. The highest biomass was found in stands of *Petasites amplus* (293.0 $t \cdot ha^{-1}$) and *R. sachalinensis* (259.2 $t \cdot ha^{-1}$). The values of leaf area index were 7.0 for *Petasites* and 21.0 for *Reynoutria*, which is more than even in the densest forests. These extreme values are reached by vigorous growth in early summer when daily height increments reach up to 15 cm. The photosynthetic active radiation is often completely absorbed due to a leaf arrangement where the upper leaves are inclined to almost vertical, the lower ones horizontal. This leads to an exploitation of the absorbed radiation which can reach as much as 22% in *R. sachalinensis*.

Sukopp and Schick (1993) have pointed out that morphological characters are important for the ability of *R. sachalinensis* to present the inflorescences in a layer above the competing vegetation. These are *(1)* a change from orthotropous to plagiotropous growth and from spiral to distichous leaf arrangement at the beginning of the reproductive phase, and *(2)* a high number of basal internodia.

Discussion

A general aim of many studies on exotic species is to contribute to an explanation as to why, among a high number of introductions only a relatively small fraction is able to spread in the new area. Features of success were sought in a great number of studies published in the SCOPE volumes (*e.g.*, Kornberg and Williamson 1986; Mooney and Drake 1986; Drake *et al.* 1989; Di Castri *et al.* 1990). A general result has been that immigration success is not generally predictable on the basis of morphological, physiological or population biological characters. It has been possible, however, to contribute to the understanding of single immigration processes to a certain degree with case studies.

One might expect *R. sachalinensis* to be a successful invader of new continents on the basis of its vegetative vigour and competitive ability. This vigour and competitive strength are clearly a preadaptation useful for its spread in Europe where the species met similar climatic conditions and at the same time weaker competitors than in its original range. Also, the fact that in both areas it can be found in near-natural vegetation and on man made substrates is an indicator of its potential to colonize new areas.

From the knowledge of its ecology in East Asia and Europe we can conclude that it has become an established member of the flora of several European countries: it is able to thrive under the climatic conditions here, there are suitable sites, it has a functioning means of dispersal (flowing water), stands can resist the immigration of trees, other potential out-competers are absent, and the impact of herbivores in Europe is minor.

In other words it is no longer dependent on anthropogenic influence on the vegetation in order to keep up populations here. According to definition of the term this makes it an agriophyte (*sensu* Schroeder 1969), and it is listed as such in Lohmeyer and Sukopp (1992).

All this, however, is not sufficient to explain its success in Europe as it is contrasted with the failure of other species. Every single species must have a strategy to survive under a given set of environmental conditions, and so it would be able to colonize new areas with similar conditions. What makes *R. sachalinensis* so special in this respect is not completely clear. Further studies on how exactly its role in the vegetation is realized may be useful.

Acknowledgments

Sincere thanks for help in the field and with literature are due to Prof. Dr. Akira Miyawaki (Yokohama), Prof. Dr. Koji Ito (Sapporo), Prof. Dr. Kazue Fujiwara (Yokohama), Dr. Masahiro Haruki (Sapporo), Dr. Hideki Takahashi (Sapporo), and Dr. Tsunehiko Nishikawa (Asahikawa). Stefan Biedermann (Berlin) has translated parts of Miyawaki (1987, 1988) from Japanese into German. Max Wade (Loughborough) kindly improved our English.

References

Alberternst, B. 1993. Zur Biologie und Bekämpfung verschiedener *Reynoutria*-Sippen. Thesis Universität Hohenheim. 106 pp.

Böcker, R. 1991. Floreninformationssystem für den Bezirk Wedding von Berlin. Thesis Technische Universität, Berlin.

Brandes, D. 1981. Neophytengesellschaften der Klasse Artemisietea im südöstlichen Niedersachsen. Braunschw. Naturk. Schr. 1: 183-211.

Buttler, K.P. and Schippmann, U. 1993. Namensverzeichnis zur Flora der Farn- und Samenpflanzen Hessens. (Erste Fassung). Botanik u. Naturschutz in Hessen, Beih. 6.

Chi, Hyung-Joon, Moon, Hee-Soo and Lee, Yong-Ju 1983. Anthraquinones from the rhizome of *Polygonum sachalinense*. Yakhak Hoeji 27: 37-43 (in Korean with English abstract).

Dettmar, J. 1991. Industrietypische Flora und Vegetation im Ruhrgebiet. Diss. Bot. 191: 1-397.

Di Castri, F., Hansen, A. J. and Debussche, M. (eds.) 1990. Biological Invasions in Europe and the Mediterranean Basin. Kluwer Academic Publ., Dordrecht.

Drake, J.A., Mooney, H.A., Di Castri, F., Groves, R.H., Kruger, F.J., Rejmánek, M. and Williamson, M. (eds.) 1989. Biological Invasions: A Global Perspective. 525 pp. John Wiley and Sons, Chichester.

Oberdorfer, E. 1983. Süddeutsche Pflanzengesellschaften. Vol.3/2a. Ed. 2. 455 pp. Gustav Fischer, Stuttgart.

Frydman, I. 1957. Mykotrofizm roslinnosci pokrywajacej i ruiny domow Wroclawia. Acta Soc. Bot. Pol. 26: 48-60.

Gödde, M. 1986: Vergleichende Untersuchung der Ruderalvegetation der Großstädte Düsseldorf, Essen und Münster. Oberstadtdirektor d. Landeshauptstadt Düsseldorf, Düsseldorf.

Haeupler, H. and Schönfelder, P. 1988. Atlas der Farn- und Blütenpflanzen der Bundesrepublik Deutschland. 780 pp. Eugen Ulmer, Stuttgart.
Hardtke, H.-J., Otto, H.-W. and Ranft, M. 1982. Zur Ausbreitung einiger Neophyten in Oberlausitz und Elbhügelland. Abh. Ber. Naturkundemus. Görlitz 56: 19-28.
Hegi, G. 1912. Illustrierte Flora von Mitteleuropa, Vol. 3, Ed. 1. 607 pp. Aufl., München.
Herberg, M. 1937. Äsung und Deckung im Jagdrevier. P. Parey, Berlin.
Herger, G., Klingauf, F., Mangold, D., Pommer, E.-H., and Scherer, M. 1988. Die Wirkung von Auszügen aus dem Sachalin-Staudenknöterich, *Reynoutria sachalinensis* (F. Schmidt) Nakai, gegen Pilzkrankheiten, insbesondere Echte Mehltau-Pilze. Nachrichtenbl. Deutsch. Pflanzenschutzd. Braunschweig 40: 56-60.
Hetzel, G. 1991. Beiträge zur Ruderalvegetation und Flora der Stadt Passau. Ber. Bayer. Bot. Ges. 62: 41-66.
Horikawa, Y. 1976. Atlas of the Japanese Flora. 2. 862 pp. Gakken, Tokyo.
Klotz, S. and Gutte, P. 1991. Zur Soziologie einiger urbaner Neophyten. Hercynia N.F. 28: 45-61.
Kornberg, H. and Williamson, M.H. (eds.) 1986. Quantitative Aspects of the Ecology of Biological Invasions. Phil. Trans. Roy. Soc. London B 314.
Krausch, H.-D. 1987. Ostasiatische Pflanzen in unseren Gärten. Gubener Heimatkalender 31: 83-90.
Klotz, S. 1984. Phytoökologische Beiträge zur Charakterisierung und Gliederung urbaner Ökosysteme, dargestellt am Beispiel der Städte Halle und Halle-Neustadt. Diss. Universität Halle.
Lohmeyer, W. and Sukopp, H. 1992. Agriophyten in der Vegetation Mitteleuropas. Schriftenr. Vegetationskd. 25.
Miyawaki, A. 1987. Vegetation of Japan. Vol. 8. 605 pp. Shibundo,Tohoku (in Japanese).
Miyawaki, A. 1988. Vegetation of Japan. Vol. 9. 563 pp. Shibundo, Hokkaido (in Japanese).
Mooney, H. A. and Drake, J.A. (eds.) 1986. Ecology of Biological Invasions of North America and Hawaii. 321 pp. Springer-Verlag, New York.
Numata, M. and Yoshizawa, N. (eds.) 1975. Weed Flora of Japan. 414 pp. Zenkoku Noson Kyoiku Kyokai Publ., Tokyo.
Oberdorfer, E. 1983. Süddeutsche Pflanzengesselschaften. Gustav Fischer Verlag, Stuttgart.
Ohwi, I. 1965. Flora of Japan. 1067 pp. Smithsonian Institute, Washington, DC.
Okuyama, S. 1983. Wild Plants of Japan. Tokyo.
Perring, F.H. and Walters, S.M. 1962. Atlas of the British Flora. 432 pp. Thomas Nelson and Sons, London.
Pyšek, P. and Prach, K. 1993. Plant invasions and the role of riparian habitats: a comparison of four species alien to central Europe. J. Biogeogr. 20: 413-420.
Reidl, K. 1989. Floristische und vegetationskundliche Untersuchungen als Grundlagen für den Arten- und Biotopschutz in der Stadt - dargestellt am Beispiel Essen. Diss. Universität GHS Essen.
Sachse, U. 1992. Invasion patterns of boxelder on sites with different levels of disturbance. Verh. Ges. Ökol. 21: 103-111.
Schroeder, F.G. 1969. Zur Klassifizierung der Anthropochoren. Vegetatio 16: 225-238.
Schubert, R. 1991. Ökologie. Ed. 3 G. 657 pp. Fischer, Jena.
Schwabe, A. 1989. Vegetation complexes of flowing-water habitats and their importance for the differentiation of landscape units. Landscape Ecol. 2: 237-253.
Slavík, B. 1986. Fytokartografické syntézy ČSR. 1. Botanicky ústav ČSAV. Průhonice u Prahy.
Stace, C. 1991. New Flora of the British Isles. 1226 pp. Cambridge University Press, Cambridge.
Starfinger, U. 1990. Die Einbürgerung der Spätblühenden Traubenkirsche (*Prunus serotina* Ehrh.) in Mitteleuropa. Landschaftsentwickl. Umweltforsch. 69: 1-119.
Starfinger, U. 1991: Population biology of an invading tree species - *Prunus serotina*. In: A. Seitz and V. Loeschke (eds.), Species Conservation: A Population Biology Approach. pp. 171-184. Birkhäuser Verlag, Basel.
Sukopp, H. and Schick, B. 1991. Zur Biologie neophytischer *Reynoutria*-Arten in Mitteleuropa. I. Über Floral- und Extrafloralnektarien. Verh. Bot. Ver. Berlin Brandenburg 124: 31-42.
Sukopp, H. and Schick, B. 1993. Zur Biologie neophytischer *Reynoutria*-Arten in Mitteleuropa. II. Morphometrie der Sprosssysteme. Diss. Bot. 196: 163-174.
Sukopp, H. and Sukopp, U. 1988. *Reynoutria japonica* Houtt. in Japan und in Europa. Veröff. Geobot. Inst. ETH, Stiftung Rübel, Zürich 98: 354-372.
Tsuyuzaki, S. 1987. Origin of plants recovering on the volcano Usu, northern Japan, since the eruptions of 1977 and 1978. Vegetatio 73: 53-58.
Tutin, T.G. *et al.* (eds.) 1993. Flora Europaea. Vol. 1. Ed. 2. 581 pp. Cambridge University Press, Cambridge.
Walter, E: 1989. Zur Ausbreitung der beiden fernöstlichen Staudenknöteriche (*Reynoutria japonica* und *R. sachalinensis*) in Oberfranken. Ber. Naturforsch. Ges. Bamberg 64: 1-17.
Walter, H. and Breckle, S.-W. 1991. Ökologie der Erde. Vol. 4: Spezielle Ökologie der gemäßigten und arktischen Zonen ausserhalb Euro-Nordasiens. 585 pp. G. Fischer, Stuttgart.
Walter, H. and Lieth, H. 1960. Klimadiagramm-Weltatlas. G. Fischer, Jena.

Walters, M. *et al.* (eds.) 1989. The European Garden Flora. Cambridge University Press, Cambridge.
Yoshioka, K. 1966. Development and recovery of vegetation since the 1929 eruption of Mt. Komagatake, Hokkaido. Ecol. Rev. 16: 271-292.
Zimmermann, K. and Topp, W. 1991. Anpassungserscheinungen von Insekten an Neophyten der Gattung *Reynoutria* (*Polygonaceae*) in Zentraleuropa. Zool. Jb. Syst. 118: 377-390.

INVASION HISTORY AND ECOLOGY OF *LYTHRUM SALICARIA* IN NORTH AMERICA

Keith R. Edwards[1], Michael S. Adams[2] and Jan Květ[3]
[1]*Institute for Environmental Studies, University of Wisconsin, Madison, Wisconsin, USA;* [2]*Botany Department, University of Wisconsin, Madison, Wisconsin, USA;* [3]*Plant Ecology Section, Institute of Botany, Academy of Sciences of the Czech Republic, Dukelská 145, CZ-379 01 Třeboň, Czech Republic*

Abstract

Lythrum salicaria, native to Europe, is an invasive species in North American wetlands. Once established, it usually becomes the dominant species and remains so over time, sometimes resulting in the formation of monocultural stands. In the present paper, its invasion history is reviewed and spread documented. The species first arrived in North America in the last years of the eighteenth century, and at present, it is found in all states and Canadian provinces in a band located between the 37th and 50th parallels. Further, the results of a natural experiment are presented that examined how various life history characteristics of *Lythrum* are affected by changing environmental conditions over a range of habitats. The study was carried out in Indiana Dunes and Sleeping Bear Dunes national lakeshores. Six sites were selected representing a natural gradient of nutrient and water conditions. Field experiments were aimed at both vegetative and reproductive characteristics. Seedling growth was followed under controlled conditions. The results show that different environmental factors, operating at different spatial and temporal scales, seem to affect *Lythrum* life history characteristics. Plants growing under nutrient-poor conditions were smaller, allocated more biomass to roots and produced fewer seeds per capsule than those growing under optimal conditions. However, the results support the idea that nutrient concentration may not be the only limiting factor as *Lythrum* appears to be stressed also by fluctuating water level. Competitive relationships among wetland species also seem to affect the pattern of *Lythrum* invasion. Applications for management are outlined and possible factors allowing the species to be a successful invader are discussed.

Introduction

The invasion and establishment of non-native species into native communities has been described as one of the four horsemen of the environmental apocalypse (Soulé 1990). Whether deliberately or by accident, plant species from all corners of the globe have the capacity to be introduced into areas far distant from their native range because of increased travel and trade between nations and other human activities. This increased movement of species is resulting in the increased homogenization of the world's communities (Lodge 1993). However, it must be remembered that colonization and change in a species' range have been, and still are, natural ecological processes (Hengeveld 1989). It is a question of the temporal and spatial scale at which one is observing these processes, and the value judgements of the observer (Allen and Hoekstra 1992), which will determine whether a particular invasion is considered to be part of a natural extension of a species range, and allowed to proceed, or a 'malignancy' which must be controlled and managed (Lodge 1993).

Numerous European plant species have been successful invaders of temperate North American communities. One such species is the wetland emergent *Lythrum salicaria* L. (purple loosestrife). *Lythrum* is an herbaceous perennial with annual

Plant Invasions - General Aspects and Special Problems, pp. 161-180
edited by P. Pyšek, K. Prach, M. Rejmánek and M. Wade

stems emerging from a perennating rootstock. The inflorescence is spike-like, covered with numerous magenta-colored flowers. The tubular flowers are pollinated by numerous bee and butterfly species, with the honey bee (*Apis mellifera* L.) being one of the most important. The fruit is a capsule, which contains many small seeds (Thompson *et al.* 1987). *Lythrum* is common along the edges of streams, lakes, ponds, and marshes (Polunin 1969).

Lythrum plants possess certain attributes of emergents described as important for increasing the probability of a species becoming a successful invader (Ashton and Mitchell 1989). Such characteristics include *(1)* a tall, clumping growth form, by which *Lythrum* can overtop native species and prevent other species from growing underneath it, *(2)* the production of a large number of small, viable seeds, and *(3)* the apparent ability for quick regeneration following cutting. *Lythrum* can grow over a wide range of nutrient and sediment conditions (Shamsi and Whitehead 1974; Keddy and Constabel 1986). In addition, the presence of a perennial rootstock allows for the storage of possibly large amounts of overwinter carbon reserves, which can be used for rapid shoot growth the following spring (Grime 1988; Tilman 1988).

There is no doubt that *Lythrum* is a successful invader of temperate North American wetland systems. It seemingly requires some disturbance in the native wetland community in order to invade and become established, although the type, size, and intensity of the disturbance have not been documented. There is also a paucity of studies concerned with the second stage of *Lythrum* invasions, namely what factors affect the growth and vigour of established *Lythrum* populations over a range of environmental conditions in which it is found. Examining these factors will provide information about *Lythrum* ecology (Nichols and Shaw 1986), information that can be used in developing a holistic and comprehensive management plan. This information will be gathered from populations undergoing significant changes in their life history characteristics. It is when these characteristics change, and the constraints affecting the population change, that one can note interesting and possibly informative phenomena (Allen and Hoekstra 1992). Such natural experiments will help to identify which environmental factors are important in affecting *Lythrum* growth and vigour, and at what spatial and temporal scales they operate, both within a certain geographic area and across the species' range.

The objective of this paper is to present results of a natural experiment that examined how various life history characteristics of *Lythrum* are affected by changing environmental conditions over a range of habitats. The results will be preceded by a brief history of the North American invasion of *Lythrum*. Lastly, we will theorize about the factors allowing *Lythrum* to be a successful invader in North America but be a benign species in its native Europe.

Invasion history and spread

The first mention of *Lythrum* in North America was in 1814, in the first edition of the flora by Torrey and Gray (Stuckey 1980). They noted scattered stands of *Lythrum* in the freshwater parts of the harbours of the major Atlantic coastal ports, including Philadelphia, New York, New Bedford, and Boston. Scattered stands and single individuals were also noticed growing along riverbanks and canals. From this citation, it has been estimated that *Lythrum* plants first arrived in North America in the last years of the eighteenth century, the seeds being transported from Europe either in sheep wool or in ship ballast (Thompson *et al.* 1987).

Lythrum spread slowly over the 130 years following this first introduction. It spread first along rivers, canals, and ditches in the northeastern portion of the United States, moving westward along the newly constructed Erie Canal (Thompson *et al.* 1987). Small and widely spaced groups of plants also became established in western Maryland and central Pennsylvania as new roads and turnpikes, such as the National Road, were constructed (Fig. 1a). By this manner of spread, it can be seen that *Lythrum* plants became established in areas that had been recently disturbed, following its behaviour pattern in Europe.

By the 1850s and 1860s, *Lythrum* was being recognized as a potential horticultural and landscape plant. It is thought that by this use *Lythrum* was first introduced into the western Great Lakes region and into the Pacific northwest. Small populations were noted at this time around Muskegon, Michigan and in Seattle, Washington and Vancouver, British Columbia.

During the second half of the nineteenth century, *Lythrum* may have been brought in with the successive waves of European immigrants. *Lythrum* was used as a medicinal herb in Europe since the first century AD, when its uses as an astringent were noted by Dioscorides. Medieval herbals also mention that *Lythrum* leaves could be used to stop the flow of blood and that snakes and flies would leave a room in which it was burned. Thus, *Lythrum* was probably part of the traditional medicine practiced by these immigrants and made up part of their herbal gardens (Thompson *et al.* 1987). In this way, *Lythrum* may have been repeatedly introduced into North America, with different European genotypes being introduced from different areas of Europe. By 1900, the greatest number of *Lythrum* populations occurred in the northeastern states, although these populations were still small and found only along riverbanks, canals and ditches. Populations in the midwest and northwest were even more widely scattered. *Lythrum* had spread further south into Virginia and West Virginia and was found further westward along the Lake Erie shoreline (Fig. 1b).

The rapid spread of *Lythrum* began in the 1930s, approximately 130 years after being first introduced into North America. *Lythrum* spread explosively into the floodplain marshes of the St. Lawrence River and along the Merrimac, Hudson, and Mohawk floodplains, with the formation of monocultural stands being reported for the first time since its introduction. This was followed by a rapid spread westward along the Great Lakes into the midwest, with additional spreading from the already extant local populations, intentionally introduced and growing for over fifty years (Fig. 1c). *Lythrum* spread rapidly from its loci in the Seattle/Vancouver area, moving quickly south into northern California and east into Idaho, Utah, and Wyoming. By 1985, *Lythrum* was found in all states and Canadian provinces in a band located between the 37th and 50th parallels (Fig. 1d).

Several hypotheses have been generated to account for this rapid spread of *Lythrum* in the last sixty years (Thompson *et al.* 1987). The 1930s was the decade of the economic depression. Many public works programs were initiated in order to put people back to work. The large scale of some of these projects, such as the Tennessee Valley Authority, construction of the Bonneville Dam along the Columbia River, and the construction of irrigation complexes and reservoirs in California and elsewhere in the west, greatly increased the area being disturbed, thereby providing large areas of available habitat into which *Lythrum* could invade. Later, the construction

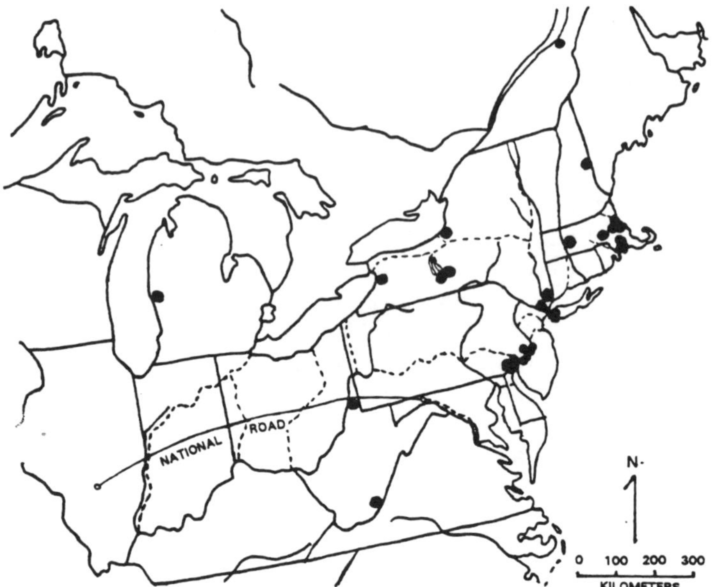

Fig. 1a.

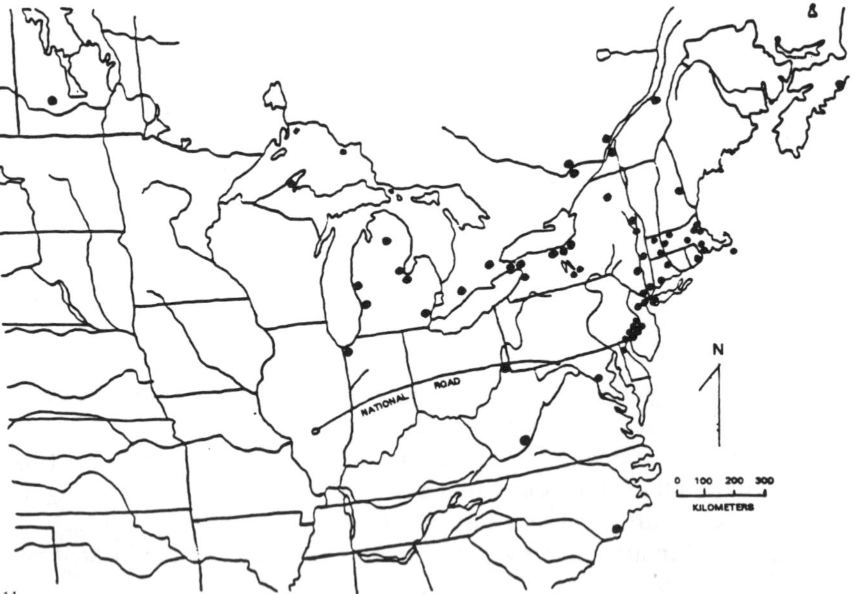

Fig. 1b.

Fig. 1. Spread of *Lythrum salicaria* in the United States (from Thompson *et al.* 1987). *(a)* By 1880. Dashed lines represent the canals in operation as of 1849. The areas of greatest *Lythrum* density are around the seaport cities where *Lythrum* is thought to have been first introduced into North America (Philadelphia, New Bedford, and Boston). *(b)* By 1900. Stands of *Lythrum* have slowly increased in number in the northeast U.S., with further spread along the Great Lakes into the midwestern states. *(c)* Distribution as of 1940, following its explosive spread in the northeast and midwest U.S., and along the floodplain of the St. Lawrence River in Canada. *(d)* By 1985. The number of new populations increased greatly in the western Great Lakes states, threatening waterfowl breeding habitat in the prairie pothole region (broken line). Also, *Lythrum* spread quickly south and east, from the Seattle/Vancouver area in the Pacific northwest, into several western states.

Fig. 1c.

Fig. 1d.

of the Interstate Highway System provided increased travel corridors by which *Lythrum* could invade areas previously cut off from invasion. The Appalachian Mountains, which had previously acted as an effective barrier to *Lythrum* spread, were now neutralized as a barrier with the construction of the highways running east-west through their interior. In addition, in the 1940s, beekeepers and honey producers began to realize the potential of using *Lythrum* as a honey source. Honey bees (*Apis mellifera*) are a major pollinator of *Lythrum* flowers (Levin 1970). There are reports of beekeepers deliberately spreading *Lythrum* seeds along streams and rivers in order to increase the density and number of *Lythrum* populations (Thompson *et al.* 1987). Arguments for retaining *Lythrum* as a honey source and against any control measures are still being made by apiculturalists (Wisconsin DATCP 1993).

Descriptions of European and North American *Lythrum* plants

Lythrum plants in Europe may grow up to 120 cm in height on average, with occasional stems attaining heights of up to two meters (Clapham *et al.* 1959; Polunin 1969; Spencer-Jones and Wade 1986). The stems have few or no branches. Plants tend to grow singly or in small stands along the edges of lakes, rivers, and in marshes. *Lythrum* seemingly requires some disturbance in order to become established in these communities (Hejný 1960), and can become locally abundant following disturbances which have produced open patches (Clapham *et al.* 1959). In these situations, *Lythrum* may be one of the first colonizing species of the newly opened patch, being a dominant for one to three years. Then, other species, such as *Juncus effusus* L., expand into the patch and become dominant, with *Lythrum* becoming a subordinate species (Ellis 1963). Hejný (1960) found *Lythrum* growing in tall sedge and degraded reed (*Phragmites australis* (Cav.) Trin. ex Steudel) communities, on drift material in the littoral zone of ponds, and diffusely in non-flooded peat meadows. *Lythrum* has been found to be a common member of floodplain fens in Estonia (Kask 1982) and as a frequent species along the littoral zone in Lake Cirisa in Latvia (Gavrilova 1986). It is also noted as being frequent in the floodplains near the mouths of the large rivers flowing into the Black Sea (Dubina and Shelyag-Sosonko 1989). Generally, *Lythrum* in Europe may be described as an early successional species that appears to require some sort of disturbance in order to colonize a wetland area. It may become a dominant or co-dominant for a few years, but then is consigned to being a subordinate species by the expansion of later successional species into the patch. Even at this initial stage, it may encounter competition from other early successional species, both native and invasive to Europe, especially among several *Epilobium* species, including *E. hirsutum* L. (Shamsi and Whitehead 1974), *E. roseum* Schreb., and recently the North American invasive *E. adenocaulon* Rafin. (= *E. ciliatum*).

Lythrum plants in North American wetlands act quite differently. They still seem to require some sort of disturbance in order to invade into a wetland. However, once it is established, *Lythrum* usually becomes the dominant species and remains so over time, sometimes resulting in the formation of monocultural stands (Rawinski 1982). *Lythrum* stems in North America grow to an average height of 2.5 m (Thompson *et al.* 1987), with particular stems reaching heights of three meters or more (K. Edwards, personal observation). By overgrowing the native wetland flora, *Lythrum* can become the top competitor in wetland systems (Gaudet and Keddy 1988). Therefore, a *Lythrum* invasion leads to a decrease in plant diversity with possibly a decrease in wildlife use of the wetland, as *Lythrum* crowds out the native plants used as food by waterfowl and other wildlife (Thompson *et al.* 1987). *Lythrum* plants tend to be multi-stemmed, with no or several branches per stem; in certain areas, larger plants may have several branches arising in the upper third of the stem, with inflorescences produced at the terminal end of each branch. This apparent lack of apical dominance did not appear to be the result of damage to the apical meristem. In other cases, damage to the apical meristem in smaller plants led to increased branching, possibly by the loss of apical dominance (personal observation). Loss of apical dominance, as a result of damage to the apical meristem caused mostly by herbivory, including that by large mammals (deer, cattle, *etc.*) is a common occurrence in European *Lythrum* plants (J. Květ, personal observation).

Methods

Study sites

Life history characteristics were measured for three years (1991-1993) in populations of *Lythrum salicaria* in two components of the U.S. national park system: Indiana Dunes (INDU) and Sleeping Bear Dunes (SLBE) national lakeshores (Fig. 2). INDU is located at the southern tip of Lake Michigan (41°50'N, 87°W), approximately 80

Fig. 2. Location of the Indiana Dunes (INDU) and Sleeping Bear Dunes (SLBE) national lakeshores. Four sites were established at INDU, with two additional sites established at SLBE. The sites range across nutrient and hydrologic gradients.

km east of Chicago. SLBE is approximately 430 km north of INDU, situated in the northwest corner of the lower peninsula of Michigan (45°N, 86°W).

Six populations were selected for the life history study; four at INDU and two at SLBE. Two of the INDU sites are located within the Tolleston ridge and swale system. This ancient beach ridge was formed at a time of higher Lake Michigan water levels following retreat of the glaciers at the end of the last Wisconsin glaciation (Doss 1993). One of these sites is within an abandoned sand mine (Tolleston sand mine site); the sediment is very sandy and nutrient-poor (organic matter <2%). The other Tolleston site (second Tolleston site) is located in a natural intradunal pond and consists of a silt-loam substrate (organic matter = 20%). Both Tolleston sites lie about 3 km inland from the present Lake Michigan shoreline. The other two INDU sites are located within a marsh complex associated with an interdunal pond (Long Lake). These sites (Long Lake Marsh East and West) are within 1.5 km of the present Lake Michigan shoreline and consist of nutrient-rich, organic substrates (organic matter = 45%). All four sites have been found to be connected hydrologically to Lake Michigan (Doss 1993).

The SLBE populations also occur in sites which have nutrient-rich, organic substrates (organic matter = 45-50%). One site is an abandoned pasture, containing two

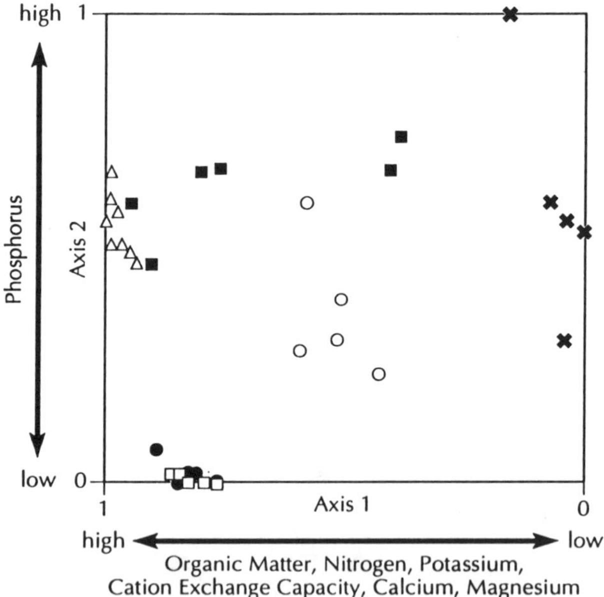

Fig. 3. The first two axes of a relativized Bray-Curtis ordination of the soil chemistry data. Axis 1 accounted for 66% of the variance while axis 2 accounted for 23%. The nutrients and chemical measures closely correlated with each axis are listed by the appropriate axis.

artificially constructed ponds, a channelized stream, and a network of shallow ditches (Aral Road site); the *Lythrum* plants grow alongside the edges of the ponds, the stream, and ditches. The second site is a sedge meadow located along the west bank of the Platte River, just before it enters into Loon Lake (Platte River site).

Soil nutrient concentrations (Ca, K, Mg, N, P, organic matter, and cation exchange capacity) were determined from five, randomly collected soil samples at each site in July 1992. Each sample consisted of 20 soil plugs, randomly collected within a 1.0 m radius circle; the plugs were taken from the root zone portion of the soil profile (the top 15 cm). A Bray-Curtis polar ordination (Ludwig and Reynolds 1988) of the soil chemistry data shows that the sites separate out along a nutrient gradient, from the nutrient-poor Tolleston sand mine site to the nutrient-rich Long Lake and two SLBE sites (Fig. 3). Except for phosphorus, all of the soil nutrients measured were highly correlated with the first axis ($r^2 > 0.50$).

The sites also differ in terms of their hydrologic characteristics. The substrates of the Long Lake and SLBE sites were always saturated during the three years of the study. However, the Long Lake sites usually contained standing water (up to 60 cm deep during parts of the 1993 growing period), while there was usually little (1 cm deep) or no standing water in the two SLBE sites. The two Tolleston sites became slightly unsaturated (6-12 cm) during the latter half of the 1991 growing period. They were slightly unsaturated for the entire 1992 growing period (1992 was a drought year in the Chicago region). In 1993, all four INDU sites contained standing water throughout the growing period, caused by the significant increase in precipitation during the spring and early summer months in the midwest region of the U.S.

Field experiments

The nutrient and hydrologic differences between the sites allowed for the establishment of a natural experiment: how are various life history characteristics of *Lythrum* populations affected by changing environmental conditions, in this case nutrient and water availability, over space and time? In order to measure these parameters, transects 30-50 m in length were randomly established in the four INDU sites and the Aral Road site in 1991. At least ten 1.0 m^2 quadrats were randomly selected along the transects in each site.

Because *Lythrum* density was significantly lower in the Platte River site, using randomly selected quadrats along line transects was an inappropriate means for insuring a large enough sample size. Therefore, four 5 x 5 m squares were established in areas of the site where *Lythrum* density was greatest. The squares were divided into 25 1.0 m^2 quadrats and, from these, ten quadrats were randomly selected. This produced a total of 40 quadrats, enough to ensure a statistically significant sample size. The squares and quadrats were established at the beginning of the 1992 field season.

Shoot biomass

All shoots in the inner 0.5 m^2 of each quadrat were tagged and measured for height and basal diameter; sampling occurred five times in the INDU sites and three times in the SLBE sites during each year of the study. Above-ground biomass was estimated in a non-destructive manner (Chiariello *et al.* 1989). Fifteen shoots from each population were selected from outside of the transect area during each sampling period and their height and diameter measured. The shoots were chosen to cover the range of shoot heights measured in the experimental quadrats. The shoots were cut at ground level, taken back to the laboratory, dried for 24 hours at 70°C, and weighed. Total shoot dry weight was then related to the height and diameter measurements by a regression equation. An equation was derived for each site during each sampling period. Shoot height and diameter were always closely correlated with total above-ground dry weight biomass ($r^2>0.77$). When height and diameter were highly correlated to each other ($r^2>0.70$), only the former was included into the regression equation. However, as these measures were not well correlated from late June 1992 onwards ($r^2<0.23$), both were included as predictors.

One-way ANOVA (Snedecor and Cochran 1967) was used to analyse the between-site differences in shoot biomass for each sampling period of 1991 and 1992. *T*-tests were run to analyze for between-year differences within each site, for sampling periods corresponding to the same time of the growing period.

Fertility

The number of stems producing inflorescences per quadrat was counted in each year of the study. One-way ANOVA was used to test for between-site differences in the number of inflorescences per unit area, after performing a square root transformation of the data to achieve normality and homogeneity of variance. As variance homogeneity was not achieved for the the 1992 and 1993 data sets, Kruskal-Wallis non-parametric ANOVA (Sokal and Rohlf 1981) was used.

Flowering phenology was noted for the 1992 and 1993 growing periods in the four INDU populations.

Fecundity

Twenty inflorescences, from outside of the transect area, were randomly collected in September 1991 from the five sites measured in that year. Fifteen capsules were randomly selected from along the length of the entire inflorescence, and the seeds contained in each capsule counted. These values were averaged to provide a mean number of seeds per capsule for each inflorescence. One-way ANOVA was used to determine whether fecundity levels differed between sites, following a natural logarithm transformation of the raw data to achieve normality and homogeneity of variance.

Seedling growth study

A seedling growth study was done in the Biotron facility at the University of Wisconsin. This facility allowed for the control of several environmental conditions. Light and air temperature levels were set to mimic daily fluctuations for July; air temperature ranged from 20-30°C, while there were 15 hours of light, with the maximum of both occurring between 1000-1500 hours. Relative humidity was kept constant at 70%. Watering of the seedlings occurred twice a day for 15 minutes each time.

Seeds from three of the INDU sites (the two Tolleston and Long Lake Marsh East sites) were sown in soil taken from those sites. These soils represented the range of the different sediment and nutrient conditions in which the study populations were growing in the field. Soil was placed into eighty 1.89 l pots for each soil type. The

Table 1. Above-ground shoot biomass (in g of dry weight/0.5 m^2) of *Lythrum salicaria* for the 1991 and 1992 growing seasons. Means ± SE are shown. The values were estimated by back transforming the natural logarithm transformations of the original shoot biomass data, which was done to achieve normality and homogeneity of variance. Shoot biomass differed significantly between sites in all sampling periods of each growing season (one-way ANOVA, degrees of freedom (df) and F-values given, $P<0.001$). Means bearing the same letter row-wise were not significantly different from each other in multiple range test ($P<0.01$). Site acronyms: TOLL: Tolleston sand mine; TORG: second Tolleston site; LOMA: Long Lake Marsh West; LOLA: Long Lake Marsh East; ARAL: Aral Road; PLRI: Platte River. ns: not sampled.

Period	TOLL	TORG	LOMA	LOLA	ARAL	PLRI	df	F
1991								
Mid June	1.09±0.30^{a}	ns	6.86±0.20^{c}	4.31±0.13^{b}	1.34±0.17^{a}	ns	3,49	48.471
Late June	1.46±0.30^{a}	1.75±0.25^{a}	11.10±0.19^{b}	ns	ns	ns	2,33	43.505
Mid July	1.47±0.30^{a}	1.11±0.20^{a}	10.37±0.17^{c}	5.45±0.12^{b}	ns	ns	3,49	36.634
Late July	1.42±0.27^{a}	2.02±0.16^{a}	7.07±0.20^{b}	10.58±0.13^{b}	2.42±0.19^{a}	ns	4,65	48.310
Mid August	2.06±0.19^{a}	2.71±0.14^{a}	11.94±0.21^{b}	14.73±0.12^{b}	2.32±0.14^{b}	ns	4,64	85.652
1992								
Early June	0.18±0.15^{a}	0.31±0.10^{b}	2.60±0.12^{c}	3.23±0.11^{c}	ns	ns	3,63	257.845
Late June	0.70±0.19^{a}	0.94±0.11^{a}	3.85±0.09^{b}	3.51±0.09^{b}	0.80±0.12^{a}	0.39±0.18^{c}	5,103	108.351
Early July	0.71±0.21^{a}	1.47±0.10^{b}	4.95±0.12^{c}	4.73±0.09^{c}	ns	ns	3,68	83.912
Late July	0.99±0.27^{a}	2.82±1.11^{b}	5.74±0.13^{c}	3.76±0.13^{b}	1.17±0.13^{a}	0.55±0.16^{d}	5,98	61.888
Late August	0.49±0.24^{a}	0.49±0.09^{a}	9.01±0.14^{b}	4.54±0.14^{c}	1.60±0.14^{d}	0.89±0.17^{e}	5,10	83.201

pots were equipped with two small polyethylene tubes to allow for drip watering of the seedlings. Four seeds were sown into each pot on February 19, 1992.

Sampling began two weeks after sowing. Five seedlings were randomly selected from each soil type, cleaned to remove soil particles, separated into root and aboveground parts, dried for 24 hours at 70°C, and weighed. Sampling occurred every four days over a 28-day period.

Root-shoot (R/S) ratios for each soil type were calculated for each sampling period. One-way ANOVA was used to analyse the data (Zar 1984), after a natural logarithm transformation was made to achieve normality and homogeneity of variance.

Results

Field experiments

Shoot biomass

During each sampling period of both 1991 and 1992, there were significant between-site differences in the average shoot biomass (Table 1). In 1991, plants in the two Long Lake populations produced significantly more shoot biomass than plants in the other sites. A similar pattern was found in 1992, except during the late July sampling period, when the shoot biomass in the second Tolleston site was similar to that in the Long Lake Marsh East site. Plants in the Aral Road population produced similar

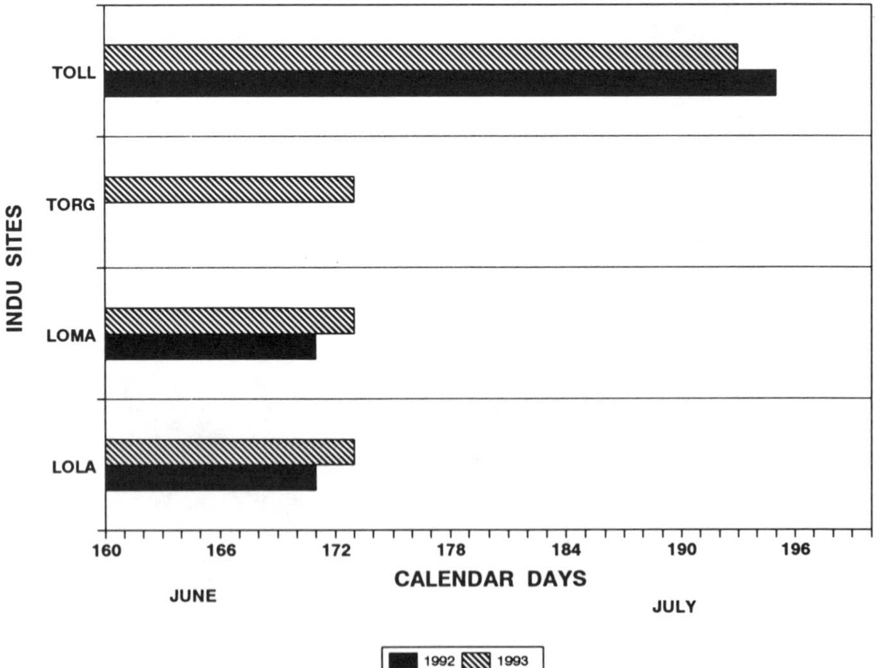

Fig. 4. Time of flowering for *Lythrum* plants in the four INDU populations for the 1992 and 1993 growing periods. No inflorescences were produced by the stems in the sample quadrats of the second Tolleston site (TORG) in 1992. Days represent the calendar days (day 1 = January 1). Site acronyms: TOLL: Tollestone sand mine; TORG: second Tolleston site; LOMA: Long Lake Marsh West; LOLA: Long Lake Marsh East.

amounts of shoot biomass as those in the two Tolleston sites, while the plants in the Platte River site produced the lowest shoot biomass of all of the populations measured. The exception was the late August 1992 sampling period, when plants in the two SLBE populations had significantly greater shoot biomass than the plants in the two Tolleston sites (Table 1).

Fertility

Differences in flowering phenology between sites were most pronounced during the 1992 growing period (Fig. 4). *Lythrum* plants began to flower on June 19 in the two Long Lake populations, but not until July 13 in the Tolleston sand mine site. No inflorescences were produced by plants in the sample quadrats of the second Tolleston site; five stems far removed from the sample quadrats flowered, but not until August 3. In 1993, the flowering time was similar to 1992 for the Long Lake and Tolleston sand mine sites. There was a substantial change in flowering phenology in the second Tolleston population, with time of flowering being similar to that seen in the Long Lake plants (Fig. 4).

Fertility, expressed as the number of flowering stems produced per 0.5 m^2 plot, differed significantly ($P<0.001$) between sites in each study year (Table 2). In 1991, the fertility in the second Tolleston site was significantly lower than in the other sites and no inflorescences were produced in the sample quadrats in this site in 1992. Average fertility levels in the Tolleston sand mine and the Aral Road sites decreased substantially between 1991 and 1992 (from 7.94 and 13.00 to 0.63 and 2.90, respectively) resulting in these populations having significantly lower fertility levels than the two Long Lake populations. The fertility in the Aral Road population increased in 1993, reaching a similar level as in 1991 (15.88 and 13.00, respectively) and this site showed the highest fertility level of all sites in those two years. Fertility levels in the Tolleston sand mine and Platte River sites remained low in 1993, while there was a substantial increase in flowering in the second Tolleston population (Table 2).

Table 2. Number of flowering stems per 0.5 m^2 plot. Means ± SD are given. The values were calculated from back transformations after using a square root transform of the raw data to achieve homogeneity of variance. Variance homogeneity was achieved only for the 1991 data: one-way ANOVA showed significant differences between sites ($F_{4,72}$ = 15.893, $P<0.001$). Kruskall-Wallis non-parametric ANOVA was used for the 1992 and 1993 data (1992: H = 71.240, $P<0.001$; 1993: H = 91.026, $P<0.001$). Means followed by the same letter columnwise were not significantly different in multiple range comparison ($P<0.05$). No inflorescences were produced in the second Tolleston site in 1992 which was not included in the analysis. ns: not sampled.

Site	1991	1992	1993
Tolleston sand mine	7.94±1.62^{bc}	0.63±0.67^{b}	0.98±1.35^{b}
Second Tolleston	0.28±0.58^{a}	0.00	6.85±0.56^{c}
Long Lake Marsh East	6.28±1.65^{bc}	8.14±0.78^{c}	13.49±1.25^{de}
Long Lake Marsh West	3.95±0.91^{b}	4.78±1.36^{c}	7.22±1.78^{cd}
Aral Road	13.00±1.74^{c}	2.90±1.11^{b}	15.88±1.25^{e}
Platte River	ns	0.06±0.27^{a}	0.09±0.34^{a}

Fecundity

Fecundity levels differed significantly between sites in 1991 (F=6.239, $P<0.001$). Fruits from the Tolleston sand mine site contained significantly fewer seeds per capsule (mean ± S.E. = 80.98±0.05, n=15) than those from the Long Lake Marsh West

(102.13±0.06) and East (102.85±0.07) populations (multiple range test, $P<0.05$). Fruits from the second Tollestone and Aral Road sites produced 92.19±0.05 and 89.65±0.07 seeds per capsule, respectively.

Seedling growth study

Seed germination began three days after sowing. Eleven days after sowing, 86% of the Long Lake seeds sown had germinated while the germination rate was only 40% and 27% for the Tolleston sand mine and second Tolleston sites, respectively. Of the seeds from the Long Lake and second Tolleston sites, the majority germinated 3-4 days after sowing whereas the germination peak for seeds from the Tolleston sand mine site occurred between days 5 and 7 (Fig. 5).

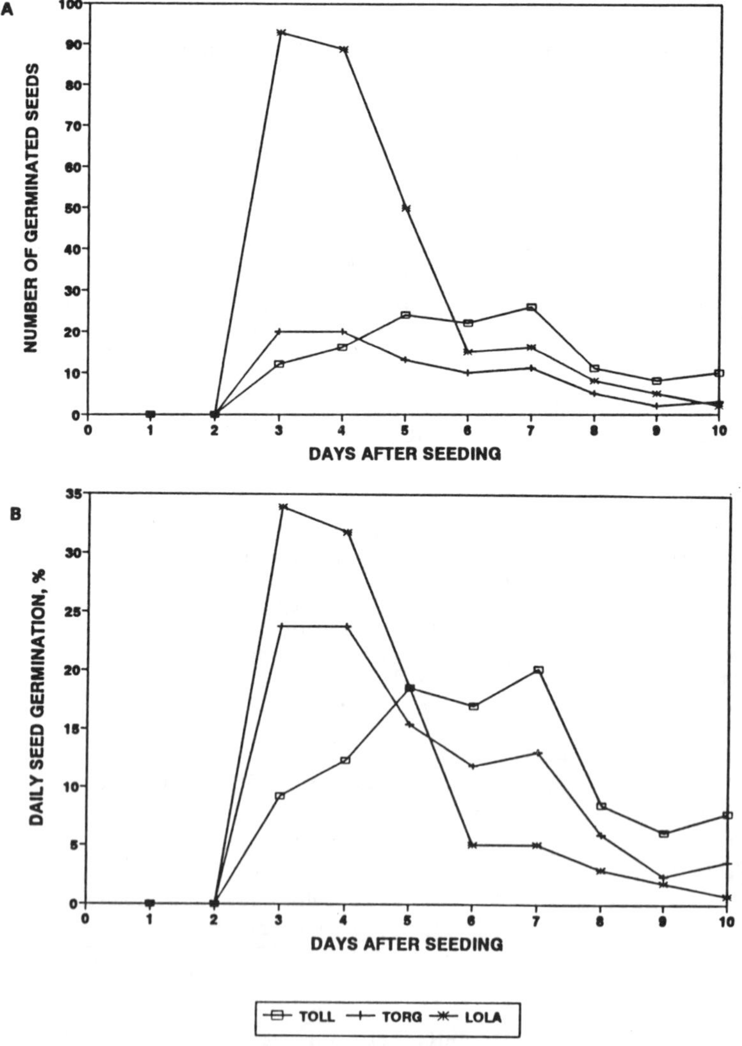

Fig. 5. Seed germination during the first 10 days from sowing. *(A)* the number of seeds which germinated on a given day; *(B)* percentage of seeds germinating per day of all of the seeds sown. Site acronyms: TOLL: Tollestone sand mine; TORG: second Tolleston site; LOLA: Long Lake Marsh East.

Table 3. Means ± SD of the total biomass (W) and root-shoot ratios (R/S) for the seedlings from three of the INDU populations grown under controlled conditions (see Methods for details). Day refers to the number of days from sowing. Total biomass is given in g of dry weight. Natural logarithms of the raw R/S ratios were made to achieve normality and homogeneity of variance. One-way ANOVA was used to test for the differences in R/S; degrees of freedom (df), F-values and significance levels (*P*) are given.

Day	Tolleston sand mine		Second Tolleston site		Long Lake Marsh East		df	F	*P*
	W	R/S	W	R/S	W	R/S			
13	0.28±0.17	0.595±1.339	0.24±0.12	0.365±2.522	0.29±0.11	0.643±1.578	2,18	4.989	<0.05
17	0.27±0.11	0.554±1.310	0.37±0.13	0.410±1.252	0.35±0.18	0.696±1.436	2,10	2.580	NS
21	0.29±0.18	0.287±2.668	0.15±0.10	0.318±3.563	1.03±0.36	0.291±1.374	2,9	0.011	NS
25	0.36±0.10	0.537±1.114	0.42±0.10	0.534±1.848	0.96±0.39	0.316±1.508	2,12	1.984	NS
29	0.33±0.10	0.444±1.601	0.44±0.20	0.375±1.408	2.99±1.66	0.239±1.356	2,12	2.836	NS
33	0.37±0.20	0.545±1.558	0.49±0.14	0.476±1.786	4.80±4.12	0.359±1.474	2,12	0.796	NS
37	0.53±0.19	0.628±1.129	0.86±0.85	0.344±1.244	7.45±6.26	0.282±1.448	2,20	16.128	<0.001

Total biomass was greater for the Long Lake seedlings than for those from the two Tolleston sites, but R/S ratios were not significantly different between the sites over most of the experimental period (Table 3). However, by the last sample (37 days after sowing the seeds), there was a highly significant difference between the three populations ($F_{2,20}$=16.128; P<0.001). The R/S ratios for the Tolleston sand mine site seedlings were significantly greater than for the seedlings from the other two sites (multiple range comparison, P<0.05). There was no significant difference between the Long Lake and second Tolleston site seedlings.

Discussion

Life history study

The results show that over the entire study period the *Lythrum* plants in the two Long Lake populations produced more aboveground biomass, and usually had higher fertility levels, than the other populations. The former populations appear to be growing under optimal conditions, while the growth and vigour of the plants in the latter are less than optimal. Different environmental factors, operating at different spatial and temporal scales, seem to affect *Lythrum* life history characteristics, especially in the two Tolleston populations.

Plants in the Tolleston sand mine site seem to be affected primarily by nutrient limitation. In addition to the smaller aboveground biomass, the greater R/S ratio of the seedlings from this site provides evidence to support this contention. Past studies have found that plants of the same species, when grown in nutrient-poor substrates, allocate more biomass to the roots relative to the shoots, when compared to plants grown in nutrient-rich substrates (Fitter and Hay 1987). In addition, the fecundity in this site was lower than that in the Long Lake populations. Nutrient limitation has been proposed as one possible explanation for increased seed abortion (Lee 1988), leading to decreased fecundity levels.

Results of the stem growth and site fertility levels support the idea that the *Lythrum* plants in the second Tolleston site may be affected more by fluctuating water levels than by nutrient limitation. This site, as well as the other INDU study sites, is connected hydrologically to Lake Michigan. Lake Michigan water levels attained their latest maximum in 1986, but since then have been decreasing (Doss 1993) which resulted in drier conditions in this site over time, with a concomitant change in community structure. The wetland was changing from a sedge meadow wetland to a wet prairie community. These drier conditions seemingly stressed the *Lythrum* plants. Evidence for this comes from remaining dead *Lythrum* stems, which represent past growing periods. *Lythrum* stems, while living for only one growing period, may remain attached to the rootstock for several years. All of the plants measured in this site had standing, dead stems attached to the rootstocks when the quadrats in this site were established. These stems attained maximum heights of 140 cm or more; this is an underestimation of their actual height because of the breaking off of most of the inflorescences and stem tops. In comparison, maximum stem height in 1991 did not exceed 120 cm, and decreased to less than 80 cm for the 1992 growing period. With increased water availability in 1993, associated with the record rainfall during the spring and early summer of that year, maximum stem height again approached 160 cm. In addition, fertility levels increased from 1991 to 1993, becoming similar to the levels found in the Long Lake sites. Also, the seedling R/S ratios from this site were similar to those from the Long Lake site. These ratios were lower than those of the Tolleston sand mine site seedling which indicates that nutrient concentrations may not be limiting.

Possible applications for management

Results from the two Tolleston sites show that *Lythrum* plants can be stressed. The environmental factors that seem to affect the *Lythrum* plants most in those two sites may not be easily employed as control agents in many other wetlands which have been invaded by *Lythrum*. The factors used that will stress *Lythrum* may have to be site specific; this requires a manager to know the wetland communities and systems under his/her jurisdiction.

However, just stressing *Lythrum* plants may not be enough to meet the goal of long-term control. For example, plants of the two Tolleston sites were stressed during the 1992 growing period, attaining much shorter stem heights and producing significantly less aboveground biomass than the plants in the two Long Lake sites. However, the plants in the Tolleston sand mine site flowered and produced seeds, while no inflorescences were produced in the second Tolleston site. The former site harbours a more open, less dense community than the latter which might have contributed to the pattern observed.

The lower fertility levels for the second Tolleston site in 1991, and the complete lack of flowering in 1992, coincided with increasing density of *Calamagrostis canadensis* (Michx.) Beauv. in 1991 and 1992, which corresponded to the decreased water levels in the site. The ecology of *C. canadensis* is similar to that of *C. canescens* (G.H. Weber ex Wigg.) Roth emend. Druce, a native European associate of *Lythrum*'s (Dostál 1989) on drier wetland sites, in contrast to *Glyceria maxima* (Hartman) Holmberg, another *Lythrum* associate but on flooded and littoral sites. These drier conditions may have allowed *C. canadensis* to increase in density, to the detriment of *Lythrum*. This implies that *C. canadensis* may have begun to outcompete *Lythrum* under the drier conditions, suggesting a shift in the competitive hierarchy; this would be contrary to what has been found in previous greenhouse competition studies. These studies (Gaudet and Keddy 1988; Johansson and Keddy 1991), performed under optimal conditions of high nutrients and water availability, always placed *Lythrum* at the top of the competitive hierarchies. Further research needs to be undertaken to determine what conditions may lead to this possible shift in the competitive hierarchy.

The results raise the possibility that control of *Lythrum* may be linked to restoration of the invaded wetland communities. In a review article, Berger (1993) lists several conditions that need to be met in order to increase the chance of restoration becoming an important component of a successful control program. Restorative plantings would be used to impede the growth of the invasive plants, restrict their fertility, deprive the invader of competitive advantages, and minimize the chance of reinvasion of the community by the target invasive species. Restorative plantings may also help to reduce the chance of successful invasion by other non-native species. The results from the two Tolleston sites suggest that *Lythrum* may be one case which meets Berger's criteria.

Possible factors allowing Lythrum to be a successful invasive species

Lythrum is an example of a successful invader; it may be informative to try and determine what factor(s) allow it to be so successful. Di Castri (1990) has listed several factors that may allow a species to successfully invade new territory. Two particular factors from Di Castri's list could be relevant to this situation.

Lythrum may have jumped ahead of its natural control agents in Europe, with no similar agents occurring in North America. It has been shown that there are several insect species in Europe that are monophagous for *Lythrum*, including *Galerucella calmariensis* L., *G. pusilla* Duftschmidt, and *Hylobius transversovittatus* Goeze (Batra 1986). Several of these species have been tested and are being released in several states in the northeast and midwest sections of the U.S. as biocontrol agents (Malecki *et al.* 1993). It appears from the studies that insect herbivory is an important agent for controlling *Lythrum* in Europe. So far, insect herbivory is the only control factor that has been tested. What effect other possible control factors may have on *Lythrum* growth and vigour, such as the role of pathogens, plant competition, and other types of herbivores, both singly and interactively, has not been studied using standard scientific methods.

Another possibility is that the *Lythrum* populations in North America are composed of a more competitive genotype than the native European populations. More competitive species should allocate more biomass and energy to growth than to reproduction, as compared to ruderal species (Harper 1977; Grime 1979; Wisheu and Keddy 1992). If the North American plants belong to a more competitive genotype, then they should allocate a significantly greater percentage of biomass to stem growth relative to reproduction, in comparison to the European plants. North American plants grow much taller on average than the European plants, a phenomenon seen with other invasive species (Crawley 1987). However, the taller growth could just be the result of *Lythrum* escaping from its natural control agents. Whatever factor(s) may most affect its growth, the increased height provides *Lythrum* with a much greater competitive advantage over native, North American species.

Acknowledgments

Many and unending thanks to Allen Tsang, Doug Luthy, and especially Barbie Lynch, for their great patience and assistance in gathering the field data. Thanks also to Kathryn McEachern for many illuminating and needed discussions, and to Noel Pavlovic and Ron Hiebert for their great cooperation and insights. This project was funded under grant CA 6000-1-8025, from the U.S. National Park Service.

References

Allen, T.F.H. and Hoekstra, T.W. 1992. Toward a Unified Ecology. 384 pp. Columbia University Press, New York.

Ashton, P.J. and Mitchell, D.S. 1989. Aquatic plants: patterns and modes of invasion, attributes of invading species, and assessment of control programmes. In: J.A. Drake, H.A. Mooney, F. di Castri, R.H. Groves, F.J. Kruger, M. Rejmánek and M. Williamson (eds.), Biological Invasions, pp. 111-154. John Wiley and Sons, Chichester.

Batra, S.W.T., Schroeder, D., Boldt, P.E. and Mendl, W. 1986. Insects associated with purple loosestrife (*Lythrum salicaria* L.) in Europe. Proc. Entomol. Soc. Washington 88: 748-759.
Berger, J.J. 1993. Ecological restoration and nonindigenous plant species: a review. Restor. Ecol. 1: 74-82.
Chiariello, N.R., Mooney H.A., and Williams, K. 1989. Growth, carbon allocation, and cost of plant tissue. In: R.W. Pearcy, J. Ehleringer, H.A. Mooney, and P.W. Rundel (eds.), Plant Physiological Ecology, pp. 327-365. Chapman and Hall, London.
Clapham, A.R., Tutin, T.G. and Warburg, E.F. 1959. Excursion Flora of the British Isles. 579 pp. Cambridge University Press, London.
Crawley, M.J. 1987. What makes a community invasible? In: A.J. Gray, M.J. Crawley and P.J. Edwards (eds.), Colonization, Succession, and Stability, pp. 429-453. Blackwell Scientific Publ., Oxford.
Di Castri, F. 1990. On invading species and invaded ecosystems: the interplay of historical chance and biological necessity. In: F. di Castri, A.J. Hansen and M. Debussche (eds.), Biological Invasions in Europe and the Mediterranean Basin, pp. 3-16. Kluwer Academic Publ., Dordrecht.
Doss, P.K. 1993. The nature of a dynamic water table in a system of non-tidal, freshwater coastal wetlands. J. Hydrol. 141: 107-126.
Dostál, J. 1989. New Flora of ČSSR. Vol. 2. 1548 pp. Academia, Praha (in Czech).
Dubina, D.V. and Shelyag-Sosonko, Yu. R. 1989. Wetlands of the Black Sea Region. 272 pp. Naukova Dumka, Kiev (in Russian).
Ellis, E.A. 1963. Some effects of selective feeding by the coypu (*Myocastor coypus*) on the vegetation of Broadland. Trans. Norfolk Norwich Natur. Soc. 20: 32-35.
Fitter, A.H. and Hay, R.K.M. 1987. Environmental Physiology of Plants. 423 pp. Academic Press, London.
Gaudet, C.L. and Keddy, P.A. 1988. A comparative approach to predicting competitive ability from plant traits. Nature 334: 242-243.
Gavrilova, G. 1986. Flora of vascular aquatic plants in the reserve 'Lake Cirisa' (Latvian SSR). pp. 118-127. In: N. Ingerpuu, T. Ksenofontova and L.-M. Laasimer (eds.), Vegetation Cover of Aquatic Macrophytes in Marshes Near the Coast of the Baltic Sea. Estonian Academy of Sciences, Tallin.
Grime, J.P. 1979. Plant Strategies and Vegetation Processes. 222 pp. John Wiley and Sons, Chichester.
Grime, J.P. 1988. The C-S-R model of primary plant strategies - origins, implications and tests. In: L.D. Gottlieb and S.K. Jain (eds.), Plant Evolutionary Biology, pp. 371-394. Chapman and Hall, London.
Harper, J.L. 1977. Population Biology of Plants. 892 pp. Academic Press, London.
Hejný, S. 1960. Ökologische Charakteristik der Wasser- und Sumpfpflanzen in den Slowakischen Tiefebenen (Donau- und Theissgebeit). 487 pp. Slovac Academy of Sciences, Bratislava.
Hengeveld, R. 1989. Dynamics of Biological Invasions. 160 pp. Chapman and Hall, London.
Johansson, M.E. and Keddy, P.A. 1991. Intensity and asymmetry of competition between plant pairs of different degrees of similarity: an experimental study on two guilds of wetland plants. Oikos 60: 27-34.
Kask, M. 1982. A list of vascular plants of Estonian peatlands. pp. 39-49. In: T. Frey, V. Masing and E. Roosaluste (eds.), Peatland Ecosystems. Estonian Academy of Sciences, Tallin.
Keddy, P.A. and Constabel, P. 1986. Germination of ten shoreline plants in relation to seed size, soil particle size and water level: an experimental study. Aquat. Bot. 27: 177-186.
Lee, T.D. 1988. Patterns of fruit and seed production. In: J. Lovett-Doust and L. Lovett-Doust (eds.), Plant Reproductive Ecology, pp. 179-202. Oxford University Press, New York.
Levin, D.A. 1970. Assortative pollination in *Lythrum*. Am. J. Bot. 57: 1-5.
Lodge, D. 1993. Biological invasions: lessons for ecology. Trends Ecol. Evolut. 8: 133-137.
Ludwig, J.A. and Reynolds, J.F. 1988. Statistical Ecology. 337 pp. John Wiley and Sons, New York.
Malecki, R.A., Blossey, B., Hight, S., Schroeder, D., Kok, L. and Coulson, J. 1993. Biological control of purple loosestrife. Bioscience 43: 680-686.
Nichols, S.A. and Shaw, B.H. 1986. Ecological life histories of the three aquatic plants, *Myriophyllum spicatum*, *Potamogeton crispus*, and *Elodea canadensis*. Hydrobiologia 131. 3-21.
Polunin, O. 1969. Flowers of Europe. 662 pp. Oxford University Press, London.
Rawinski, T.J. 1982. The ecology and management of purple loosestrife (*Lythrum salicaria*) in central New York. M.S. Thesis. New York Cooperative Wildlife Research Unit, Cornell University, Ithaca, NY.
Sculthorpe, C.D. 1967. The biology of Aquatic Vascular Plants. 610 pp. Edward Arnold Press, London.
Shamsi, S.R.A. and Whitehead, F.H. 1974. Comparative eco-physiology of *Epilobium hirsutum* L. and *Lythrum salicaria* L. I. General biology, distribution, and germination. J. Ecol. 62: 279-290.
Shamsi, S.R.A. and Whitehead, F.H. 1977. Comparative eco-physiology of *Epilobium hirsutum* L. and *Lythrum salicaria* L. III. Mineral nutrition. J. Ecol. 65: 55-70.
Snedecor, G.W. and Cochran, W.G. 1967. Statistical Methods. 593 pp. Iowa State University Press, Ames.
Sokal, R.R. and Rohlf, F.J. 1981. Biometry. 859 pp. W.H. Freeman, San Francisco.
Soulé, M.E. 1990. The onslaught of alien species, and other challenges in the coming decades. Conserv. Biol. 4: 233-239.
Spencer-Jones, D. and Wade, M. 1986. Aquatic Plants. 169 pp. ICI Professional Products, Farnham, Surrey.

Stuckey, R.L. 1980. Distributional history of *Lythrum salicaria* (purple loosestrife) in North America. Bartonia 47: 3-20.

Thompson, D.Q., Stuckey, R.L. and Thompson, E.B. 1987. Spread, Impact, and Control of Purple Loosestrife (*Lythrum salicaria*) in North American Wetlands. 55 pp. Fish and Wildlife Research Report 2. U.S. Department of Interior, Fish and Wildlife Service, Washington, D.C.

Tilman, D. 1988. Plant Strategies and the Dynamics and Structure of Plant Communities. 360 pp. Princeton University Press, Princeton.

Wisconsin Department of Agriculture, Trade, and Consumer Protection. 1993. Reviewer Comments to the Environmental Assessment for a Permit to Release Biological Control Agents of Purple Loosestrife. Division of Agricultural Resource Management, Madison, Wisconsin.

Wisheu, I.C. and Keddy, P.A. 1992. Competition and centrifugal organization of plant communities: theory and tests. J. Veget. Sci. 3: 147-156.

Zar, J.H. 1984. Biostatistical Analysis. 718 pp. Prentice Hall, Englewood Cliffs, NJ.

THE DISTRIBUTION OF *LEMNA MINUTA* WITHIN THE BRITISH ISLES: IDENTIFICATION, DISPERSAL AND NICHE CONSTRAINTS

John L. Bramley, Jo T. Reeve and Georges B.J. Dussart
Ecology Research Group, Canterbury Christ Church College, Canterbury, Kent CT1 1QU, United Kingdom

Abstract

The distribution of an alien species *Lemna minuta*, native to temperate regions of North and South America, is increasing in the British Isles. Identification, distribution and dispersal of *L. minuta* are described. The invasive nature of the species and ecological constraints imposed by water chemistry were studied at Vauxhall Lakes near Canterbury, UK. Further, the abundance of each duckweed species present in 62 different sites was regressed on water chemistry characteristics. It appears that, unlike in other species of *Lemnaceae*, the abundance of *L. minuta* is not significantly controlled by the water chemical conditions. Hence in relation to water chemistry, *L. minuta* has a wider tolerance of habitat range than native *Lemnaceae* in the south-east region of the United Kingdom.

Introduction

Geographical boundaries between communities and the isolation of gene pools can be one mechanism by which speciation occurs (allopatric speciation). Over evolutionary history, these boundaries are occasionally removed or overcome and the biota readjusts to a different dynamic state, as organisms invade areas from which they had previously been excluded. This picture of natural spread is becoming more complicated under human influences; transactions and journeys within the 'global village' are commonplace, increasing the likelihood of the accidental and deliberate spread of organisms into novel habitats. At the same time, 'sophisticated' societies have become more reliant upon the extensive management of the environment, such as intensive farming, which has, in turn, created a number of new and/or disturbed habitats. Within this framework of anthropogenic influences, invasions occur and in Europe there are several notable examples, particularly in the freshwater macrophyte communities. Perhaps the most widely known are the invasions of *Elodea canadensis, Azolla filiculoides* and *Crassula helmsii*. Unfortunately another species to add to this list is *Lemna minuta*, which is a small floating macrophyte originating from the temperate Americas.

Study species

Identification

The *Lemnaceae* (duckweeds) are aquatic monocotyledons and show extreme reduction in morphology: they are usually found floating on or just beneath the surface of rivers, lakes and canals. There are four genera and 34 species within the family, eight

Plant Invasions - General Aspects and Special Problems, pp. 181-185
edited by P. Pyšek, K. Prach, M. Rejmánek and M. Wade

of which have been identified in Europe (Table 1).

Numerous taxonomic treatments (Brooks 1940; Daubs 1965; Rich and Rich 1988; Stace 1991) have been produced which differentiate *Lemnaceae* species by various morphological features, *e.g.*, size, root length, gibbosity and colour. Perhaps, though, the most reliable feature for identifying *L. minuta* is vein (nerve) number, it has only one; mature fronds of native European species either have no veins or more than three (see Table 1).

Table 1. Identified European *Lemnaceae* (after Landolt 1986). Species native *versus* alien status in the British Isles is indicated.

Species	Distribution	Number of veins on mature fronds
Indigenous species:		
1. *Spirodela polyrhiza* (L.) Schleiden	Distributed all over the world except the eastern and southern parts of South America, and New Zealand.	7-16, occasionally up to 21
2. *Lemna minor* L.	Native to the oceanic regions of North America, Europe, Africa and western Australia.	generally 3, rarely 4-5
3. *Lemna gibba* L.	Distribution is mainly centred around regions with Mediterranean climates, which include areas of North and South America, much of Europe and parts of Africa.	usually 4-5, infrequently up to 7
4. *Lemna trisulca* L.	Inhabits cooler climate zones throughout the world, but is absent from South America.	generally 3
5. *Wolffia arrhiza* (L.) Horkel ex Wimmer	Distributed in Europe, Africa, western Asia and Brazil.	absent veination
Alien species:		
1. *Spirodela punctata* (G.F.W. Meyer) Thompson	Originally distributed throughout South America, southern Asia and Australia.	3-7
2. *Lemna aequinoctialis* Welwitsch.	Originally distributed throughout warmer zones of the world, including much of Africa.	generally 3
3. *Lemna minuta* Kunth (*L. minuscula* Herter nom. illeg.)	Originally distributed in temperate zones of North and South America.	1

Distribution

Two of the three alien lemnids, namely *Spirodela punctata* and *Lemna aequinoctialis*, appear limited in distribution to their sites of introduction (Landolt 1986); *L. minuta* however does not. For example in 1982 in Britain *L. minuta* was recognised at 23 sites; by 1993 this figure had risen to 89 (J. Bramley, unpublished data). Not only has the number of pond and lake sites increased but it has also been invading new types of habitat such as flooded peat and mine workings (J. Bramley, unpublished data).

Dispersal

L. minuta originates from temperate regions of North and South America (Landolt 1986) and it would, perhaps, be reasonable to assume that the species has been transported accidentally to Europe in assignments of exotic fish or other water plants. It is certainly present at several aquatic nurseries (Leslie and Walters 1983; J. Bramley, unpublished data) where it is sometimes sold as *L. minor*. Once introduced into semi-natural environments, further spread by birds, anglers, flood water or even deliberate release is easily imaginable. This, coupled with its floating habit and the influence of currents on dispersal, indicate that once released, its spread down a river or drainage system is largely assured. Distribution and abundance, thereafter, will then relate more to competitive ability within various niche constraints.

Study site

In some circumstances *L. minuta* has been seen to outcompete indigenous *Lemnaceae* (Leslie and Walters 1983; Oliver 1991): one of the most striking examples in Britain has occurred at Vauxhall Lakes near Canterbury, Kent, United Kingdom (Dussart *et al.* 1993). This three hectare nature reserve consists largely of an irregular shaped meso-eutrophic shallow lake which, until 1989, contained several native duckweed species. These species had a combined water surface area cover of approximately one or two percent. *L. minuta* was first identified at Vauxhall in that year, possibly being introduced accidentally by anglers. Before the end of 1990, it had become the dominant aquatic plant species and, by early 1991, almost the whole surface of the lake was covered by a dense mat, which in places was 12 cm deep. The lake below the surface became anoxic and there were high fish mortalities. Aquatic invertebrate diversity also decreased, from 32 to two species (T. Harman, personal communication). The situation now seems to have stabilised in an alternating boom-bust cycle. Indigenous *Lemnaceae* are still present, though in reduced numbers, and appear marginalised.

Prompted by the devastating effects of *L. minuta* on Vauxhall Lakes, a case study was set up to investigate the invasive nature of this species, the results of which are summarised below.

Methods

Field investigations were carried out between 1991-1993 and comprised 74 site visits, to 62 different sites, in the south-eastern region of the United Kingdom. At every

Table 2. Methods used in water chemistry analysis. Analysis carried out using test kits followed the manufacturers' instructions, and the results for all tests were calibrated against known standards.

Ion	Method of analysis	Detection limit (mg/l)
Silica	Palintest kit AP 181	0.1
Sulphate	Palintest kit AP 154	2.5
Magnesium	Palintest kit AP 193	0.5
Nitrate	Palintest kit AP 163	0.05
Nitrite	Palintest kit AP 109	0.05
Ammonia	Ciba-Corning chroma kit (Ammonia)	0.01
Phosphate	Ciba-Corning chroma kit (Phosphate)	0.05
Sodium	Gallenkamp FGA 330 flame analyser	1.0
Potassium	Gallenkamp FGA 330 flame analyser	1.0
Calcium	Gallenkamp FGA 330 flame analyser	1.0
Chloride	Merck chloride test kit 11106	2.0

site visit the abundance of each species, in terms of percentage cover, was estimated using the following scale: 0% cover scored 1; less than 1%: 2; 1-5%: 3; 5-25%: 4; 25-50%: 5; 50-75%: 6; 75-100%: 7; greater than 100%: 8.

A one litre water sample was collected, outside of any developed mat, in an acid-washed glass flask, and chemical analysis was carried out later in the laboratory (Table 2 outlines the procedures used for water chemistry analysis).

The abundance of each species, where present, was then regressed on water chemistry measurements using a backward step-wise multiple regression.

Results and discussion

The results of multiple regression analysis are summarized in Table 3. Multiple regression analysis, as used here, allows synergistic effects within the water chemistry parameters measured to be accounted for but does not directly relate abundance to any one single ion; it is therefore difficult to extrapolate data from experiments which do. It may, however, give a more realistic picture of the dynamic chemical habitat in which aquatic plants have to survive. Thus, although it would be very tempting to speculate over why some ions appear to be limiting factors, the importance of other coexisting ions must also be considered. A detailed summary of the nutrient requirements for individual *Lemnaceae* species, including known toxic levels, can be found in Landolt (1986). The results do, however, indicate that the abundance of *L. minuta*, in the investigation sites at which it was present, is not significantly controlled by the water chemical conditions. This situation is in contrast to native lemnids where in each case there do appear to be chemical constraints.

The apparent plasticity, with respect to water chemistry parameters, of *L. minuta* may characterize it as a 'good' invader, as would a high vegetative reproductive potential (Ashton and Mitchell 1989). Laboratory work now in progress indicates a general, though not significant, trend in which *L. minuta* exhibits a faster growth rate in monoculture, in terms of frond number increase, than any of the native British *Lemnaceae* (J. Bramley, unpublished data). If plasticity and relatively high reproductive capacity are true attributes of *L. minuta*, they may account, at least partially, for the continued and sometimes aggressive invasion of the species (Oliver 1991; Dussart *et al.* 1993).

Table 3. Multiple regression analysis of field results. The abundance of each species was related to water chemistry measurements. Each observation in the table is the F-value from a backwards stepwise multiple regression for that species. Significant predictors are indicated: *$P<0.05$; **$P<0.01$; ***$P<0.001$.

Ion	*S. polyrhiza*	*L. minor*	*L. minuta*	*L. gibba*	*L. trisulca*
Silica	19.97**	-7.66*	- 4.08	0.86	- 2.06*
Sulphate	1.77**	0.27	-1.26	-0.20**	-0.05
Magnesium	-0.82*	-0.30	-1.18	0.30	1.05*
Nitrate	13.38**	-13.02	-11.89	2.70**	-1.56
Nitrite	-260.14*	366.57*	206.40	8.46	-38.56
Ammonia	-17.23***	-20.37	0.11	0.08	-1.85
Phosphate	-12.42*	4.50	17.13	5.58***	0.95
Sodium	0.03	-0.24	1.37	-0.08	-0.76*
Potassium	4.16	0.42	-6.20	-0.53	4.18**
Calcium	-4.53***	2.66*	2.70	-0.46*	0.09
Chloride	0.28	-0.02	-0.23	3.48*	-0.07*

Crawley (1986) found that the best correlate of high invasibility in plant communities is the average degree of plant cover; low cover is associated with high invasibility. This statement may apply directly to the surface of freshwaters, which could, perhaps, be seen as a large underexploited niche. On a world stage this may partially account for the 'success' of two of the most noxious tropical and subtropical weeds, *Eichhornia crassipes* and *Salvinia molesta,* whose floating habit has caused problems to human populations in several continents (Sculthorpe 1967; Barrett 1989). Perhaps, *L. minuta* with its wider tolerance and possible reproductive advantage over indigenous *Lemnaceae,* is a temperate equivalent.

Conclusions

The recorded distribution of *L. minuta* within Britain has been increasing (J. Bramley, unpublished data); to what extent this is new colonization, as opposed to that which may have been previously overlooked, is unclear. Records from other parts of Europe are even more sparse. Nevertheless, this alien can, under certain conditions, be an aggressive invader and there may be important ecological consequences at numerous sites.*

References

Ashton, P.J. and Mitchell, D.S. 1989. Aquatic plants: Patterns and modes of invasion, attributes of invading species and assessment of control programmes. In: J.A. Drake, H.A. Mooney, F. di Castri, R.H. Groves, F.J. Kruger, M. Rejmánek and M. Williamson (eds.), Biological Invasions: A Global Perspective, pp. 111-154. John Wiley and Sons, Chichester.

Barrett, S.C.H. 1989. Waterweed invasions. Scient. Am. October 1989.

Brooks, J.S. 1940. The Cytology and Morphology of the *Lemnaceae.* Thesis. Cornell University.

Crawley, M.J. 1986. The population biology of invaders. Phil. Trans. Roy. Soc. London B 314: 711-731.

Daubs, E.H. 1965. A monograph of the *Lemnaceae.* The University of Illinois Press, Urbana.

Dussart, G., Robertson, J. and Bramley, J. 1993. Death of a lake. Bio. Sci. Rev. 5(5): 8-10.

Landolt, E. 1986. The family of *Lemnaceae* - a monographic study. Vol. 1. Veröff. Geobot. Inst. ETH, Stiftung Rübel 71: 1-566.

Leslie, A.C. and Walters, S.M. 1983. The occurrence of *Lemna minuscula* Herter in the British Isles. Watsonia 14: 243-248.

Oliver, J.E. 1991. Spread of *Lemna minuscula* in Wiltshire. Bot. Soc. British Isles News 58: 10.

Rich, T.C.G. and Rich, M.D.B. 1988. Plant Crib. Botanical Society of the British Isles, London.

Sculthorpe, C.D. 1967. The Biology of Aquatic Vascular Plants. 610 pp. Edward Arnold, London.

Stace, C. 1991. New Flora of the British Isles. Cambridge University Press, Cambridge.

*This is part of an ongoing study into the ecophysiology of *Lemna minuta* and the authors would be pleased to receive correspondence concerning the species.

INTERSPECIFIC INTERACTIONS AND PERSISTENCE OF AN INVASIVE MOSS, *ORTHODONTIUM LINEARE*, IN INVADED COMMUNITIES OF INDIGENOUS CRYPTOGAMS

Tomáš Herben
Institute of Botany, Academy of Sciences of the Czech Republic, CZ-25245 Průhonice, Czech Republic

Abstract

Orthodontium lineare, a Southern Hemisphere moss of a circum-Antarctic temperate distribution, was found in Europe for the first time in 1911 in the British Isles. In the following fifty years it has been spreading throughout Western, Northern and Central Europe. It grows on organic substrata, such as rotting wood, raw humus and bare peat; it spreads both in artificial spruce forests and more natural communities of mixed forests and epilithic communities. *O. lineare* regularly produces vast numbers of spores. Permanent plots of two types of communities invaded by this species demonstrate that *O. lineare* behaves as a fugitive species. It is able to overgrow some smaller, mainly hepatic, species, but this relationship is always symmetrical in that other species also overgrow *O. lineare*. It is always overgrown both by taller mosses and *Cladonia* lichens. In contrast, *O. lineare* is able to easily establish in the uncolonized substratum in suitable sites, both in undisturbed areas and in artificially disturbed sites. The data presented show that its presence in communities is temporary and in the long term it is replaced by other species.

Introduction

Recently, a number of invasive species have been identified among bryophytes (Söderström 1993). They are species of diverse habitats and a significant proportion are more or less able to establish in natural habitats, where they should be able to persist in communities of indigenous species (in Europe, *e.g.*, *Orthodontium lineare* and *Campylopus introflexus*). The mechanisms of the invasion of natural habitats are the most difficult subject within the field of biological invasions, since (apart from the population parameters of the invasive species) it depends on the interaction structure within the invaded community, which is, as a rule, difficult to work out. However, compared with vascular plants, the community structure of terrestrial cryptogams (bryophytes and lichens) is relatively simple and the dynamics of interspecific interactions can sometimes be studied more easily.

These plants are in most cases essentially two-dimensional due to their very small stature; their compact growth form making them strong space monopolizers (During 1990, 1992). Further, their underground structures (rhizoids) extend only short distances from the parent shoots. Because of these features, interactions in bryophyte communities are essentially local at the contact zone between colonies and pairwise; the role of higher order interactions and diffuse competition is considered to be negligible (Van Tooren and During 1988; During 1990). Many approaches developed for the study of marine sessile invertebrates are applicable for the study of these communities; for example, community dynamics and the interactions between plants are easily studied by sequential mapping (Connell 1961; Chornesky 1989).

The aim of this study is to determine how this successful moss, *Orthodontium lineare*, persists in the communities of other bryophytes in Europe.

Plant Invasions - General Aspects and Special Problems, pp. 187-192
edited by P. Pyšek, K. Prach, M. Rejmánek and M. Wade

The species

Orthodontium lineare is a moss of the southern hemisphere, distributed in a circum-Antarctic temperate zone (Meijer 1951; Ochyra 1982). In Europe, it was found for the first time in 1911 in the British Isles, though it was identified as a new species only a decade later (Burrell 1940), and the final conclusion about its identity was reached as late as the fifties (Meijer 1951; Ochyra 1982). In the following fifty years it has been spreading throughout Western, Northern and Central Europe; being reported from Britain, the Netherlands, Germany (throughout almost the whole country), Belgium, Luxembourg, France, Poland, the Czech Republic, Slovakia, Denmark and Sweden (Söderström 1992). It is a species growing on organic substrata, such as rotting wood, raw humus and bare peat. It spreads extensively in artificial spruce forests of Europe (DeZuttere and Schumacker 1980), and in acid sandstone areas of central Europe it invades more natural communities of mixed forests and epilithic bryophyte and lichen communities. A more detailed sociological analysis shows that it invades primarily three types of bryophyte communities: *(i)* species rich communities of small hepatics and *Tetraphis pellucida*, confined mainly to rotting wood and sand; *(ii)* species poor communities of taller mosses (*Pohlia nutans*, *Dicranum scoparium*), occurring on various organic and sandy substrata; and *(iii)* species poor lichen communities (mainly *Cladonia* spp.) on rotting wood and sandy substrata (Herben 1990).

O. lineare is known to produce vast numbers of spores (Hedenäs *et al.* 1989a) and the spore production is not variable over time (unpubl. data). The species is able to produce a persistent protonema, which probably enables it to survive unfavourable weather conditions and to spread vegetatively (Hedenäs *et al.* 1989a). At the boundary of its current distribution area, it may be restricted by unfavourable climate, summer drought being the most likely factor. The synchronous mortality at an array of sites was demonstrated (Hedenäs *et al.* 1989a; Herben 1994 and unpubl. data).

Methods

A set of permanent plots was established in communities invaded by *O. lineare* in sandstone areas in central Bohemia, Czech Republic (Polomené hory and Prachovské skály). Eight plots were established in the species rich communities with small hepatics, *Tetraphis pellucida* and *Pohlia nutans*, fifteen plots were established in species poor communities with *Cladonia* spp., and two plots were in intermediate communities. The plots were of variable size, from 100 cm^2 up to 400 cm^2. The plots were recorded twice a year in spring and autumn. Since not all plots were established at the same time, the mean number of recordings for the species rich and species poor plots was 5.5 and 4.5, respectively. The plots were recorded by drawing the contours of plant colonies on transparencies attached to the ground at fixed positions.

Different approaches were used to analyze the species rich and species poor communities. In species poor *Cladonia*-communities, the cover of all colonies was measured planimetrically; further the length of contacts with neighbouring species was measured. The effects of length of contact with other species on the change in the colony size was determined using the multiple regression. Only colonies completely within the plot were included into the analysis. Colonies increasing in size and colonies decreasing in size of *O. lineare* and *Cladonia* spp. were analyzed separately.

In the species rich plots, the change of the position of lateral contact zones between species colonies were recorded on squared millimeter paper. The movement of a contact zone was recorded in a qualitative way, *i.e.* counting the number of recordings of the given plot where the particular movement type occurred. The figures were summed for all the plots. Additionally, the appearance of new colonies (not physically attached to the older ones) was recorded.

In the same sandstone area (Prachovské skály) a set of experimentally disturbed plots 500 cm^2 in size were established in an area with abundant *O. lineare*. These plots were located within an area with *O. lineare* present. The plots were selected so that no *O. lineare* would be immediately outside the disturbed plot. The plots were carefully cleared of the all bryophytes and lichens exposing the bare sandy soil. Twelve such plots were established. The plots were recorded in the same way as the permanent plots. The simple change in the cover of individual species was recorded planimetrically.

Results

In the species rich communities, *O. lineare* behaved as a competitively inferior species (Table 1). It was spreading freely over the unoccupied substratum, but its ability to invade established colonies of other species was rather restricted. It did invade growths of hepatics (*Calypogeia integristipula*, *Bazzania trilobata*, and partly also *Lepidozia reptans*) and of *Tetraphis pellucida*. On the other hand, all these species showed the ability to overgrow *O. lineare*. With the exception of *Tetraphis pellucida*, this occurred with at least the same frequency as that of *O. lineare*. In contrast, *O. lineare* only rarely overgrew larger mosses such as *Pohlia nutans* and *Leucobryum juniperoideum*.

Table 1. Patterns of lateral overgrowth of species in species rich communities. The numerator, *e.g.* 12, indicates the number of cases for all plots in which the species in the row was advancing; the denominator, *e.g.* 0, indicates the number of cases where the species in the column was advancing. The species pairs for which no contacts were observed are indicated by a dash (-).

	Uncolonized substratum	*O.l.*	*L.r.*	*L.j.*	*C.i.*	*T.p.*	*D.d.*
Orthodontium lineare	12/0						
Lepidozia reptans	1/0	14/5					
Leucobryum juniperoideum	5/0	10/1	6/2				
Calypogeia integristipula	5/0	7/4	1/4	0/2			
Tetraphis pellucida	3/0	5/8	5/7	-	2/6		
Dicranodontium denudatum	1/0	4/1	8/7	-	5/4	5/1	
Bazzania trilobata	3/0	3/3	-	-	1/0	-	0/1
Pohlia nutans	3/0	7/2	3/2	2/2	1/0	1/0	-

The ability of *O. lineare* to establish new colonies within the established growths of other bryophytes in species rich communities shows similar pattern (Table 2). The species established colonies well in the uncolonized substratum and within growths of hepatics (*Calypogeia integristipula*, *Bazzania trilobata*, and *Lepidozia reptans*) and *Tetraphis pellucida*. These species all show a reciprocal ability. *O. lineare* was never observed to establish new colonies within growths of larger mosses (*Pohlia nutans* and *Leucobryum juniperoideum*).

Table 2. Patterns of new colony establishment in species rich communities. The numerator, *e.g.* 9, indicates the number of cases for all plots where the species in the row established at least one new colony in the growth of the species in the column; the denominator, *e.g.* 0, indicates the number of cases where the species in the column established at least one new colony in the growth of the species in the row.

	Uncolonized substratum	*O.l.*	*L.r.*	*L.j.*	*C.i.*	*T.p.*	*D.d.*
Orthodontium lineare	9/0						
Lepidozia reptans	1/0	7/5					
Leucobryum juniperoideum	2/0	5/0	0/3				
Calypogeia integristipula	0/0	0/2	1/1	0/1			
Tetraphis pellucida	3/0	2/2	0/2	0/0	2/1		
Dicranodontium denudatum	0/0	0/0	0/0	0/0	0/0		
Bazzania trilobata	2/0	2/1	0/0	0/0	0/0	0/0	
Pohlia nutans	1/0	4/0	0/0	0/0	0/0	0/0	0/0

In the species poor *Cladonia* communities, the increase of *O. lineare* is strongly correlated with the contact zone between *O. lineare* and the uncolonized substratum; the effect of *Cladonia* spp. is marginal (Table 3). Increase in *Cladonia* shows a similar pattern. In contrast, the decrease of *O. lineare* is strongly correlated with the length of the contact zone with *Cladonia*, indicating suppressive effect of *Cladonia* on *O. lineare*. No such phenomenon was shown by *O. lineare* on *Cladonia*.

Table 3. Stepwise multiple regressions of change in colony size on lengths of contact zones with other species (species poor plots). I - initial colony size, F - length of the contact with uncolonized substratum, C - length of the contact with the competitor (*Cladonia* spp. for *O. lineare*, *O. lineare* for *Cladonia* spp.). The initial colony size was forced into the equation as the first variable; further variables were included in the order of their F-to-enter values. F-to-enter values refer to the step when the particular variable was entered to the equation (not to the initial values). Absolute values of the decrease were taken.

Species	Type of change	N	Independent variable	F-to-enter	R^2	Regression coefficient sign
O. lineare	increase	21	1. I	2.53	0.09	-
			2. F	39.37	0.67	+
			3. C	3.98	0.72	+
	decrease	14	1. I	4.35	0.15	-
			2. C	9.86	0.40	+
			3. F	1.20	0.43	-
Cladonia spp.	increase	55	1. I	15.48	0.17	+
			2. F	5.18	0.22	+
			3. C	0.06	0.22	-
	decrease	54	1. I	6.89	0.09	+
			2. F	8.59	0.18	-
			3. C	0.37	0.19	-

At the artificially disturbed sites the recolonization is very fast. In eight plots it took less than one year for the first bryophytes to appear; only in one plot there was no establishment after two years. *O. lineare* is invariably in the first species to establish. Mean bryophyte cover after one year was 22.4%, after two years 66.2%. The same figures for *O. lineare* cover only were 14.5% and 45.5%, respectively. In only two plots were species other than *O. lineare* more abundant after the two years, namely *Tetraphis pellucida* and *Dicranella cerviculata*.

Discussion

Results of both species rich and species poor plots indicate that *O. lineare* is not a competitively dominant species. It is able to overgrow some smaller, mainly hepatic, species, but this relationship is not deterministic and may reverse. *O. lineare* is always overgrown both by taller mosses and *Cladonia* lichens. This is also reported by Hedenäs *et al.* (1989b). Its success in established communities is therefore limited. Importantly, the interspecific interactions between cryptogamic plants do not seem to be fixed for each species pair; instead, there is a large proportion of pairs where the overgrowth may run in both directions, making the long term dynamics of bryophyte communities difficult to predict. Similar phenomena are known for sessile animals (Chornesky 1989 and references therein).

On the other hand, *O. lineare* is able to establish easily in uncolonized substratum in suitable sites, both in undisturbed areas (Table 2) and in artificially disturbed sites, and is able to spread effectively over this uncolonized substratum. This accounts for its very high ability to spread, both locally (Herben 1994) and regionally (DeZuttere and Schumacker 1980), and, given its relative competitive inferiority, for its fugitive behaviour within the communities studied. Although the way of its establishment is unknown, in some other fugitive mosses, *e.g.*, *Funaria hygrometrica* and *Bryum argenteum*, spore establishment was shown to take place regularly under field conditions (Miles and Longton 1992). The high and regular spore output in *O. lineare* may play a role here. Simulation results showed that given the high spore output, even a very low establishment probability of the order 10^{-6} spores falling onto favourable substratum would account for the observed dynamics in the sandstone sites (Herben *et al.* 1991). *O. lineare* is, however, also known to produce rhizoidal gemmae (Whitehouse 1987); since gemmae are bigger and establish with higher probability (Kimmerer 1991), they can also play a role in invasions by the moss species.

The ability of *O. lineare* to invade established communities of indigenous plants should be reconsidered. The data presented show that its presence in both species rich and *Cladonia* communities is temporary and in the long term it is replaced by other species. Its ability to establish itself permanently in undisturbed natural communities is due to its fugitive strategy, not to the competitive superiority over indigenous species.

References

Burrell, W.H. 1940. A field study of *Orthodontium gracile* (Wilson) Schwaegrichen and its variety *heterocarpum* Watson. Naturalist 785: 295-301.

Chornesky, E.A. 1989. Repeated reversal during spatial competition between corals. Ecology 70: 843-855.

Connell, J.H. 1961. The influence of competition and other factors on the distribution of the barnacle *Chthamalus stellatus*. Ecology 42: 710-723.

DeZuttere, P. and Schumacker, R. 1980. L'extension d'*Orthodontium lineare* Schwaegr. subsp. *lineare* en Belgique et au Grand-Duché du Luxembourg. Dumortiera 14-15: 15-22.

During, H.J. 1990. Clonal growth patterns among bryophytes. In: J. van Groenendael and H. de Kroon (eds.), Clonal Growth in Plants: Regulation and Function, pp. 153-176. SPB Academic Publ., The Hague.

During, H.J. 1992. Ecological classifications of bryophytes and lichens. In: J.W. Bates and A.M. Farmer (eds.), Bryophytes and Lichens in Changing Environments, pp. 1-31. Oxford University Press, Oxford.

Hedenäs, L., Herben, T., Rydin, H. and Söderström, L. 1989b. Ecology of the invading moss species *Orthodontium lineare* in Sweden: substratum preference and interactions with other species. J. Bryol. 15: 565-581.
Hedenäs, L., Herben, T., Rydin, H. and Söderström, L. 1989a. Ecology of the invading moss species *Orthodontium lineare* in Sweden: spatial distribution and population structure. Holarctic Ecol. 12: 163-172.
Herben, T. 1990. Sociology of communities invaded by *Orthodontium lineare* (*Bryophyta*) in Europe (excl. British Isles). Preslia (Praha) 62: 215-220.
Herben, T. 1994. Local rate of spreading and patch dynamics of an invasive moss species, *Orthodontium lineare*. J. Bryol. 18: 115-125.
Herben, T., Rydin, H. and Söderström, L. 1991. Effect of spore establishment probability on the persistence of the fugitive invading moss, *Orthodontium lineare*: a spatial model. Oikos 60: 215-221.
Kimmerer, R.W. 1991. Reproductive ecology of *Tetraphis pellucida*. II. Differential success of sexual and asexual propagules. Bryologist 94: 284-288.
Meijer, W. 1951. The genus *Orthodontium*. North-Holland Publ. Co., Amsterdam.
Miles, C.J. and Longton, R.E. 1992. The role of spores in reproduction of mosses. Bot. J. Linn. Soc. 104: 149-173.
Ochyra, R. 1982. *Orthodontium lineare* Schwaegr. - a new species and genus in the moss flora of Poland. Bryol. Beitr. (Duisburg) 1: 23-36.
Söderström, L. 1992. Invasions and range expansions and contractions of bryophytes. In: J.W. Bates and A.M. Farmer (eds.), Bryophytes and Lichens in Changing Environments, pp. 131-158. Oxford University Press, Oxford.
Van Tooren, B. and During, H.J. 1988. Early succession of bryophyte communities on Dutch forest earth banks. Lindbergia 14: 40-46.
Whitehouse, H.L.K., 1987. Protonema-gemmae in European mosses. Symp. Biol. Hung. 35: 227-232.

IV
CONTROL AND MANAGEMENT OF INVASIVE PLANTS

TREATMENT OF *CRASSULA HELMSII* - A CASE STUDY

Lois E. Child[1] and David Spencer-Jones[2]
[1]*International Centre of Landscape Ecology, Loughborough University, Loughborough, Leicesteshire LE11 3TU, United Kingdom;* [2]*Water Wise Consultancy, Plowmans, Park Road, Forest Row, Sussex RH18 5BX, United Kingdom*

Abstract

Two sites infested with the aggressively invasive aquatic and marginal plant *Crassula helmsii* were selected for herbicide trials over the summer of 1992. All three forms of the plant were treated, the submerged form with diquat alginate and dichlobenil, the emergent form with glyphosate and diquat and the floating form with diquat. A reasonable level of control of the emergent form of *C. helmsii* was achieved using glyphosate, with a 75% reduction in plant height recorded. Neither dichlobenil, diquat nor diquat alginate controlled any of the forms of the plant.

Introduction

Crassula helmsii (*Crassulaceae*), a succulent perennial herb, is a native of Australia and New Zealand (see Dawson and Warman 1987 for distribution). The submerged form has erect stems up to 1.3 m in length when growing fully submerged or in shallow water. The emergent form is initially prostrate, rooting up to 5 cm on wet mud and forming dense stands or swards of ca. 10 cm in height. Stems are simple to once or twice branched and bear linear to ovale-lanceolate leaves, 4-15(-20) mm wide. The species is probably self pollinating (Clapham *et al.* 1987; Dawson and Warman 1987) but is not thought to produce viable seeds in the British Isles.

The species' adventive distribution in Eurasia was believed to be limited to the British Isles (Casper and Krausch 1981), but Asovsky (1981) records the species in the Baikal region of the former USSR. It was introduced to the British Isles in the early twentieth century as an oxygenating plant for ornamental use (Swale and Belcher 1982). Since its first recorded escape into the wild in 1956 (Lousley 1957), this invasive aquatic and marginal plant has extended its distribution and now presents severe localised problems throughout the British Isles with the frequency of new sites doubling about every two years (Dawson and Henville 1991).

Previous studies on the herbicidal control of *C. helmsii* have identified diquat alginate as an effective herbicide for treatment of the submerged form of the plant (Dawson and Henville 1991). Spencer-Jones (1994) reported a complete kill when this treatment was applied late in the season to an area heavily infested with the submerged form of the plant.

This study, to investigate the effects of chemical treatment, presents data collected from field trials conducted over the summer of 1992 (Wade *et al.* 1994). Two treatment sites were selected, both were privately owned lakes in the south of England, and various chemicals were used to control particular growth forms of *C. helmsii.*

Plant Invasions - General Aspects and Special Problems, pp. 195-202
edited by P. Pyšek, K. Prach, M. Rejmánek and M. Wade

Site descriptions

Site 1

The first site, located ca. 10 km north of Worthing, West Sussex, is an artificial ornamental lake with an area of approximately 1500 m^2 comprising two main parts. The larger triangular area (800 m^2) with a mean depth of approximately 0.5 m was excavated in 1975 with a smaller (700 m^2) deeper addition in 1986-1987, the latter being approximately 1.0 m depth. The lake is stream-fed, although the flow can be diverted to enable the lake to become completely isolated. The owner has observed that the lake does not freeze over in winter, even in extreme cold. This is probably due to its source which is pumped into the feed stream from a nearby sand quarry.

The lake is surrounded by overhanging trees (*Salix* sp.) on the northern banks, with a lawned garden and a seasonally grazed meadow along the south bank. The bankside is colonised by emergent *Iris pseudacorus*, *Sparganium erectum* and *Juncus* spp. Floating and submerged aquatic vegetation includes *Glyceria fluitans* and *Callitriche* spp. Very few areas of bare hydrosoil exist. The lake is well stocked with fish.

C. helmsii had been observed at this site for more than 5 years prior to this study. The lake has been the subject of former trials, with various herbicidal treatments having been applied since 1989 (Spencer-Jones 1994). The final application of diquat alginate in October 1991 had, on visual assessment, left the lake almost completely clear of *C. helmsii*. At the time of the present trials however, the plant was once again growing aggressively in its submerged form (approximately 85% cover of hydrosoil). It was also present but to a much lesser extent, in small patches amongst the marginal vegetation in its emergent form (approximately 5% of bankside covered).

Site 2

The second site, an ornamental lake, is situated on a large well managed estate ca. 3 km northeast of East Grinstead, West Sussex. This lake, slightly under 11,000 m^2 in area and 2 m depth, is one of a series of three lakes which are spring-fed. The lake margins are surrounded by emergent vegetation including *Iris pseudacorus*, *Menyanthes trifoliata*, *Sparganium erectum*, *Typha latifolia*, *Scirpus lacustris*, *Eleocharis palustris*, and *Juncus* spp. On an initial visit to the site, the lake was populated by a small flock of Canada geese (*Branta canadensis*).

C. helmsii was definitely not present at this site in 1990, but was recorded early in 1992 in its emergent form along the bank margins. No treatment for control of *C. helmsii* had been carried out at this site prior to this study. At the time of the current trials, the plant was present in its emergent form as a dense turf around the lake margins (approximately 50% cover), and in the shallower parts of the lake there were patches of the submerged rooted form extending into the water body for approximately 2-3 m in places (approximately 5% cover).

Methods

Visits were made to both sites for treatment and monitoring throughout the summer of 1992. Water samples were collected on each visit and were analysed on site for pH and conductivity. Laboratory analyses were conducted for total oxidised nitrogen (as NO_3), phosphate (as PO_4) and chloride content using 'Palintest'. A visual assessment of the extent of *C. helmsii* was made, based on percentage ground/hydrosoil cover and height of emergent plants.

Site 1

The first treatment over the whole water surface (*i.e.*, the submerged form) was carried out on 3 August from the bank, using diquat alginate applied at a rate of 10 l/ha (containing 100 g/l active ingredient as diquat cation in a viscous gel formulation).

Re-treatment was carried out on 15 September using the liquid formulation of diquat applied at rates of 25 and 50 l/ha (containing 200 g/l active ingredient as dibromide) from a boat as a surface spray. The dose was based on the respective water volumes of the shallow and deep areas and the whole of the water surface was treated. This treatment was also considered suitable for the marginal areas in order to control the emergent form of the plant which by this stage was increasing around the lake margins.

Site 2

Treatment of the emergent form was carried out on 3 August on three plots each of 1 x 6 m, using glyphosate at 6 l/ha product in 250 l/ha water (2.46 kg/ha active ingredient as isopropylamine salt). The herbicide was applied as a foliar spray using a small hand-held pressure sprayer. A control plot of similar size was left untreated.

Treatment was also carried out on a small (1 x 4 m) plot of the submerged plant on 20 July using dichlobenil granules at 150 kg/ha (10.1 kg/ha active ingredient as dichlobenil).

Results

Water chemistry

At both sites the water was slightly basic with a low conductivity. Water entering the lake at Site 1 is not subject to agricultural run off and there is no industrial or domestic input to either water body. Nutrient levels were generally low which is to be expected from isolated water bodies (see Tables 1 and 2).

Site 1

Monitoring the effects of diquat alginate showed that poor control had been achieved (see Table 1 for the summary of results). The plant showed some initial die-back of the lower leaves but growth had continued and the apical tips of each stem were green and healthy. By September, the submerged form was covering up to 95% of the hydrosoil area. Some submerged *C. helmsii* had grown up to the water surface and had begun to float. One of the effects of the treatment with diquat alginate was to cause brittleness of the plant. This resulted in fragmentation causing large masses of fully viable free-floating matter with many plant fragments showing vigorous growth

Table 1. Visual assessment of cover of all forms of *Crassula helmsii* at Site 1 over the trial period. Results of water analyses are also presented (conductivity in µS, other values in mg/l). Number of days after the treatment follows the date in brackets).

	Water analysis					Visual assessment		
Date	pH	Conductivity	NO_3	PO_4	Cl	Submerged form	Floating form	Emergent form
4.6.1992	6.9	410	4.40	0.22	40	85% cover of hydrosoil; plant rooting and shooting, regenerating particularly in shallower part of lake	none	5% cover of bank margins; small patches of dense turf, localised around lake margins, particularly amongst *Iris pseudacorus*
20.7.1992	-	-	-	-	-	virtually 100% cover of hydrosoil in shallower area, many large areas, bottom rooted with bright green new growth; 60-75 % in deeper area, large amount of new growth	none	10% cover of bank margins; patches of dense turf, some new areas infested
3.8.1992	-	-	-	-	-	TREATMENT: diquat alginate		
14.8.1992 (11 days)	7.0	410	1.88	0.14	40	95-100% cover of hydrosoil; plants in centre of lake, bright green and healthy, plants in shallow areas, brittle and stripped of lower leaves	10% cover of lake surface, large mats of broken stems, still green and shooting	10% cover of lake margins as dense patches of turf
7.9.1992 (35 days)	7.4	360	4.80	0.12	20	95% cover of hydrosoil; roots broken and disintegrated, lower leaves decayed, upper leaves very green, thriving	70% cover of water surface, large mats of floating stems with new shoots proud of the water surface and fine white roots 3-5 cm long	15-20% cover of lake margins in patches of dense turf, flowering, reddish/pink colouration to stems
15.9.1992	7.7	360	5.40	0.18	40	TREATMENT: diquat	TREATMENT: diquat	TREATMENT: diquat
1.10.1992 (16 days after diquat, 59 days after diquat alginate)	7.3	360	6.00	0.34	40	95% cover of hydrosoil; some lower leaves stripped, the majority of submerged material shooting and rooting	70% cover of water surface, fragmented into small lengths, broken/stripped lower leaves but healthy green shoots at apices	15% cover of lake margins, dense patches of turf, some brown/dead material but new regenerating material underneath

Table 2. Visual assessment of cover of both forms of *Crassula helmsii* at Site 2 over the trial period. Results of water analyses are also presented (conductivity in µS, other values in mg/l).

	Water analysis					Visual assessment		
Date	pH	Conductivity	NO_3	PO_4	Cl	Submerged form	Emergent form	Control plot
4.6.1992	7.5	230	0	0.24	20	5% cover of lake hydrosoil; blackened stems and lower leaves with bright green shoots and white roots from the nodes	50% cover of lake margins; 100% plot cover; dense turf	100% cover; dense turf
20.7.1992	-	-	-	-	-	15-20% cover of treatment plot, patchy TREATMENT: dichlobenil	60% cover of lake margins; 100% plot cover	100% cover; dense turf
3.8.1992	-	-	-	-	-		TREATMENT: glyphosate	
14.8.1992	7.3	210	0.003	0.06	20	patchy cover 1-2 m from the bank (26 days after treatment)	60% plot cover; turf 4 cm in height, 15-20% vigour (11 days after treatment)	100% plot cover; turf 9 cm in height
7.9.1992	8.3	230	0.006	0.46	20	patchy cover 1-2 m from bank	turf 4 cm in height (35 days after treatment)	100% plot cover; turf 15 cm in height, flowering
1.10.1992	7.9	210	0.028	0.28	20	patchy cover 1-2 m from bank	treatment band still distinguishable but new green shoots appearing (59 days after treatment)	100% plot cover; dense turf in flower

both as rooting from the nodes and as shoots emerging from the water surface. When assessed, floating material (which had not been present prior to the treatment) was covering 70% of the water surface.

After re-treatment with diquat, the growth of *C. helmsii* was not affected, either in its rooted submerged, floating or its emergent form. The other bankside vegetation including *Iris*, *Sparganium* and *Juncus* had however been very severely affected and was considerably damaged.

Site 2

When the site was visited 11 days after treatment with glyphosate, the water level of the lake had risen due to heavy rainfall, leaving some of the treated material under the water. The treated emergent plants which were still above the water were significantly yellowed compared to the control plot (see Table 2 for a summary of the results).

The glyphosate-treated plots after 35 days showed considerable browning of the leaves and stems with reduced growth. Below this mat of rather brittle material, the plant was beginning to throw up green and healthy shoots. Some of the treated plots were still under water and consisted of apparently dead, dark brown material. However, some of the stems which appeared to be dead, on closer inspection had fresh green shoots at the apices and were rooting from the nodes. Although this material was gaining in vigour, it had been severely affected by the treatment. By comparison, the control plot was now a luscious turf and was flowering profusely.

On the final visit to the site, the plots were again under water although a treatment band could still be distinguished across the glyphosate-treated plots. The growth had been severely checked but new growth with bright green shoots was appearing underneath the mat of dead material. The plant was still in flower on the control plot.

The submerged form treated with dichlobenil granules showed no change throughout the observation period.

Discussion

Analysis of the water samples shows that the values obtained from the trial sites were within the limits observed at other *C. helmsii* sites (Dawson and Henville 1991) in terms of nutrient availability for plant growth.

A previous trial carried out at Site 1 with diquat alginate (Spencer-Jones 1994) resulted in a reasonable level of control of *C. helmsii*. However, at the time of the first diquat alginate treatment in 1992, the flow of water through the lake had not been stopped, with the result that during treatment and for some hours after, diquat-containing water continued to flow out of the lake, thus diluting the herbicidal effect. It is possible that the poor control of submerged growth was due to this.

Dawson and Henville (1991) suggest that the typical biomass of submerged *C. helmsii* is considerably greater during the herbicide application period than that of many other aquatic plants. This low ratio of herbicide to weed biomass could further explain why poor results have been observed in the herbicidal treatment of submerged *C. helmsii*. If this is a valid assumption, then treatment of the plant early in the season before the biomass becomes excessive is indicated.

One effect of the diquat alginate treatment was to cause stem brittleness of the submerged plant. This caused fragmentation of the treated biomass which in turn resulted in large free-floating masses of vegetation. The lack of containment of this floating material probably resulted in some increase in the cover of emergent *C. helmsii* around the lake margins, with previously unaffected sites becoming colonised following treatment. Dawson and Henville (1991) recommend containment of free-floating matter following diquat alginate application and it is suggested that had this procedure been followed, the subsequent spread of the plant in this trial would have been limited.

The second treatment at Site 1 with liquid diquat had no effect on any of the forms of *C. helmsii*, despite controlling filamentous algae present in the lake and emergent vegetation, such as *Iris*, *Sparganium* and *Juncus* which colonised the lake margins. It is difficult to explain why this was so.

At Site 2, immediately after treatment with dichlobenil a period of heavy rainfall resulted in a marked increase in the water level of the lake. The mode of action of this herbicide is by soil uptake *via* the plant roots. As no discernible effect of this treatment was noted, it is suggested that either the herbicide granules had been washed away before becoming fully absorbed by the hydrosoil or the heavy biomass of the vegetation prevented the granules from reaching the hydrosoil resulting in negligible uptake by the roots.

Glyphosate at Site 2 achieved some measure of control over the trial period. Although a second treatment may have affected a greater level of lasting control, this was prevented by the rising water level of the lake which covered the emergent plots during much of the trial.

A factor in the lack of adequate control of *C. helmsii* from some herbicides could result from the prolific growth rate of the plant and its ability to regenerate from single nodes. Total elimination of the plant appears to be necessary to ensure that there is no subsequent colonisation at a site following treatment.

Acknowledgments

The authors wish to thank the land owners for allowing the use of their lakes for trials and also ICI Professional Products for provision of the diquat alginate, dichlobenil and diquat used in these trials. This study was supported by the National Rivers Authority under R and D Project 294.

References

Asovsky, M.G. 1981. The finds of rare for the east Siberia littoral-aquatic and aquatic plants, discovered on the Baikalo-Amur route. Bot. Zh. 66: 1218-1219.

Casper, S.J. and Krausch, H.D. 1981. Süsswasserflora Mitteleuropas. Vol. 2. Gustav Fischer, Stuttgart.

Clapham, A.R., Tutin, T.G. and Moore, D.M. 1987. Flora of the British Isles. 688 pp. Cambridge University Press, Cambridge.

Dawson, F.H. and Henville, P. 1991. An Investigation of the Control of *Crassula helmsii* by Herbicidal Chemicals (with Interim Guidelines on Control). Final report. Report to the Nature Conservancy Council. 107 pp.

Dawson, F.H. and Warman, E.A. 1987. *Crassula helmsii* (T. Kirk) Cockayne: is it an aggressive alien aquatic plant in Britain? Biol. Conserv. 42: 247-272.

Lousley, J.E. 1957. Botanical records for 1956. London Naturalist 36: 11.
Spencer-Jones, D.H. 1994. Some observations on the use of herbicides for the control of *Crassula helmsii.* In: L.C. de Waal, L.E. Child, P.M. Wade and J.H. Brock (eds.), Ecology and Management of Invasive Riverside Plants, pp. 15-18. John Wiley and Sons, Chichester.
Swale, E. and Belcher, H. 1982. *Crassula helmsii,* the swamp stonecrop, near Cambridge. Nature Cambs. 25: 59-62.
Wade, P.M., De Waal, L.C., Child, L.E., Dodd, F.S. and Darby, E.J. 1994. Control of Invasive Riparian and Aquatic Plants. 292 pp. R & D Project Record 294/7/W. National Rivers Authority, Bristol.

TREATMENT OF *FALLOPIA JAPONICA* NEAR WATER - A CASE STUDY

Louise C. de Waal
School of Applied Sciences, Division of Environmental Science, University of Wolverhampton, Wolverhampton WW1 1SB, United Kingdom

Abstract

A site with a dense monoculture of the invasive and vigorously growing *Fallopia japonica* was selected for herbicide trials. The site was chosen for regular treatments at different times throughout the growing seasons of 1991, 1992 and 1993. Plots within the site were treated with either 5 l/ha glyphosate or 5 l/ha glyphosate with an adjuvant (2% Mixture-B). Both treatments were made using a low water volume (80 l/ha) and were applied using a knapsack sprayer through a specifically recommended very low volume nozzle. A good level of control, up to 99% kill, was achieved with both treatments. A separate treatment using a long lance sprayer was carried out at the height of the growing season to apply glyphosate at a rate of 5 l/ha with a normal water volume of 180-200 l/ha. This method achieved good control for up to two years after treatment.

Introduction

Fallopia japonica (Houtt.) Ronse Decraene (Japanese knotweed), also known as *Reynoutria japonica* Houtt. and *Polygonum cuspidatum* Sieb. and Zucc., is a native plant of Japan, Taiwan and northern China. It was introduced into Europe as a fodder and ornamental garden plant in around 1825. Only 60 years later in 1886, *F. japonica* was naturalised in the British Isles (Conolly 1977). Since then it has spread rapidly throughout the British Isles, especially in Wales and around conurbations (Child *et al.* 1992). The species has become a serious problem in many habitats such as river corridors, road verges, railway embankments, waste ground, cemeteries, parks and gardens (Conolly 1977; De Waal *et al.* 1994).

In spring *F. japonica* grows vigorously from an extensive underground rhizome system. True *F. japonica* seeds are not found in the Bristish Isles as only male sterile plants have been recorded to date (Bailey 1990). The spread therefore is by vegetative means, primarily by rhizome fragments but also through sections of stem material (Brock and Wade 1992; Richards, Moorehead and Laing Ltd. and International Centre of Landscape Ecology 1993).

Although there is no legal requirement to control *F. japonica* in the U.K., it is an offence to plant or otherwise cause the species to grow in the wild under the Wildlife and Countryside Act 1981, Section 14, Schedule 9 (Part II). The success of a control programme is largely dependent upon the elimination of the large rhizome system and therefore the use of a translocated herbicide is essential. From previous studies on the chemical treatment of *F. japonica* two translocated, non-persistent herbicides, glyphosate and 2,4-D amine, both currently approved for use in or near water under the Control of Pesticides Regulations 1986 (Department of the Environment 1992), have been shown to achieve some control. Roblin (1988) states that spraying with 6

Plant Invasions - General Aspects and Special Problems, pp. 203-212
edited by P. Pyšek, K. Prach, M. Rejmánek and M. Wade

l/ha of glyphosate early in the year will arrest the plant's growth, however, is unlikely to kill the plant. Spraying the plants during the peak of their activity (July) at a lower concentration (4 l/ha) should be sufficient to kill the plants, but the size of the plants will make spraying difficult. Beerling (1990) has studied the effects of 2,4-D amine applied twice a year at a rate of 2790 g a.i./ha and glyphosate applied once and twice a year at a rate of 2154 g a.i./ha. Both herbicides achieved good control, *i.e.,* reduced biomass and leaf area by the end of the season. However, in neither case has long term treatment and/or monitoring been undertaken. Therefore, the main objective of this research is to study the long-term effects of chemical treatment with glyphosate on the growth of *F. japonica.* Further specific aims are to study the effects of timing and the practicalities of the different treatments and the effects of the combination of mechanical and chemical treatments on the control of *F. japonica.*

Site location

To undertake this research an experimental area was selected riparian to a river corridor and highly infested with the invasive and vigorously growing *F. japonica* which had not been treated before. The site is approximately 10 km north-east of Swansea, Wales, in between Glais and Pontardawe, National Grid Reference SN 712 021. The site, on the floodplain of the River Tawe, lies between a grassland area to the south-east and the river channel to the north-west. The main part of the site, approximately 0.2 ha, is densely covered with *F. japonica*. Associated species recorded at this part of the site are *Alnus glutinosa, Arum maculatum, Calystegia sepium, Cirsium* spp., *Filipendula ulmaria, Heracleum sphondylium, Impatiens glandulifera,* and *Rubus fruticosus* agg. At the north-western part of the site towards the river, there is a mixed vegetation of *F. japonica* with predominantly *Ulex europaeus, Cytisus scoparius* and *Carex* spp. The whole of this site is privately owned and is not readily accessible to the public.

Methods

The site was visited for treatment and monitoring during the growing seasons of 1991-93. An assessment of the state of the plants was made at each visit. Percentage cover was recorded by visual assessment of three randomly chosen 1 m^2 plots. If percentage cover was smaller than approximately 5%, density of stems was recorded by counting the number of shoots per m^2. The height of the stems in each randomly chosen 1 m^2 plot was measured (in metres) and the average over the three plots calculated. The state of the leaves, *i.e.* colour and size, was assessed visually as well as the presence/absence of flowers of *F. japonica.* Furthermore, percentage cover of other native species colonizing the plots was recorded as described above. Tables 1 and 2 give an account of all the site visits made, the types of treatment undertaken in 1991-93 and monitoring details obtained.

In July 1991 herbicide trials were carried out on fully grown *F. japonica* plants with an average height of 2.5 m. In order to gain access through the area and to make treatment possible, paths were cut through the dense and tall *F. japonica* stand using a brush cutter with a metal blade. An area of approximately 100 m^2 was treated with

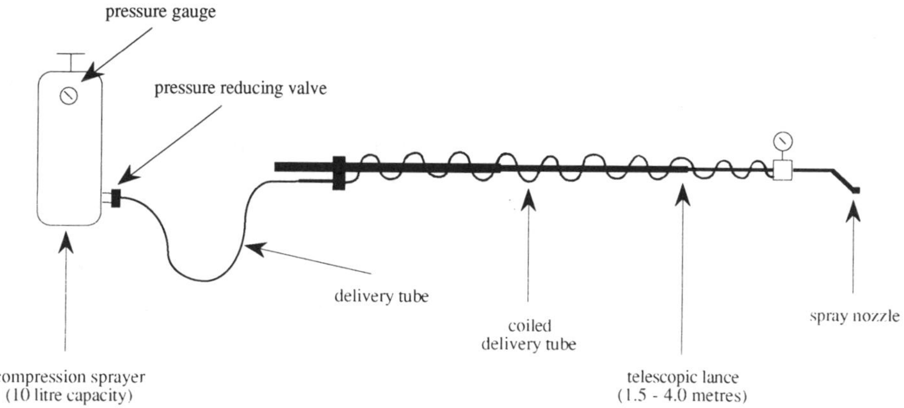

Fig. 1. Long lance sprayer fitted with a polijet nozzle.

glyphosate (as Roundup® Pro) applied by means of a long lance sprayer fitted with a polijet nozzle (Fig. 1). The herbicide was applied at a rate of 5 l/ha (2.4 kg/ha active ingredient as isopropylamine salt) diluted with a normal water volume (180-200 l/ha).

In March 1992, experimental plots were established in the area highly infested with *F. japonica*. The dead, semi-woody stems from the previous years were cut to ground level using a brush cutter with a metal blade. This enabled the spray to be directly targeted onto the green leaves rather than being intercepted by the dead stems and also made movement through the stands easier when spraying was carried out later in the growing season. The area was divided, using hazard tape, into three major plots (plots 1 to 3) each with an approximate area of 100-120 m^2 and subdivided into two equal subplots (A and B) (Fig. 2). Plots 1 and 2 were subsequently treated to compare the practicalities and effectiveness of treatments at different times throughout the growing season. The subplots, A and B, were treated with either glyphosate (as Roundup® Pro) or with glyphosate and an adjuvant (Mixture-B^{TM}). Both treatments were undertaken by means of a conventional knapsack sprayer fitted with a very low volume nozzle. The herbicide was applied at the same rate as for the long lance sprayer, however, in a low water volume (80 l/ha). The adjuvant was added at a rate of 2% of the spray volume. The first applications were made with the plants at a

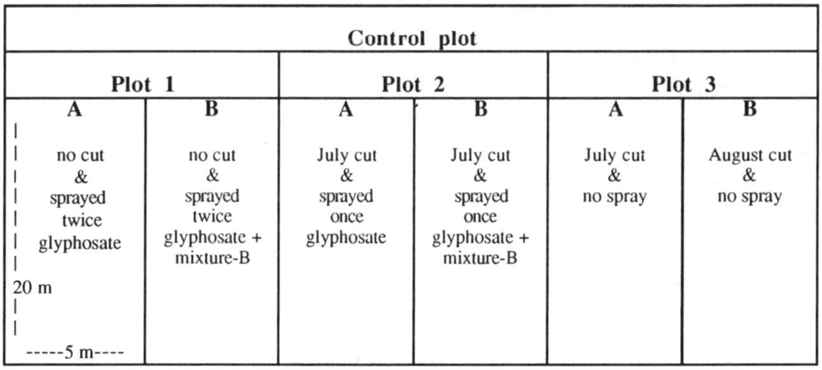

Control plot					
Plot 1		**Plot 2**		**Plot 3**	
A	**B**	**A**	**B**	**A**	**B**
no cut & sprayed twice glyphosate 20 m -----5 m----	no cut & sprayed twice glyphosate + mixture-B	July cut & sprayed once glyphosate	July cut & sprayed once glyphosate + mixture-B	July cut & no spray	August cut & no spray

Fig. 2. Plan showing the experimental plots with an indication of the various treatments.

height of 1.6 m and with sufficient leaf area available to allow effective uptake of the herbicide. Where possible a second application was carried out later in the season. The timing of spraying was accommodated according to weather conditions. When the plants exceeded a height of 1.6 m they were cut with the metal blade of a brush cutter to reduce the height and vigour of the plant. All the biomass was removed from the site, stacked in piles and left to dry out. Re-growth in the plots was treated as described above. Plot 3 was cut at different times throughout the growing season to compare the effectiveness of herbicide treatment against cutting only.

In the summer of 1993, plots 1 to 3 were monitored to assess the effectiveness of the various treatments undertaken in the summers of 1991-92.

Results

Long lance sprayer

- *1991:* During an initial visit in spring 1991 a dense monoculture of *F. japonica* was recorded. Seventy-three days after the first treatment, the plants were still alive and flowering (Table 1). Approximately 70% of the leaves were green and appeared to be healthy. The remaining 30%, mainly the top leaves, were visually affected by the treatment. These affected leaves appeared burned and had a yellow and brown discolouration, whereas the mature plants in the control plot were healthy throughout the growing season.
- *1992:* Although the treatment initially appeared to be unsuccessful, its long-term effects were more evident in 1992 and 1993. In May 1992, monitoring showed that the plants in the treated area were set back in growth in both percentage cover and height, in comparison to the untreated control plot (Table 1). They were also clearly deformed with smaller and more pointed leaves. In July 1992, two separate areas could be distinguished within the treated area. In area 1, *F. japonica* plants appeared to be deformed with long and narrow leaves appearing in small bunches growing from the axils of the branches. They were often shrivelled with yellow and brown edges. The upper leaves were approximately 10 mm long by 2-4 mm wide and the lower leaves 25-30 mm long by 10-20 mm wide. In area 2, both percentage cover and height were only slightly set back compared to the control plot, however, the leaves were still deformed. The upper leaves were 25-30 mm long by 10-20 mm wide. The lower leaves were of a normal shape, but small compared to the control plot. The differences between areas 1 and 2 were probably due to the number of leaves being wetted during the herbicide application in 1991. It is likely that in area 1 more leaves were properly wetted due to the more accurate direction of the spray, which facilitated the plant to take up herbicide more effectively.
- *1993:* In mid-May 1993, the treated plants were still deformed with a maximum height of 0.2 m. The leaves were long and narrow and the stem purplish-red coloured rather than speckled.

Table 1. Record of site visits and monitoring details (1991-93) for *Fallopia japonica* plots with long lance sprayer treatment. Mean values from three 1 m² plots are given. See text for details on area 1 and area 2 data. Cover in %, height in metres. Key to activities: A: treatment with 5 l/ha glyphosate, normal water volume; M: monitoring.

Date		Control	Treated area	
15.06.91	treatment	M	M	
	cover	100	100	
	height	2.5	2.5	
21.07.91	treatment	M	A	
	cover	100	100	
	height	2.5	2.5	
02.10.91	treatment	M	M	
	cover	100	100	
	height	2.5	2.5	
	flowering	yes	yes	
08.05.92	treatment	M	M	
	cover	40	5	
	height	0.8	0.3	
			Area 1	Area 2
07.07.92	treatment	M	M	M
	cover	100	10	80
	height	2-2.5	0.3-0.4	1.5
17.05.93	treatment	M	M	
	cover	80-90	5-10	
	height	1.5	0.2	

Knapsack sprayer low water volume

In early May 1992 the percentage cover of the whole *F. japonica* stand was approximately 20% with an average stem height of 0.8 m (Table 2). Most stems had only six leaves, though not fully grown, and the top leaves were still rolled back.

Plots 1A and 1B

- *1992:* In early June, the first herbicide treatment was carried out. Thirty-five days after this treatment plant height had not increased (Table 2 and Fig. 3). In plot 1A approximately 50% of the leaves were affected by the treatment. These leaves were partly yellow and/or red-brown and some of them were shrivelled. Re-growth was present from the axils of the branches. In plot 1B approximately 60% of the leaves were affected, they were either discoloured like those in plot 1A or had fallen off. No re-growth appeared in this plot.
  In July 1992, 57 days after the first treatment, 50% of the leaves had fallen off and the remaining leaves appeared pale lime-green and brown in both plots. Re-growth was present from the axil of the branches, however, these leaves appeared brown and unhealthy. During this visit the second treatment was applied to both plots.
  In August, 83 days after the first and 26 days after the second treatment, most leaves had fallen off and the remaining leaves appeared unhealthy and discoloured in both plots.
  In September, 48 days after the second treatment, 100% kill was recorded and no re-growth was observed in either of the plots (Table 2 and Fig. 4).
- *1993:* In May during the following growing season hardly any *F. japonica* re-growth had appeared in either plot 1A or 1B. The cover was approximately 1% and this was still the case in September. The plants were deformed reaching a height

Table 2. Record of site visits and monitoring details (1992-93) for *Fallopia japonica* plots with knapsack sprayer treatment. Mean values from three 1 m^2 plots are given. Cover in %; height in metres; density in shoots/m^2. Key to activities: A: cutting with brush cutter to remove semi-woody stems after winter period; B: treatment with 5 l/ha glyphosate, low volume, special nozzle; C: treatment with 5 l/ha glyphosate + 2% Mixture-B, low volume, special nozzle; D: cutting with brush cutter to reduce vegetation height; M: monitoring. Number of days after the last treatment is indicated.

Date		Control plot	Plot 1		Plot 2		Plot 3	
			A	B	A	B	A	B
30.03.92	treatment	M	A	A	A	A	A	A
08.05.92	treatment	M	M	M	-	-	-	-
	cover	20	20	20				
	height	0.8	0.8	0.8				
02.06.92	treatment	M	B	C	M	M	-	-
	cover	100	100	100	100	100		
	height	1.6	1.6	1.6	1.6	1.6		
	days after	-	0	0	-	-		
07.07.92	treatment	M	M	M	D	D	-	-
	cover	100	75	30-50	biomass removed	biomass removed		
	height	2.25	1.6	1.6				
	days after	-	35	35	0	0		
29.07.92	treatment	M	B	C	M	M	D	-
	cover	100	60	60	1	1	biomass removed	
	density				10	10		
	height	2.4	1.6	1.6	0.25	0.25		
	days after	-	0	0	22	22	0	
24.08.92	treatment	M	M	M	M	M	M	D
	cover	100	30	20	60	60	1	biomass removed
	density						15	-
	height	2.4	1.6	1.6	0.5	0.5	0.2	
	flowering	yes	no	no	no	no	no	-
	days after	-	26	26	48	48	26	0
15.09.92	treatment	M	M	M	M	M	M	M
	cover	100	0	0	75-80	75-80	2	<1
	density						20	1
	height	2.4	-	-	0.6	0.6	0.45	0.15
	flowering	yes	-	-	no	no	no	no
	days after	-	48	48	70	70	48	22
25.09.92	treatment	M	M	M	B	C	M	M
	cover	100	0	0	75-80	75-80	2	<1
	density						20	1
	height	2.4	-	-	0.6	0.6	0.45	0.15
	flowering	yes	-	-	no	no	no	no
	days after	-	58	58	0	0	58	32
10.11.92	treatment	M	M	M	M	M	M	M
	cover	10	5	5	5	5	5	-
	height	2.4	1.45	1.45	0.75	0.75	0.5	-
	days after	-	104	104	46	46	104	78
17.05.93	treatment	M	M	M	M	M	M	
	cover	80-90	1	1			60-70	
	density				20	10		
	cover others	-	<1	<1			-	
	height	1.5	0.8	0.8	0.2	0.05	1.1	
07.09.93	treatment	M	B	B	C	C	M	
	cover	100	<1	<1	5-10	5	30-40	
	density				20	10		
	cover others	-	80	80	60	50	10	
	height	2.5	0.3	0.25	0.15	0.1	1.3	
	flowering	yes	no	no	no	no	yes, but less than control plot	

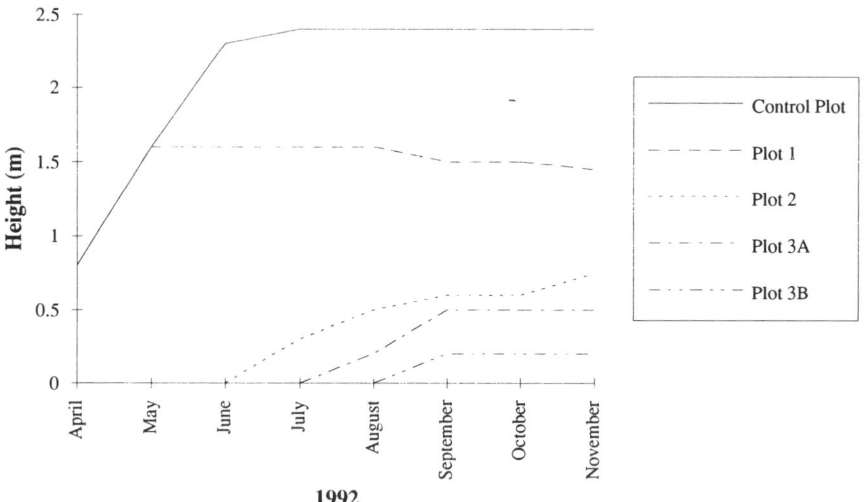

Fig. 3. Comparison of height of *Fallopia japonica* in plots 1 to 3 and the control plot for 1992. Values represent means from three 1 m^2 plots. Subplots of plots 1 and 2 did not differ in height. Complete winter kill is indicated by the end of the line.

of up to 0.5 m in September with narrow leaves of 30-50 mm long by 10-30 mm wide and hardly any flowers. Overall, a 99% kill was achieved in both plots 1A and 1B. Other vegetation started to regenerate with a cover of less than 1% in May and up to 80% in September (Table 2). Species included *Chenopodium polyspermum, Cirsium* spp., *Filipendula ulmaria, Galium aparine, Heracleum sphondylium, Hyacinthoides non-scripta, Impatiens glandulifera, Ranunculus* spp., *Rubus fructicosus* agg., *Rumex* spp., *Urtica dioica* and some grasses and mosses.

Plots 2A and 2B

- *1992:* In July, 22 days after cutting, the re-growth of the *F. japonica* stands was minimal (Table 2 and Fig. 4). Each new shoot had only produced 6-7 leaves, which were not fully grown.
  In August, 48 days after cutting, the height of the re-growth in both plots 2A and 2B was 0.5 m and the cover 60%. The stems produced plenty of leaves, most of which were fully grown.
  In early September, 70 days after cutting, a cover of 75-80% with an average height of 0.6 m was reached. The leaves were sometimes bored and/or nibbled by insects and none of the plants were flowering. In late September the plots were treated as describe above.
  In November, 46 days after the herbicide treatment, all the *F. japonica* plants appeared to be dead. The leaves were brown and the majority of them had fallen off the stems.
- *1993:* In May during the following growing season hardly any *F. japonica* re-growth had appeared in either plots 2A or 2B (Table 2). The re-growth in plot 2A was generally deformed with long and narrow leaves which were nibbled. In plot 2B the re-growth had just started to regenerate with deformed and red-purplish coloured stems and leaves.
  In September, at the end of the growing season, the plants were still deformed in

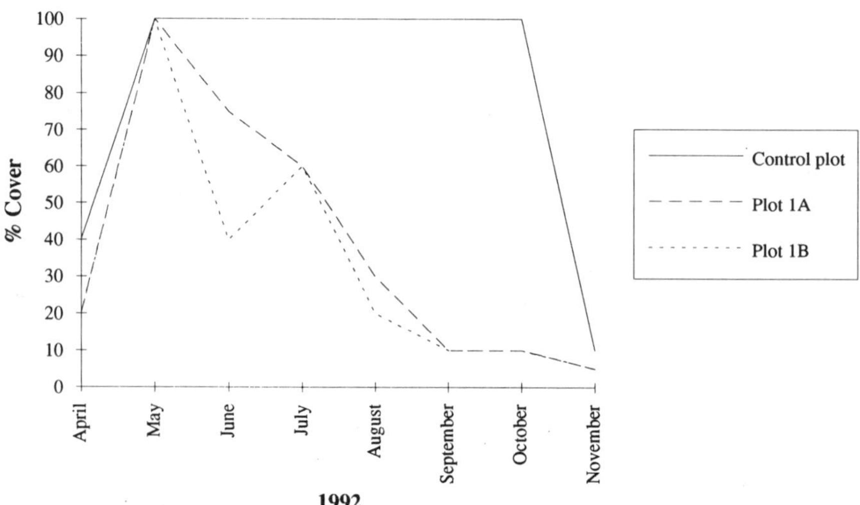

Fig. 4. Comparison of percentage ground cover of *Fallopia japonica* in plots 1A and 1B and the control plot for 1992. Means from three 1 m^2 plots are shown.

both plots with narrow leaves of 30 mm long by 10 mm wide. Overall, a 90% kill was achieved. Other species, starting to re-establish in both plots with a cover of 50-60%, included such species as *Chenopodium polyspermum, Cirsium* spp., *Filipendula ulmaria, Impatiens glandulifera, Rubus fructicosus* agg., *Urtica dioica* and grasses.

Plots 3A and 3B

- *1992:* Plot 3A was cut in July 1992 and plot 3B in August at a height of 2.4 m with a cover of 100% (Table 2).
  In August, 26 days after cutting, and in September, 48 days after cutting, the new shoots had produced very few leaves. Most of the leaves were still rolled back and some of which were damaged by insects. None of the plants was flowering.
  In November, the plants died back naturally.
- *1993:* In May, plots 3A and 3B were set back in growth, both in height and percentage cover. Approximately 20% of the leaves were brown, which was probably due to frost as a result of the exposed situation at the time. The plants were also grazed by cattle to a height of 1 m.
  In September, the stands were still set back in growth. The overall cover in plots 3A and 3B was 50%, with 30-40% *F. japonica* and approximately 10% grass cover, and a *F. japonica* plant height of up to 1.3 m. The plants were flowering, although with lower numbers than on the control plots (Table 2).

Discussion

This research was undertaken to study the long-term effects of chemical treatment with glyphosate and the effects of combined chemical and mechanical treatment on the growth of *F. japonica* and to study the effects and practicalities of timing of the different treatment methods. As a result of the nature of research, *i.e.,* using herbicides on *F. japonica* stands in the wild, it was not possible to follow an appropriate statistical design with several replicates. Neither was it possible to arrange the treatment plots randomly. Firstly, the size of the area necessary for treatment would be too large, and secondly, the spread of the herbicide through the rhizome system is very unpredictable. However, the following qualitative conclusions can be drawn from this research.

The study has shown that the herbicide glyphosate (as Roundup® Pro) applied at the top of the growing season using a long lance sprayer, at a rate of 5 l/ha with a normal water volume of 180-200 l/ha can achieve effective control of *F. japonica.* After the first year, follow-up treatments can be applied using conventional methods, such as a knapsack sprayer or even a weed wiper, as the re-growth will not exceed 0.5 m in height. However, the treatment of large and dense stands is very labour intensive and therefore expensive, as paths need to be cut before spraying is possible. Although, this method could be very effective on narrow and linear stands, such as river banks.

When using a knapsack sprayer for the application of herbicides to *F. japonica* stands, a pre-treatment is necessary to remove dead stems from the previous year, *e.g.*, using a brush cutter with a metal blade. The most effective timing of herbicide treatment appears to be early in the growing season, *e.g.*, in May, when sufficient leaf area has developed, but before the plant has reached an unworkable height, with a second application later in the season, *e.g.*, in July. Using this method, glyphosate applied at a rate of 5 l/ha with a low water volume has shown to achieve up to 99% control. The knapsack sprayer, however, needs to be fitted with a specifically recommended very low volume nozzle. Additionally, adjuvants may improve the activity of the herbicide and can slightly increase the effectiveness of the treatment.

If weather conditions do not permit spraying to be carried out before growth has exceeded an unworkable height (approximately 1-1.5 m), the plant can be cut to ground level using a brush cutter with a metal blade. After approximately seven weeks the re-growth should be treated as described above. However, the later the plants are cut the longer it takes to produce sufficient re-growth and leaf area to enable the plants to take up the herbicide effectively. The latest date for cutting to provide sufficient re-growth for treatment in the same growing season is late July-early August.

This study has also shown that cutting can be an effective, short term method of control. It reduces height and vigour of the *F. japonica* plants considerably in the same and the following growing season. Cutting the plants to ground level late in the growing season, *e.g.*, after mid August, produces little or no re-growth in the remaining period of the season and sets growth back in the following year. Beerling (1990) states that cutting might increase lateral rhizome growth and hence may encourage the spread of the plant. However, this has not been confirmed in this research.

If there is a risk of any plant material accidentally falling into a watercourse, the cut biomass should be removed from the site, as stems floating in water can produce new shoot growth and adventitious roots (Brock *et al.* 1995). The plant material should be stacked in piles at a safe site and left to dry out. The piles need careful monitoring throughout the growing season as healthy new shoot growth can also be produced at the bottom of the piles of plant material, potentially producing another new stand of *F. japonica.*

Regardless which control method is used, re-treatment will be necessary on an annual basis until no new shoots appear. After one or two years of applying glyphosate a substantial cover of grasses and herbs can be developed in the treated area. At this time, the use of a selective herbicide, *e.g.*, 2,4-D amine, and the use of a more precise applicator, *e.g.*, a weed wiper, should be considered to avoid damaging the established vegetation.

Acknowledgments

This research on the treatment of *F. japonica* was undertaken in 1991 and 1992 as a part of the National Rivers Authority's Research and Development project on the control and management of invasive riparian and aquatic plants. The author wishes to thank Monsanto plc. for supplying the chemicals used and for providing funding to continue this research in 1993. The author also wishes to thank the Aquatic Weeds Research Unit for the use of their long lance sprayer and the landowner for allowing the site to be used for this research.

References

Bailey, J.P. 1990. Breeding behaviour and seed production in alien giant knotweed in the British Isles, pp. 121-129. Proceedings of the Conference of the Industrial Ecology Group of the British Ecological Society.

Beerling, D.J. 1990. The Ecology and Control of Japanese Knotweed (*Reynoutria japonica* Houtt.) and Himalayan Balsam (*Impatiens glandulifera* Royle.) on River Banks in South Wales. 138 pp. Thesis, University of Wales, Cardiff.

Brock, J.H. and Wade, P.M. 1992. Regeneration of *Fallopia japonica*, Japanese knotweed, from rhizome and stems: observations from greenhouse trials. IXe Colloque International sur la Biologie des Mauvaises Herbes, pp. 85-94. Dijon.

Brock, J.H., Child, L.E., De Waal, L.C. and Wade, M. 1995. The invasive nature of *Fallopia japonica* is enhanced by vegetative regeneration from stem tissues. In: P. Pyšek, K. Prach, M. Rejmánek and M. Wade (eds.), Plant Invasions - General Aspects and Special Problems, pp. 131-139. SPB Academic Publ., Amsterdam.

Child, L.E., De Waal, L.C., Wade, P.M. and Palmer, J.P. 1992. Control and management of *Reynoutria* species (knotweed). Aspects Appl. Biol. 29: 295-307.

Conolly, A.P. 1977. The distribution and history in the British Isles of some alien species of *Polygonum* and *Reynoutria*. Watsonia 11: 291-311.

Department of the Environment. 1992. Guidance for Control of Weeds on Non-Agricultural Land. HMSO, London.

De Waal, L.C., Child, L.E., Wade, P.M. and Brock, J.H. 1994. Ecology and Management of Invasive Riverside Plants. 217 pp. John Wiley and Sons, Chichester.

Richards, Moorehead and Laing Ltd and International Centre of Landscape Ecology. 1993. Japanese Knotweed Control: Model Specifications and Knotweed Fragment Regeneration Study. Welsh Development Agency, Cardiff.

Roblin, E. 1988. Chemical control of Japanese knotweed (*Reynoutria japonica*) on river banks in South Wales. Aspect Appl. Biol. 16: 201-206.

Wildlife and Countryside Act. 1981. HMSO, London.

RHODODENDRON PONTICUM AND SOME OTHER INVASIVE PLANTS IN THE SNOWDONIA NATIONAL PARK

Rod H. Gritten
Snowdonia National Park, Penrhyndeudraeth, Gwynedd, LL48 6LS, United Kingdom

Abstract

Rhododendron ponticum is a very successful invasive species found growing in a large variety of different habitats covering a significant area of the Snowdonia National Park, North Wales. Its distribution and the reasons for its success as an invasive species are discussed. A scheme, now in its third year, to control an extensive area of *R. ponticum* is described and an assessment made to predict the ultimate outcome of the present control strategy. Finally, a general overview is given describing several other invasive species troubling the National Park and the reasons for their success are discussed.

Introduction

The Snowdonia National Park is a mountainous area of exceptionally wild character covering 2173 km^2 of North Wales, UK; its description has already been given by Gritten (1992).

Rhododendron ponticum L. was introduced into Britain in 1763 (Elton 1958), though when it arrived in Snowdonia is not known. Its natural distribution is disjunct, with *R. ponticum* ssp. *baeticum* in Spain and Portugal and *R. ponticum* ssp. *ponticum* in Bulgaria, Turkey and the Lebanon (Shaw 1981). It was probably first planted in Snowdonia as an ornamental shrub in large estates. It was also used extensively as rootstock for grafting ornamental varieties. Subsequent neglect of grafted bushes led to the phenomenon of reversion, where the rootsock developed shoots from the base and the scion became overwhelmed quite quickly (Shaw 1984). It was also extensively used as a low cover in shelter belts and game coverts, the latter principally in woodland. Its biology and the reasons for its success are fully described by Cross (1973, 1975), Gritten (1987, 1988), Rotherham (1983) and Shaw (1984), though it may be appropriate to summarise some of these here.

Some aspects on the biology of *Rhododendron ponticum*

It is not known when *R. ponticum* first became perceived as a problem in Snowdonia. Indeed, it is unclear as to when it first began to be invasive. It is a well documented phenomenon, however, (see Elton 1958, and numerous papers in Drake *et al.* 1989) that there is often a considerable lag between the time a plant species is first introduced into a different country and when it first becomes truly invasive there (Kowarik 1995). The core distribution of *R. ponticum* in Snowdonia appears to be associated with large estates, many of which were the homes of successful slate quarry owners. As the slate industry waned around the turn of the century, however,

Plant Invasions - General Aspects and Special Problems, pp. 213-219
edited by P. Pyšek, K. Prach, M. Rejmánek and M. Wade

gardeners were probably the first casualties of declining fortunes and the prolific seed production of *R. ponticum* allowed the shrub to spread rapidly into the surrounding countryside. High humidity, mild winters and acid soils all favour the growth of *R. ponticum*. Its spread is aided by the fact that its tiny seeds are easily wind dispersed, it is unpalatable to grazing animals, it is remarkably shade tolerant, it is long lived and is apparently free from pests and diseases. Heavy sheep grazing and burning may well be factors contributing to its spread. It is also able to spread by layering. Allelopathy may also be a factor in its competitive success and so too mycorrhizal association (Rotherham, 1983; Rotherham and Read, 1988) though the latter assertion is refuted by Cross (1973, 1975) and Tester (1988).

Distribution of *Rhododendron ponticum* in Snowdonia

The first survey of *R. ponticum* in the Park was carried out in 1984 by asking approximately 100 selected local people if they knew where the shrub was growing in areas they knew well. This survey was carried out on a presence/absence per km basis. It was revealed that 28% of the kilometre squares covering the park had stands of *R. ponticum*. A much more thorough ground survey was carried out during the winter of 1985-1986, largely on horseback and this not only confirmed the accuracy of the first survey but showed that *R. ponticum* covered some 3407.5 ha of the Park or 1.6% by area (Gritten 1987). Although no surveys have been carried out since then, anecdotal evidence from many sources indicates that the plant is still spreading. A number of different bodies are controlling stands in a variety of habitats but it is considered that, overall, a larger area is now affected.

Significantly, the habitat most affected is coniferous woodland where 49.8% of the stands were found. Conditions of high acidity, high humidity and a measure of ground disturbance are found in abundance in forests and it is under such conditions that *R. ponticum* thrives (see Shaw 1981 and 1984 for a detailed discussion on the ecology of *R. ponticum*). It is of interest to note that *R. ponticum* is one of the few plants that is able to grow in abundance in the very low light intensities found under commercial coniferous plantations, though flowering may be suppressed. It has been suggested, however, that *Tsuga heterophylla* (Raf.) Sang. may cast a shade of sufficiently low light intensity to destroy *R. ponticum* (Forest Enterprise, personal communication). By their very nature, coniferous forests were the least well surveyed and it is likely the problem is a great deal more extensive in this habitat.

The Snowdonia National Park Authority (SNPA) has a presumption against further large scale commercial coniferous afforestation in the area (National Park Plan Review 1986). Of more significance on landscape grounds, therefore, are the stands to be found outside of coniferous plantations.

14.7% of stands were found in broadleaved woodland. Many of these stands were originally planted as understorey for pheasant rearing and have been spreading ever since. These woodlands, which are composed largely of *Quercus petraea* (Mattuschka) Lieblein and *Betula pubescens* Ehrh. are considered by the SNPA to be of great landscape and wildlife significance. Considerable staff time and resources are spent in conserving them. Some 80-90% of these woodlands are currently grazed by sheep (internal SNPA report) and so little or no natural seedling regeneration is occurring. A combination of allelopathy and the deep shade cast by *R. ponticum* stands also prevents tree seedling regeneration. Indeed, the moss *Isopterygium elegans* (Brid.) Lindb. appears to be one of the few plants that can tolerate conditions found under dense *R. ponticum*.

Of even greater significance in the Snowdonia landscape are the stands found on the open mountain (23.5%). It is not uncommon to find entire hillsides covered by dense and impenetrable stands. Whilst very considerable quantities of seed are produced by bushes growing in these open conditions every year, conditions for successful germination are fortunately only achieved irregularly. A combination of wet springs and the availability of suitable seedbeds - which in turn appear to be created by a certain level of sheep grazing and perhaps burning - limit successful germination to about one year in seven. However, though seedling invasion of areas peripheral to established flowering colonies is confined to certain years, the infilling of existing stands of scattered bushes by natural growth and by layering is a rapid, continuous and readily observable phenomenon in many areas.

The part played by sheep grazing in the spread of *R. ponticum* is not clearly understood. Thomson *et al.* (1992) using aerial photography and multiple regression techniques have shown that *R. ponticum* appears to favour areas with lower sheep grazing intensities. However, it has been suggested in the past (Shaw, personal communication) that seed germination is more successful on a closely-nibbled turf or one that has been recently burned. It would be very useful to establish what part is played by grazing (and burning) in the spread of *R. ponticum* in open mountain habitats.

7.8% of stands were found growing in private gardens. This represented 633 sites and, significantly, 33% of these sites contained stands which were deemed to be invasive.

Control and management of *Rhododendron ponticum*

Control methods

A variety of methods has been and is currently being used to control *R. ponticum*. In habitats where herbicide use is considered undesirable *and* where resources allow, bushes are cut down, burned and the stumps winched out. Often, where herbicide use is favoured, cut stumps have been treated with Amcide or Glyphosate, though success using this method is often variable. Increasing the surface area of the cut stumps by drilling or scoring with a chainsaw prior to herbicide application has resulted in little additional success.

In practice, the most effective method of treatment has been to cut and burn bushes and to spray resultant regrowth arising from the cut stumps up to a year later with a mixture of Glyphosate and Mixture 'B' (Tabbush 1987). Timing of the spray is important, a warm day in late July-August being the most effective. Where herbicide treatment has been considered to be inappropriate, continual slashing of the regrowth over a several year period has been found to be time-consuming but effective.

Shaw (1984) has discussed in depth various ecological reasons for the success of *R. ponticum* as an invasive species but it is worth highlighting a few additional points here.

The control of *R. ponticum* is an extraordinarily labour-intensive task. All but the tiniest of bushes have to be cut down and their regrowth sprayed. *R. ponticum* has very poor tangential translocation so that for herbicide treatment to be effective almost every leaf has to receive a dose of chemical. Only three quarters of a bush will be destroyed if only three quarters' is sprayed and clearly it is impossible and undesirable to spray the entire leaf surface of bushes some of which can reach up to five metres in height. Spraying a season's regrowth thoroughly has been shown to be the most effective method of treatment. Such labour intensive methods, whilst effective, are nevertheless extremely expensive and one might argue that one of the successes of this plant as an invasive is the astronomical costs of control on a Park wide basis, estimated at £45 million at 1992 prices.

A second reason for the success of this plant is the rapidity with which cleared areas are recolonised by seedling re-invasion. Once an area has been cleared of large bushes, a highly acidic and, often, homogeneous litter remains which is an ideal seedbed for *R. ponticum* germination. Preliminary (unpublished) results from the analysis of permanent quadrats laid down by the author in areas cleared of dense stands in 1993 show that *R. ponticum* seedlings are amongst the first species to recolonise. This would appear to be in part due to lack of competition from other colonising species which are perhaps inhibited by allelopaths remaining in the soil. Apart from re-invading *R. ponticum* seedlings, there is evidence to suggest that mosses are the first species to appear on ground cleared of large bushes.

Where volunteer labour is available, re-invading seedlings can be removed by hand. This is only effective, however, if seedlings are over five years old since early development is slow and seedlings younger than this are tiny. Whilst herbicide control of seedlings has been employed, this can be undesirable since other successful colonising species are also affected. In principle, it would be best to control large areas of infestation rapidly so that seedling re-invasion from neighbouring flowering bushes can be kept to a minimum. Cleared areas do not possess a persistent seedbank since the viability of *R. ponticum* seed is only seven months.

Now that so many coniferous forests in Snowdonia are ready for felling, there is increasing concern over the potential for *R. ponticum* to spread into those areas since the litter formed under conifers and the resultant increase in light intensity creates an ideal seedbed for germination.

Current control initiatives and their success

The SNPA, forestry bodies, National Trust, the Countryside Council for Wales, the Royal Society for the Protection of Birds and private landowners are all controlling stands of *R. ponticum* in Snowdonia but their actions are largely uncoordinated and control successes are probably of little overall significance. It is quite likely that the plant is spreading faster through the Park than it is being controlled.

The initial stages of one control scheme involving volunteer labour largely provided by the Army was described by Gritten (1992). In this, teams of army personnel are sporadically cutting and burning a 90 ha stand covering some rugged mountainous terrain near Beddgelert. This programme has now been stepped up by the addition of a permanent team of Employment Training personnel partially funded by the SNPA and jointly administered by the SNPA and the National Trust. This approach is slow, clearance of approximately 4-5 ha being achieved annually. Private contractors, though expensive, are motivated to clearing much bigger areas much more quickly.

Recent trials within the same programme, however, using a Menzi 6000 walking excavator have shown considerable promise. This Swiss-made machine, able to operate successfully on rugged slopes of up to 45°, incorporates a powerful hydraulically-driven flail mower attachment which macerates *R. ponticum* bushes of even 150 mm stem diameter into small pieces. The machine is able to destroy bushes at a rate considerably faster than by hand and the effective maceration process does away with the need for burning. The resultant heterogeneous litter appears to inhibit seedling re-invasion. It has been observed by the author that successful seedling germination appears to be confined to areas where the sward is closely grazed, recently burnt or consisting of simple bryophyte mats. Evidently, *R. ponticum* seeds require high light intensities for germination and are rarely found germinating under existing stands of the shrub or in a dense three-dimensional sward. The highly heterogeneous litter generated by the flail mower may well not allow adequate light intensities to penetrate between the interstices where seeds are likely to fall. Regrowth from the macerated stump is adequate for future herbicide treatment and the impact by the machine on the other elements of the environment, at least in woodland and on the open mountain, is minimal.

The current scheme, now in its third year, is using a variety of comparative control methodologies and involves a variety of land ownerships. It is becoming increasingly clear which methodologies are the most successful though the cost-effectiveness of some of them is still in some doubt. The failure of this scheme, as is so often the case, is the lack of long-term financial commitment. Recent attempts to gain EEC funding for this operation, or indeed a much more comprehensive Park-wide scheme, may be more successful now that there is a possibility of speeding up the control process by mechanisation.

Meanwhile, seedling re-invasion into areas already cleared under the scheme has become firmly established.

Other invasive plants in Snowdonia

Whilst *R. ponticum* is considered to be the most potentially devastating invasive plant in Snowdonia, there are several other species of note.

Spartina anglica C.E. Hubb

Spartina anglica is found growing on sand and mud in most of the estuaries in the National Park where it is rapidly suppressing the few native species which are able to colonize this habitat. Its ability to trap silt suspended in tidal waters is legendary and the grass has been planted extensively in other countries, notably The Netherlands, to reclaim marginal intertidal estuarine habitats. The invasion of mudflats which are an important feeding ground for birds has also provoked considerable concern and control programmes of variable success have been undertaken. The SNPA undertook a control programme in one estuary in Snowdonia but this has subsequently been seen to be of limited success and it is likely that further control will not be undertaken.

Fallopia japonica Houtt.

Fallopia japonica is found growing in a wide variety of ruderal habitats throughout the Park, notably in roadside verges and riverbanks to where it is largely confined by livestock grazing. Whilst control in roadside verges with Tordon 22K has met with some success, this herbicide has not been approved for use by water. Control programmes have been undertaken on several stands growing next to rivers using an approved formulation of glyphosate but these have only been effective with repeated applications spread over three or four years. A Park-wide survey has been initiated but it will be some time before the extent of the problem can be assessed. However, it is clear that numerous stands of varying sizes are spreading throughout lowland areas of the Park and that the cost of a concerted control programme would, therefore, be considerable.

Spiraea x *rosalba* Dippel

Little attention has been given to this plant until recently when a survey (unpublished internal SNPA report) showed that a large number of farms centred around the Bala area in the east of the Park are affected. The plant is typically found spreading longitudinally along hedges and ditches where it is apparently confined by cattle and sheep grazing. However, in areas where grazing has been excluded, *S.* x *rosalba* quickly spreads out from ditch and hedge lines. No control programmes have been undertaken in Snowdonia to date largely because it is not perceived as a problem by local farmers. Indeed, several farmers interviewed claimed that it was an ideal hedging (barrier) plant and that they wished they had more of it. It is considered to be an increasingly serious problem in at least one ungrazed wetland Site of Special Scientific Interest in the area (Countryside Council for Wales, personal communication).

Conclusions

There has been considerable speculation in the literature, *e.g.,* Elton (1958), Drake *et al.* (1989) and Usher (1987) as to why some communities are more susceptible to plant invasions than others. Most of the discussion has been centred on the ecological aspects of plant invasions. Many of these criteria might be used to address the vexing question as to why an area such as the Snowdonia National Park has such manifest problems with invasive species.

However, there are several other non-ecological criteria that need to be addressed and two reasons are abundantly clear. The area is of low agricultural productivity (and low population) and whilst it has received, and rightly deserves, the highest landscape protection status that can be bestowed on an area in the U.K. - National Park designation - it is nevertheless perceived as a low priority area for Government expenditure. The extent to which invasive plants have spread in Snowdonia would never have been tolerated had they occurred in the productive south of England. The tragedy is that inevitably both Government and senior land managers in the UK will one day fully appreciate the extent of the devastation being caused by invasive plants in landscapes such as Snowdonia. By then the costs of control programmes will be astronomical.

Meanwhile, it is interesting to speculate that reduction in sheep grazing regimes inherent in such environmentally based farm schemes as Tir Cymen, Countryside Stewardship and Environmentally Sensitive Areas (ESA) may cause an increase in the invasiveness of plants such as *R. ponticum, Spiraea* x *rosalba* and *Fallopia japonica*. The very schemes being initiated to try and improve the environment whose nature conservation and landscape interest has been so reduced by agricultural intensification may perversely cause degradation by invasive species.

At the Rio Earth Summit in 1992 more than 150 states signed the Biodiversity Convention, part of which required each country to produce a national action plan for biodiversity. In relation to this plan, it has been stated that "...the diversity of species within the UK is relatively modest...of importance at the species level is the ecological control of invasive introduced species which can have a devastating effect on the diversity of native communities, *e.g., R. ponticum...*" (Walton 1993).

Maybe there is light at the end of the tunnel.

References

Cross, J.R. 1973. The Ecology and Control of *Rhododendron ponticum* L. Thesis. University of Dublin.

Cross, J.R. 1975. Biological flora of the British Isles: *Rhododendron ponticum* L. J. Ecol. 63: 345-364.

Elton, C.S. 1958. The Ecology of Invasions by Animals and Plants. 181 pp. Methuen, London.

Drake, J.A., Mooney, H.A., Di Castri, F., Groves, R.H., Kruger, F.J., Rejmánek, M. and Williamson, M. (eds.) 1989. Biological Invasions: A Global Perspective. 525 pp. John Wiley and Sons, Chichester.

Gritten, R.H. 1987. The Spread of *Rhododendron ponticum* - a National Problem. Report of Discussion Conference. Snowdonia National Park Authority, Gwynedd.

Gritten, R.H. 1988. Invasive plants in the Snowdonia National Park. Ecos 9(1): 17-22.

Gritten, R.H. 1992. The control of *Rhododendron ponticum* in the Snowdonia National Park. Aspects Appl. Biol. 29: 279-286.

Kowarik, I. 1995. Time lags in biological invasions with regard to the success and failure of alien species. In: P. Pyšek, K. Prach, M. Rejmánek and M. Wade (eds.), Plant Invasions - General Aspects and Special Problems, pp. 15-38. SPB Academic Publ., Amsterdam.

National Park Plan Review 1986. Snowdonia National Park Authority, Gwynedd, LL48 6LS.

Rotherham, I.D. 1983. The Ecology of *Rhododendron ponticum* with Special Reference to its Competitive and Invasive Capabilities. Thesis. University of Sheffield.

Rotherham, I.D. and Read, D.J. 1988. Aspects of the ecology of *Rhododendron ponticum* with reference to its competitive and invasive properties. Aspects Appl. Biol. 16: 327-335.

Shaw, M.W. 1981. *Rhododendron ponticum*. An Assessment of the Current Knowledge about this Species in Relation to its Invasive Behaviour and Possible Control Methods. Institute of Terrestrial Ecology Report. Bangor Research Unit, Gwynedd.

Shaw, M.W. 1984. *Rhododendron ponticum* - ecological reasons for the success of an alien species in Britain and features that may assist in its control. Aspects Appl. Biol. 5: 231-241.

Tabbush, P.M. 1987. The Use of 'Mixture B' to Enhance the Effect of Glyphosate Herbicide on *Rhododendron ponticum* and Coarse Grasses. Research Information Note, Forestry Commission.

Tester, M. 1988. Putting down roots. New Scientist 1624: 94.

Thomson, A.G., Radford, G.L., Norris, D.A. and Good, J.E.G. 1992. Monitoring and Modelling *Rhododendron* Invasion in Wales. Institute of Terrestrial Ecology Report, Project No. TO9054m5. Bangor Research Unit, Gwynedd.

Usher, M.B. 1987. Invasibility and wildlife conservation: invasive species on nature reserves. Phil. Trans. Roy. Soc. London B 314: 695-710.

Walton, D.W.H. 1993. U.K. National action plan for biodiversity. Bull. Brit. Ecol. Soc. 24: 140-147.

V

RESEARCH IN PLANT INVASIONS: PRESENT AND FUTURE

RECENT TRENDS IN STUDIES ON PLANT INVASIONS (1974-1993)

Petr Pyšek
Institute of Applied Ecology, University of Agriculture Prague, CZ-281 63 Kostelec nad Černými lesy, Czech Republic

Abstract

Studies on plant invasions (*i.e.*, those dealing with any aspect of ecology of non-native species) indexed in Ecological Abstracts 1974-1993 were analysed at the global scale. In total, 872 studies were recorded, with a steep increase in the number per year beginning in the middle of the 1980s. At present, about 100 new papers appear each year. The role of compendia and proceedings volumes is fairly important and may be related to the international SCOPE project on biological invasions launched in 1982. The initial period of data accumulation and registration of the occurrence of alien species was followed by an increasing interest in species biology and ecology, interactions at higher hierarchical levels and general aspects of plant invasions. The research appears to be most intensive in areas where aliens contribute significantly to the flora and impose a serious threat to native vegetation (Australia, New Zealand, South Africa, west coast of United States, Hawaii). Global distribution is rather disproportionate and some heavily affected areas are underestimated. The research focus differs between geographical areas and reflects the level of practical problems associated with alien species. The available data are scattered in at least 189 journals but among these, the 20 being most frequently used as publication media account for almost a half of the studies published. The proportion of papers indexed in Current Contents has been steadily increasing from the middle of 1980s which also indicates the increasing importance of the field. The question whether or not to launch an independent journal devoted to biological invasions is discussed.

Introduction

Since Elton's classical study (Elton 1958), research into biological invasions has yielded an immense number of papers and the attention paid to this field has established it as a respectable branch of contemporary plant ecology (see, *e.g.*, Lodge 1993). The increasing importance of biological invasions has been widely recognized and the research in the last decade yielded several important milestones which have the potential to direct further development in this field (Mooney and Drake 1986; Groves and Burdon 1986; Kornberg and Williamson 1986; Drake *et al.* 1989; Di Castri *et al.* 1990; Groves and Di Castri 1991; Stone *et al.* 1992).

The present paper attempts to analyse in quantitative terms the papers on plant invasions published in the last twenty years and addresses a simple question: what kind of information is available and how is it distributed over time and space? It also aims to document the increasing importance of the field.

Data sources

Ecological Abstracts 1974-1993 (EA, Jarvis 1974-93) were scanned for the following keywords: ADVENTIVE, ALIEN, ARCHAEOPHYTE, COLONIZATION, (BIOLOGICAL, WEED) CONTROL, EXOTIC, (PLANT) INTRODUC (-TION, -ED),

Plant Invasions - General Aspects and Special Problems, pp. 223-236
edited by P. Pyšek, K. Prach, M. Rejmánek and M. Wade

(BIOLOGICAL, PLANT) INVASION, NATURALIZATION, WEED. As some of these terms can be used in a sense which does not match the purpose of the present study, each entry was checked for the content to select those relating to alien plants. Only higher plants were considered. Alien species are understood as those deliberately or accidentally released into an area to which they are not native (see, *e.g.*, Binggeli 1994a, and Pyšek 1995 for definition). Two compendia of principle importance (Mooney and Drake 1986 and Drake *et al.* 1989, the former not indexed, the latter covered only as an entire title without separate entries) were added. These are considered in analyses based on the year of publication but not in those based on the volume of EA.

Each study was classified according to its focus. The following topics, subsequently termed 'study types', were distinguished:

a. historical studies on particular species: history of introduction, analyses of origin, phytogeographical analyses;
b. floristic records: reports on the occurrence of new aliens, local floras with the role of aliens mentioned or stressed;
c. population level: autecological studies on biology and ecology of introduced species, comparison of native and alien species;
d. community and ecosystem levels: impact of aliens on native flora, vegetation studies;
e. conservacy, control;
f. problems of general interest: general properties of invaders and invaded communities, theoretical studies, discussion papers, reviews.

For each study, the area of interest (*i.e.,* the geographical area to which the respective invasion was related) was recorded. The classification of geographical areas adopted for the analyses follows that used in the Regional Index of EA.

The following publication media were distinguished: *(a)* books and compendia (considering each chapter indexed as a separate entry), *(b)* periodicals and *(c)* others which mostly referred to internal reports, theses, *etc.* Periodicals were further classified with respect to the coverage by Current Contents/Agriculture, Biology and Environmental Sciences.

Clearly, the data set used for the analysis is far from being complete (see Binggeli 1994b). Some studies certainly escaped inclusion because the alien status of a species was not stressed (hence, they did not appear in the keywords scanned), others would have not been covered by EA because of having been published in non-indexed journals. Similarly, some books were not covered by EA or only some chapters were indexed. Nevertheless it may be assumed that a significant proportion of papers published on the topic in the period assessed appeared in the record. At least, the data set analysed appears to be a reasonably representative sample of the literature available in the field. Another important aspect is simply that this data set represents what one can trace in one of the most widely recognized on-line databases.

Temporal trends in the number of studies

Considering the total number of studies on plant invasions indexed in Ecological Abstracts, a steady and conspicuous increase is evident from the middle of the eight-

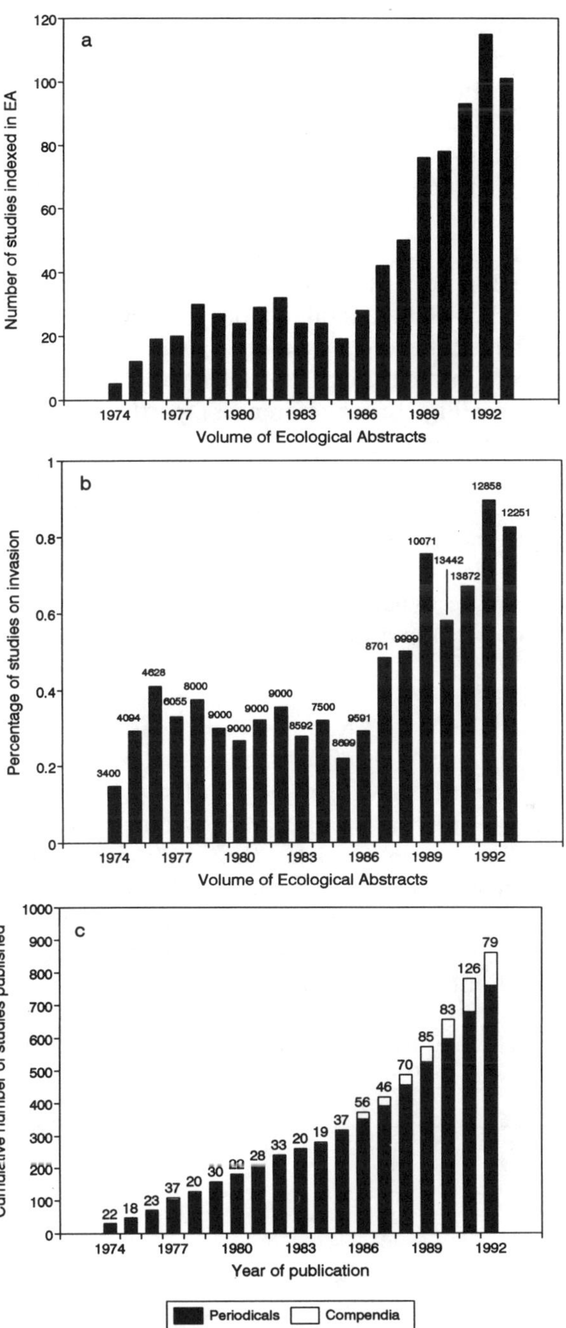

Fig. 1. Dynamics of studies on plant invasions over time. *(a)* Number of studies indexed in Ecological Abstracts. *(b)* Percentage of studies on plant invasions of the total number of papers (shown on the top of bars) indexed in the respective volume of Ecological Abstracts. *(c)* Cumulative number of studies on plant invasions published up to the respective year. Number of studies published in the respective year is shown on top of the bar. Papers published in compendia are shown separately; the following were considered: Kruger *et al.* 1977, Groves and Burdon 1986, Mooney and Drake 1986, Kornberg and Williamson 1986, Duffey and Usher 1988, Drake *et al.* 1989, Di Castri *et al.* 1990, Groves and Di Castri 1991, Agriculture, Ecosystems and Environment Vol. 1991 (1-3), Ramakrishnan 1991.

ies (Fig. 1a). This is still true when it is taken into account that the absolute number of entries in EA was gradually increasing from 1974 up to the present, so that the probability of studies on plant invasions to be covered was also increasing (Fig. 1b).

As to the year of publication, the cumulative number of studies shows two distinct periods of linear increase: up to 1985 ($Y = 23.97 X - 4729.2$, $r = 0.9926$) and from 1986 onwards ($Y = 81.17 X - 160888.0$, $r = 0.9912$). Slopes of both regressions were highly significantly different ($F = 260.35$, $P<0.0001$, tested according to Snedecor and Cochran 1967) indicating that the rate of increase in the total number of studies was higher in the latter period (Fig. 1c). In the last few years, about one hundred new studies per year have been published and the total number recorded up to 1992 is 872.

This represents the interest of ecologists in biological invasions in the last decade which has been widely recognized (Lodge 1993). During this period, several specialized volumes appeared (see Fig. 1c for references), partly as a consequence of the international SCOPE project launched at the beginning of the eighties (Drake *et al.* 1989). Although the contribution of these volumes is significant (Fig. 1c), the interest in the field they have helped to stimulate is also reflected by the increase of studies published in periodicals. Another reason may be simply that ecologists realized how intriguing a subject biological invasions is and this fact, together with the gradual shift of attention to the more dynamic aspects of plant ecology, lead to the more intensive research in biological invasions.

Research priorities

The distribution of studies differs with respect to the particular research focuses (Table 1). Papers on control and conservation are most numerous and constitute more than a quarter of the studies available which reflects the practical implications of the field. On the other hand, historical studies are quite rare. Reconstruction of spreading dynamics on the long-term temporal scale represents valuable information important for our understanding of the invasion process; their low representation reflects difficulties associated with limited availability of such data.

Table 1. Number of studies published at 5-year intervals classified according to the type of the study. The total number of studies exceeds 872 as some of them cover more than one type.

	<1977	1978-82	1983-87	1988-92	Total	%
Historical	8	16	19	35	78	7.72
Floristic	32	52	24	84	192	18.99
Species biology and ecology	7	23	39	80	149	14.74
Community level	12	13	47	94	166	16.42
Control, conservancy	54	44	39	144	281	27.79
General papers	2	10	41	92	145	14.34

Changes in the importance of particular study types are shown in Fig. 2. At the beginning, floristic studies and those on control were by far the most frequent. Their relative importance (assessed as the percentage contribution to the total number of studies available up to the respective year) was gradually decreasing, although namely in the latter the absolute number of studies is maintained at a high level (Table 1). An increase in the proportion of papers focused upon biology and ecology

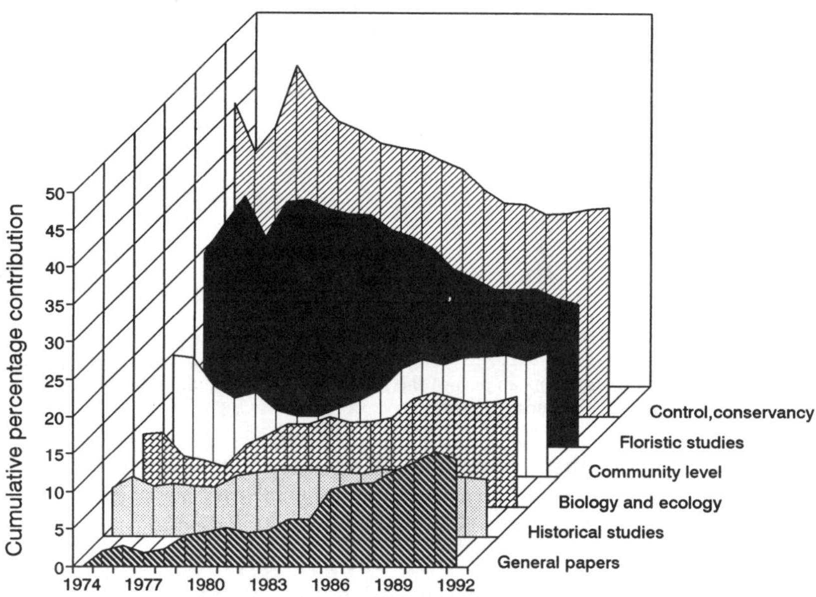

Fig. 2. Trends in the cumulative percentage contribution of studies focused on particular topics to the total number of studies on plant invasions.

of invasive species and interactions at the community level is typical of the 1980s. The same trend can be seen for the papers of general interest. Clearly, the period concentrating on data accumulation and registration was followed by an increasing interest in basic scientific reserach (*i.e.,* the mechanisms underlying the invasion process, impact of alien species on native flora, a search for general properties of invasive species and invaded communities) and reviewing and summarizing the results achieved (Crawley 1986, 1989; Newsome and Noble 1986; Rejmánek 1989; Noble 1989; Roy 1990).

Geographical pattern

When classified according to the continents (Table 2), the highest numbers of studies have been published from the Australian region (27.3%) Northern America (25.1%) and Europe (18.5%). On the other hand, rather low numbers were found for Asia (5.1%), South America (2.2%) and Central America (1.3%). The contribution of Africa is relatively high (11.3%) but 78.4% of studies related to this continent are accounted for by the Republic of South Africa alone.

A more detailed insight (Fig. 3) reveals considerable differences in particular areas of the world and clearly indicates those in which the research is most intensive: Australia, New Zealand, South Africa and the west coast of the United States of America. Comparison of the temporal trends in the number of studies available for these areas indicates that the research in Australia started to be intensified about five years earlier than in other areas, *i.e.,* at the very beginning of the 1980s. Also, a very conspicuous increase at the end of the 1980s was detected for the Republic of South Africa (Fig. 4).

Table 2. Total number of studies on plant invasion focused upon particular areas of the world shown at 5-year intervals. Total number of studies does not equal 872 as some of them were not geographically focused and some covered more than one area.

	<1977	1978-82	1983-88	1989-92	Total	%
North America	36	34	32	93	195	25.06
Central America[1]	4	2	1	3	10	1.29
South America	3	2	3	9	17	2.19
Africa	10	9	18	51	88	11.31
Europe[2]	23	25	28	68	144	18.51
Asia[3]	4	5	4	27	30	5.14
Australia[4]	8	22	33	61	124	15.94
New Zealand	7	9	21	51	88	11.31
Pacific Islands[5]	3	3	9	23	38	4.88
Global	0	1	4	27	32	4.11

[1]including Carribean; [2]including European part of the former USSR; [3]including Asian part of the former USSR; [4]including New Guinea; [5]including Hawaii

These figures show the absolute number of studies related to particular areas and thus represent a good measure of how the available information is distributed at the global scale. On the other hand, Fig. 5, showing the 'relative research intensity' (the number of studies on plant invasions expressed as a percentage of the total number of studies related to the area) may be considered a rough measure of how a large proportion of funds spent on ecological science in a particular area goes to research into plant invasions. Hence, it expresses better how important the field is considered to be in particular areas. The results basically correspond to those obtained by considering the absolute number of studies, with Australia, New Zealand and Republic

Fig. 3. World distribution of studies on plant invasions indexed in Ecological Abstracts 1974-1992. Number of studies: 0: empty areas; 1-10: dotted; 11-30: hatched; 31-50: cross-hatched; >50: full. Note that particular areas were considered as a whole and the following were distinguished (classification according to EA): Canada (30 studies), USA-west coast (63), USA-midwest (13), USA-south (26), USA-east coast (22), Central America incl. Carribean (10), South America (17), Pacific Islands incl. Hawaii (38), Europe-west (incl. British Isles, 43), Europe-north (30), Europe-central, east (43), Europe-south (16), Africa-north, west (10), Africa-central, east (6), Africa-south (1, excluding Republic of South Africa 69), former USSR-European part (11), Asian part (4), Asia-Middle East (3), Asia-south east (23), Asia-Far East (6), Australia (115, excl. New Guinea 9), New Zealand (88). Studies related to entire continents are not considered; the number of these was negligible except of North America (41).

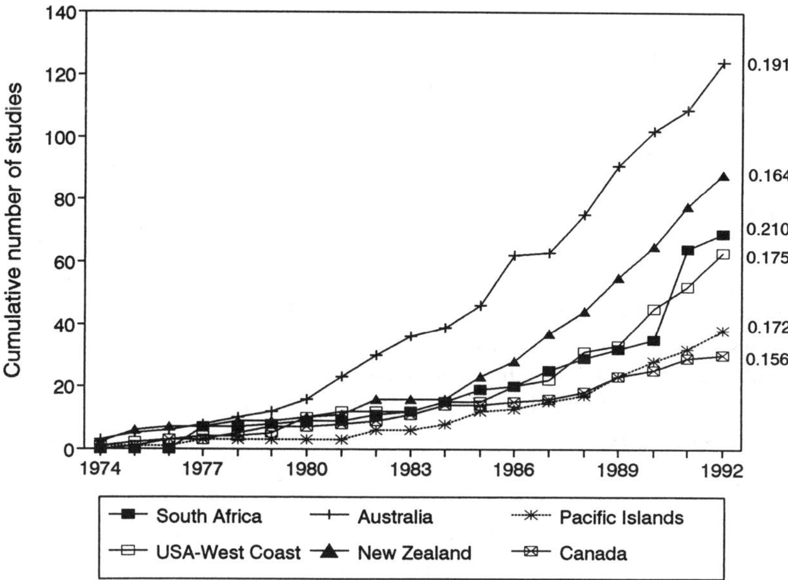

Fig. 4. Cumulative increase in the number of studies on plant invasions shown for the areas studied most intensively. Rate of increase in the number of studies, expressed as the regression coefficient of CUMULATIVE NUMBER OF STUDIES = exp (a + b × YEAR) regression equation, is shown on the right.

of South Africa showing the highest values. However, it reveals that the relative research intensity in North America does not exceed that in other parts of the world. This is still true if the differences between particular areas of United States as displayed in Fig. 3 are taken into account: for none of them do the values exceed 0.5%.

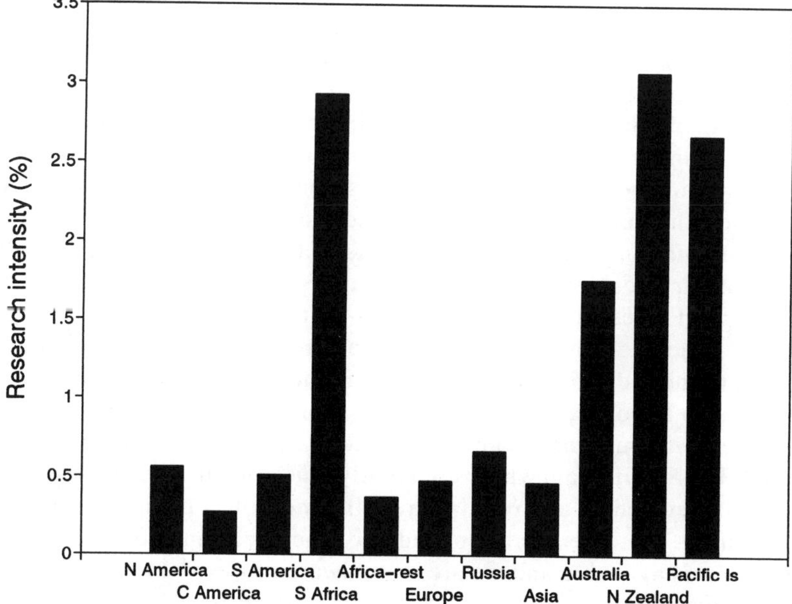

Fig. 5. Relative research intensity expressed as a percentage of the number of studies on plant invasions of the total number of studies related to the area (as indexed in Ecological Abstracts). Republic of South Africa is displayed separately from the rest of the continent because of considerably different research intensity. N: North; S: South; C: Central; Is: Islands.

On the contrary, plant invasion research, viewed in relative terms, is very important in the Pacific Islands, namely Hawaii.

It can be assumed that the attention paid to plant invaders in a particular area of the world would increase with *(a)* the number and proportion of alien species in the local flora, and *(b)* practical problems imposed by these species. Simply, the more invading species are present in the local flora, the higher the probability they happen to become a subject of ecologists' interest. This is further stressed by the 'attractiveness' of alien species, both from the floristic (botanists tend to publish the records of plants new to the area they work in), ecological (dynamic changes brought about by an invader) and practical (implications for management) viewpoints.

To a large extent, the intensity of research over particular areas of the world appears to be broadly correlated with the global pattern of extent of plant invasions. The number of studies is high for the areas of large scale transformation of original vegetation by man such as regions with Mediterranean climate and island areas where the high number and/or proportion of alien species in local floras has been reported: Australia (400-850 species in the southern part of the continent, *i.e.,* 21.0-23.6% of the total flora, Heywood 1989), New Zealand (1570 - 46.7%, Heywood 1989), Hawaii (800 - 45.1%, Loope and Mueller-Dombois 1989), California (650-750 naturalized species - 11.4-12.9%, Rejmánek *et al.* 1991), South Africa (503 species of naturalized exotic weeds, Wells and Stirton 1982). Research intensity also seems to be reasonably high in most of Europe where the proportion of aliens in the floras of particular countries is being estimated to range between 10-20% with absolute numbers comprising several hundreds (Heywood 1989; Pyšek 1989). However, it appears that, seen from the global viewpoint, the research intensity is rather disproportionate as some heavily affected areas have been only poorly studied (*e.g.,* parts of Africa, South America, and south-east Asia, see Drake *et al.* 1989). Undoubtedly, lower research intensity in some areas of the world reflects the fact that the extent of invasions or the level of problems they cause are less significant (North of Europe and North America, some relatively undisturbed tropical forests of South America, less populated areas of the Asian part of USSR). Nevertheless, the research intensity depends principally on the overall level of science in a given area, *i.e.,* on the wealth of particular countries and their scientific traditions. This may explain the low research intensity in some developing countries (majority of Africa, part of Asia and South America, the European part of the former Soviet Union). It cannot be assumed that the representation of aliens in those parts of these areas that are heavily transformed by man (tropical grasslands of Africa and South America, mediterranean area of Chile, tropical forests of Africa and south-east Asia) is lower than, for example in temperate regions of Europe or North America where the research is much more developed. For these developing countries, not only are the studies on invasive species rare but also the overall botanical information is incomplete (Heywood 1989).

This problem with missing data brings about a danger of circular reasoning: unless the research is proportional (which apparently it is not), it tends to appear that the more effort is devoted to a particular area, the more it is being invaded by aliens simply because a more detailed knowledge of the occurrence of alien species and their behaviour is available and we are more aware of the problems these can cause in that particular area.

Apparently, the magnitude of occurrence of alien species corresponds to how intensively they are studied but cannot explain the global pattern of research intensity

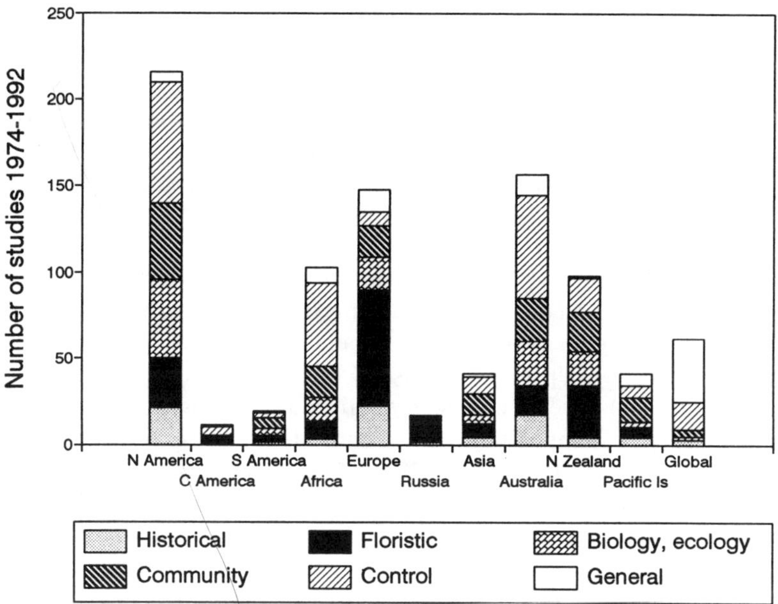

Fig. 6. Structure of studies on plant invasions in particular areas of the world. Total number of studies as indicated by bar height does not correspond exactly to the figures given in Table 2 as some of the studies cover more than one topic.

entirely. The seriousness of practical problems caused by alien invaders also appears to significantly affect the pattern of research intensity. This is not surprising as the resources spent on research are being more and more conditioned by its practical implementation (see, *e.g.*, Lawton 1994).

The progress in studies on plant invasion does not depend only on how intensively they are studied but also on which are the research priorities (Fig. 6). Not surprisingly, a substantial part of studies in those areas suffering most from plant invasions are focused upon control and conservation aspects (Africa 47.6%, Australia 38.2%, North America 32.4%). A high proportion of floristic papers among European studies (46.0%) reflects a strong floristic tradition in the region and indicate that the practical problems associated with aliens are minor compared to other intensively studied areas of the world. The work in the field here is, to a large extent, in the phase of registering their occurrence. Floristic records are in fact almost the only type of research that has been carried out so far in the former Soviet Union (88.2%). Historical studies are almost completely restricted to Europe, North America and Australia and these four areas account for 75.0% of the total number of historical studies published. There are only very few historical studies, if any, from Central America, South America, Africa and Asia. This is particularly unsatisfactory as historical studies represent especially valuable information. By considering species spreading dynamics on a longer time scale, they make it possible to relate species historical performance to changes in such factors as human impact and climate (see Kowarik 1995; Beerling 1995). General problems are, not surprisingly, addressed mostly in those studies adopting a global point of view of plant invasions accounting for 59.7% of the total.

Publication media

The available information on plant invasions is scattered (not including compendia, book chapters, internal reports and theses) in at least 189 journals. However, nine journals in which the papers on the topic have been published most frequently account for 28.0% of the total number of indexed studies and the 20 most frequently used journals cover almost 50% of studies (Fig. 7).

Of the 872 papers recorded in total, 136 (15.6%) have been published in books or separate proceedings, 36 (4.1%) were classified as internal reports or theses and 701 (80.4%) appeared in periodicals. The increasing importance of the field is further documented by an increasing proportion of papers published in journals covered by Current Contents (Fig. 8). Up to 1992, 452 (64.5% of the total number of papers published in periodicals) papers have been published in journals covered by Current Contents and 249 (35.5%) appeared in those not indexed there.

The structure of papers differs remarkably with respect to the type of publication media (Fig. 9). A high proportion of papers focusing upon general problems (47.5%) is typical of compendia. The papers published in Current Contents-covered journals often focus upon the biology and ecology of invasive species and the ecology of their communities (42.4%) and upon control and conservancy problems (33.1%). Among periodicals not indexed in Current Contents, there is a high proportion of floristic papers (38.9%) and also that of historical studies (12.1%) exceeds the figure obtained for other types of publication media.

The journals that may be considered as most important for the development of the field of plant invasions are shown in Fig. 10.

Obviously, the results of the research in plant invasions are considerably scattered (in fact even more than the present analysis indicates, given that the data set used

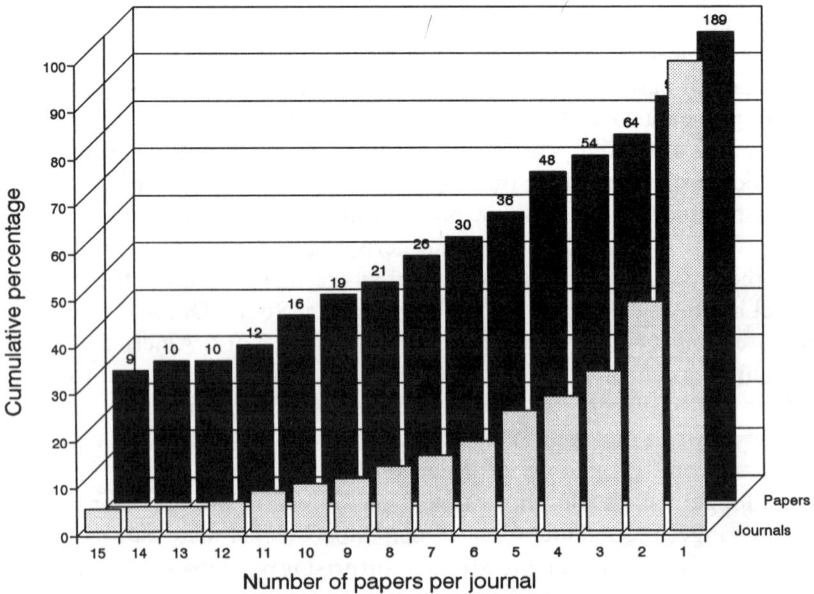

Fig. 7. Cumulative percentage of journals ranked according to the number of studies published in it up to 1992 and corresponding cumulative percentage of studies accounted for. Cumulative number of journals is given on top of the bar for papers.

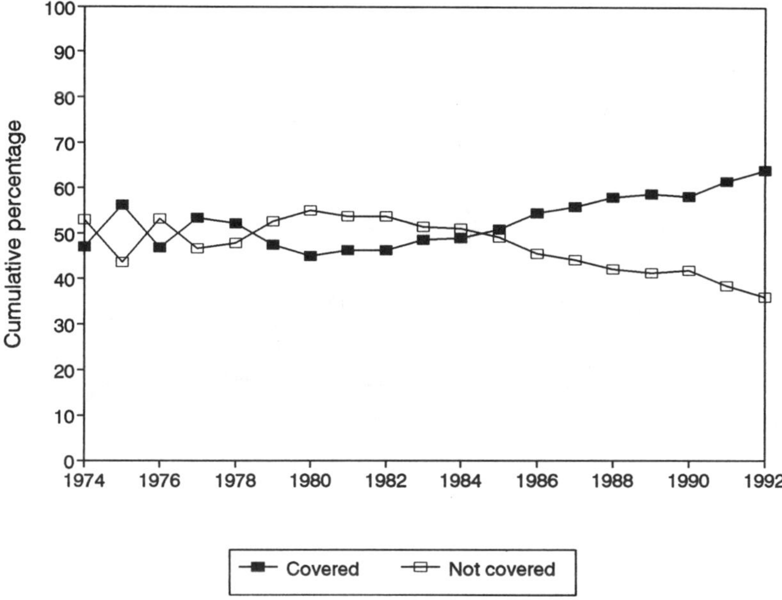

Fig. 8. Trends in the cumulative percentage of journals covered and not covered by Current Contents.

here is by no means complete), with a significant proportion of the information published in conference proceedings, see also Binggeli 1994b). On the basis of the results obtained by EA indexed papers, it seems reasonable to assume that at present, there are, as a very conservative estimate, at least about 100 new papers published per year and of these a large proportion meets an international standard. Considering that the

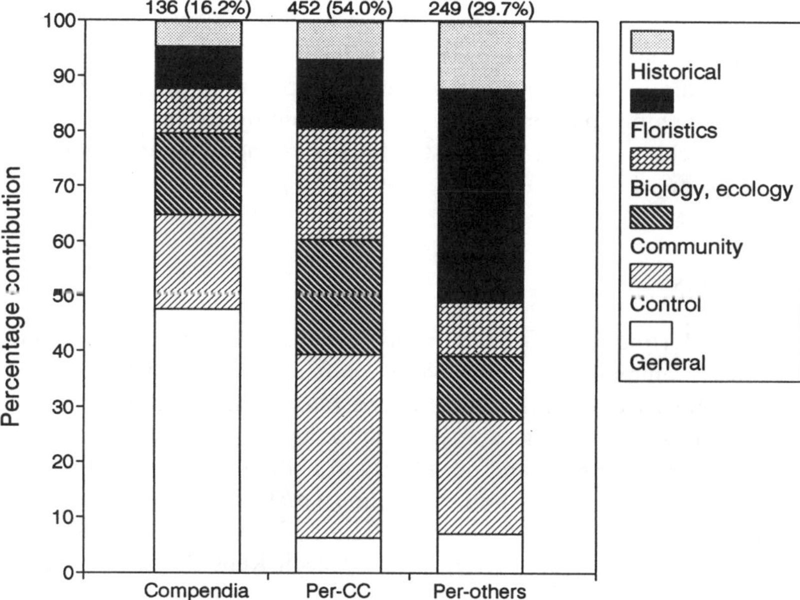

Fig. 9. Structure of studies according to the focus on particular topics shown for compendia and periodicals (Per) covered and not covered by Current Contents (CC). Total numbers and percentages are shown on top of the bars. The total number does not equal 872 as internal reports were not considered.

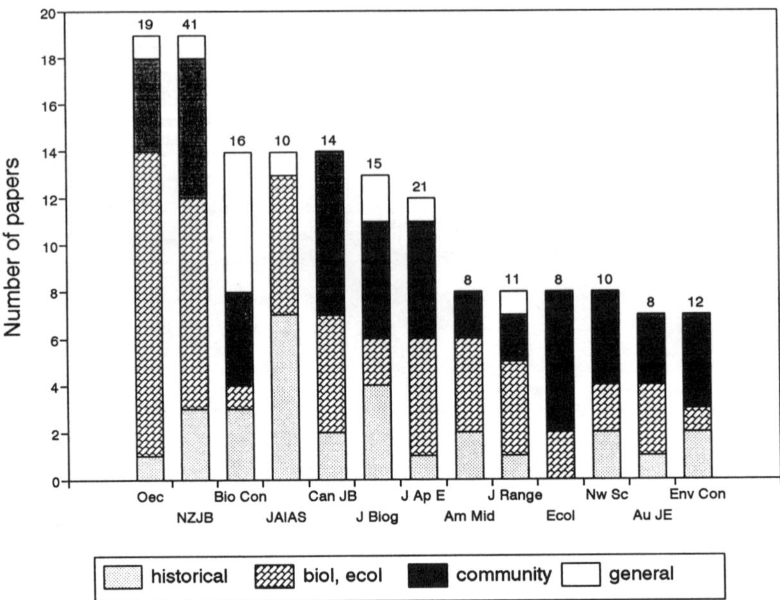

Fig. 10. Leading journals in plant invasions research. Floristic studies and those focused exclusively on control were excluded to stress the contribution of journals associated rather with the basic research and show thus those journals that are most important for the development of the field. Total number of all studies (*i.e.*, including those on floristics and control) is shown on top of the bar. Dotted area: historical studies; Cross-hatched : species biology and ecology; Filled: community level; Empty: general papers; Oec: Oecologia; NZJB: New Zealand Journal of Botany; Bio Con: Biological Conservation; JAIAS: Journal of Australian Institute of Agricultural Research; Can JB: Canadian Journal of Botany; J Biog: Journal of Biogeography; J Ap E: Journal of Applied Ecology; Am Mid: American Midland Naturalist; J Range: Journal of Range Management; Ecol: Ecology; Nw Sc: Northwest Science; Au JE: Australian Journal of Ecology; Env Con: Environmental Conservation.

number of studies on animal invasions is at least the same (and probably even higher), consideration should be given to whether the development in the field has not come to the point of launching an international journal of biological invasions. Concentration of the major achievements (or at least of the part of them) would certainly further accelerate the progress in the studies on plant invasions.

Acknowledgments

I thank David Beerling, Sheffield, Tomáš Herben, Průhonice, and Karel Prach, Třeboň, for their comments on the previous draft of the paper and to Roger Hall, Oxford, for support. Roger Hall, Oxford, and Max Wade, Loughborough, kindly improved my English. My thanks are also due to the Department of Plant Sciences, University of Oxford for space and possibility of using its facilities. The study was made possible by funding from The Leverhulme Trust Foundation. Thanks are also due to Dana Pavelčíková, British Council Prague, for her kind assistance. It was partly supported by the Grant Agency of the Czech Republic (grant No. 204/93/2440).

References

Beerling, D.J. 1995. General aspects of plant invasions: an overview. In: P. Pyšek, K. Prach, M. Rejmánek and M. Wade (eds.), Plant Invasions: General Aspects and Special Problems, pp. 237-247. SPB Academic Publ., Amsterdam.

Binggeli, P. 1994a. The misuse of terminology and anthropometric concepts in the description of introduced species. Bull. Brit. Ecol. Soc. 25(1): 10-13.

Binggeli, P. 1994b. Invasive woody plant species database: early trends with special reference to the tropics. In: C.E. Hughes (ed.), Tropical Trees as Invasive Species. Proceedings of the Tropical Forest Forum Workshop, Oxford Forestry Institute, 17 May 1994 (in press).

Crawley, M.J. 1986. The population biology of invaders. Phil. Trans. Roy. Soc. London B 314: 711-731.

Crawley, M.J. 1989. Chance and timimg in biological invasions. In: J.A. Drake, H.A. Mooney, F. di Castri, R.H. Groves, F.J. Kruger, M. Rejmánek and M. Williamson (eds.), Biological Invasions: A Global Perspective, pp. 407-437. John Wiley and Sons, Chichester.

Di Castri, F., Hansen, A.J. and Debussche, M. (eds.) 1990. Biological Invasions in Europe and the Mediterranean Basin. 463 pp. Kluwer Academic Publ., Dordrecht.

Drake, J.A., Mooney, H.A., Di Castri, F., Groves, R.H., Kruger, F.J., Rejmánek, M. and Williamson, M. (eds.) 1989. Biological Invasions: A Global Perspective. 525 pp. John Wiley and Sons, Chichester.

Duffey, E. and Usher, M.B. (eds.) 1988. Biological invasions of nature reserves. Biol. Conserv. 44: 1-135.

Elton, C. 1958. The Ecology of Invasions by Animals and Plants. 181 pp. Methuen, London.

Gray, A.J. 1986. Do invading species have definable genetic characteristics? Phil. Trans. Roy. Soc. London B 314: 675-693.

Groves, R.H. and Burdon, J.J. (eds.) 1986. Ecology of Biological Invasions: An Australian Perspective. 166 pp. Australian Academy of Sciences, Canberra.

Groves, R.H. and Di Castri, F. (eds.) 1991. Biogeography of Mediterranean Invasions. 485 pp. Cambridge University Press, Cambridge.

Heywood, V.H. 1989. Patterns, extents and modes of invasion by terrestrial plants. In: J.A. Drake, H.A. Mooney, F. di Castri, R.H. Groves, F.J. Kruger, M. Rejmánek and M. Williamson (eds.), Biological Invasions: A Global Perspective, pp. 31-55. John Wiley and Sons, Chichester.

Jarvis, P. (ed.) 1974-1993. Ecological Abstracts. Vol. 1-20. Elsevier Science Publ., The Hague.

Kornberg, H. and Williamson, M.H. 1986. Quantitative Aspects of the Ecology of Biological Invasions. 240 pp. Phil. Trans. Roy. Soc., London.

Kowarik, I. 1995. Time lags in biological invasions with regard to the success and failure of alien species. In: P. Pyšek, K. Prach, M. Rejmánek and M. Wade (eds.), Plant Invasions: General Aspects and Special Problems, pp. 15-38. SPB Academic Publ., Amsterdam.

Kruger, F.J. *et al.* 1977. Proceedings of the 2nd National Weed Conferrence of South Africa, Stellenbosch. A.A. Balkema, Cape Town.

Lawton, J.H. 1994. Peer review, co-evolution and tortoises. Oikos 69: 361-363.

Lodge, D.M. 1993. Biological invasions: lessons for ecology. Trends Ecol. Evolut. 8: 133-137.

Loope, L.L. and Mueller-Dombois, D. 1989. Characteristics of invaded islands, with special reference to Hawaii. In: J.A. Drake, H.A. Mooney, F. di Castri, R.H. Groves, F.J. Kruger, M. Rejmánek and M. Williamson (eds.), Biological Invasions: A Global Perspective, pp. 257-280. John Wiley and Sons, Chichester.

Mooney, H.A. and Drake, J.A. (eds.) 1986. Ecology of Biological Invasions of North America and Hawaii. 321 pp. Springer-Verlag, New York.

Newsome, A.E. and Noble, I.R. 1986. Ecological and physiological characters of invading species. In: R.H. Groves and J.J. Burdon (eds.), Ecology of Biological Invasions: An Australian Perspective, pp. 1-20. Australian Academy of Sciences, Canberra.

Noble, I.R. 1989. Attributes of invaders and the invading process: terrestrial and vascular plants. In: J.A. Drake, H.A. Mooney, F. di Castri, R.H. Groves, F.J. Kruger, M. Rejmánek and M. Williamson (eds.), Biological Invasions: A Global Perspective, pp. 301-313. John Wiley and Sons, Chichester.

Pyšek, P. 1989. Archaeophytes and neophytes in the ruderal flora of some Czech settlements. Preslia 61: 209-226 (in Czech).

Pyšek, P. 1995. On the terminology used in plant invasion studies. In: P. Pyšek, K. Prach, M. Rejmánek and M. Wade (eds.), Plant Invasions: General Aspects and Special Problems, pp. 71-81. SPB Academic Publ., Amsterdam.

Ramakrishnan, P.S. (ed.) 1991. Ecology of biological invasion in the tropics. 206 pp. International Scientific Publications, New Delhi.

Rejmánek, M. 1989. Invasibility of plant communities. In: J.A. Drake, H.A. Mooney, F. di Castri, R.H. Groves, F.J. Kruger, M. Rejmánek and M. Williamson (eds.), Biological Invasions: A Global Perspective, pp. 369-388. John Wiley and Sons, Chichester.

Rejmánek, M., Thomsen, C.D. and Peters, I.D. 1991. Invasive vascular plants of California. In: R.H. Groves and F. di Castri (eds.), Biogeography of Mediterranean Invasions, pp. 81-101. Cambridge University Press, Cambridge.

Roy, J. 1990. In search of the characteristics of plant invaders. In: F. di Castri, A.J. Hansen and M. Debussche (eds.), Biological Invasions in Europe and Mediterranean Basin, pp. 335-352. Kluwer Academic Publ., Dordrecht.

Snedecor, G.W. and Cochran, W.G. 1967. Statistical Methods. 593 pp. Iowa University Press, IA.

Stone, C.P., Smith, C.W. and Tunison, J.T. (eds.) 1992. Alien Plant Invasions in Native Ecosystems of Hawaii: Management and Research. 887 pp. University of Hawaii Press, Honolulu.

Wells, M.J. and Stirton, C.H. 1982. South Africa. In: W. Holzner and M. Numata (eds.), Biology and Ecology of Weeds, pp. 339-345. Dr. W. Junk Publ., The Hague.

GENERAL ASPECTS OF PLANT INVASIONS: AN OVERVIEW

David J. Beerling
Department of Animal and Plant Sciences, University of Sheffield, PO Box 601, Sheffield S10 2UQ, United Kingdom

Abstract

Viewed in a broader context, studies of plant invasions can be used to address specific issues relating to fundamental ecological theories. This chapter considers how future studies of plant invasion may be used to relate ecosystem structure to function, explore the concept of keystone species and how invasions by alien species can alter, in a measurable fashion, the biogeochemistry of a system. Given that a general effect of plant invasions is invariably to reduce trophic levels within a system, the net result of invasions parallel those of a reduction in species diversity at the global scale as a result of man's disturbance of the terrestrial biosphere and global climate system. Therefore both converge to the single question: what is the effect of a reduction in species diversity on ecosystem stability? Here, a number of case studies are presented which have begun to use plant invasions to address all these issues. It is concluded that, with the advent of global change, alien invasions will be more frequent and so play an increasingly important role in affecting ecosystems. We require the initiation of large scale detailed experiments to begin to adequately quantify and therefore be in a position to predict the effects of an increased frequency of invasions on ecosystems.

Introduction

Plant invasions have increasingly attracted attention from all regions of the globe (Pyšek 1995b), but many of these studies are concerned, understandably, with the necessarily important issues of control (see Section IV) and have not sought generalities applicable to other systems. This chapter considers how we may direct future studies of plant invasions to address firstly the possibility of the existence of a set of unifying properties which may be used *a priori* to identify whether a particular species will become invasive and secondly how invasions may be used to investigate fundamental issues of ecology relating ecosystem structure to function. Specific examples of where these types of study have begun to be undertaken are reviewed in the context of the present volume together with a consideration of the possible impact of global change on invasions.

Many documented examples of invasions by organisms involve the introduction of species into new areas beyond what is considered their 'native range'. Species' introductions have subsequently led to the production of a wide variety of terms to describe these organisms; these issues have been extensively addressed by Pyšek (1995a). In considering the question: "what constitutes an introduced species?" emphasis must be placed on the use of the fossil record. In general the fossil record has been much neglected, yet this must be a key consideration (Betancourt *et al.* 1984), and in some instances the presence of a species in the fossil record has led to its alien status being overturned (Byrne and McAndrews 1975).

Plant Invasions - General Aspects and Special Problems, pp. 237-247
edited by P. Pyšek, K. Prach, M. Rejmánek and M. Wade

Physiological and ecological studies

Elton (1958) recognised early on that certain types of organisms appeared to be more successful invaders than others (in the sense that they achieve dominance within an ecosystem). Several chapters in the present volume investigate the characteristics permitting plants to become successful invaders (Rejmánek 1995; Pyšek *et al.* 1995; Vogt-Andersen 1995), and it is clear that common features emerge, especially rapid, early season growth and vegetative reproduction (Brock *et al.* 1995; Bailey *et al.* 1995) and efficiently dispersed propagules (Bramley *et al.* 1995; Edwards *et al.* 1995; Pyšek *et al.* 1995; Vogt Andersen 1995). These studies and others (summarized by Drake *et al.* 1989; Di Castri *et al.* 1990; Kornberg and Williamson 1986) all highlight the fact that our ability to identify *a priori* potential invaders by screening for these types of ecological traits is extremely limited (Noble 1989). The *post hoc* identification of why species became successful invaders in a particular area provides us with little predictive capacity to identify potential future invaders.

The elusive nature of any general predictive properties of invaders is apparent, in part, because in most instances we lack knowledge of the underlying physiological mechanisms. Comparative ecological approaches are undoubtedly useful but are invariably limited in that they are not applicable to systems beyond those in which they were developed. Similar constraints apply to a consideration of the geographical amplitude of an invader and its past history of success (Daehler and Strong 1993). Where mechanisms can be identified at the physiological level, stronger rules of prediction may begin to emerge (Austin 1982). An example of this approach is the study of changes in the dominance of a native and introduced grass species in North American prairies (Harris 1967). Over a large area of its former distribution the native grass *Agropyron spicatum* has become replaced by the introduced Eurasian grass *Bromus tectorum* despite it being a very successful competitor against other native species. The physiological mechanism for this replacement centres on differ-

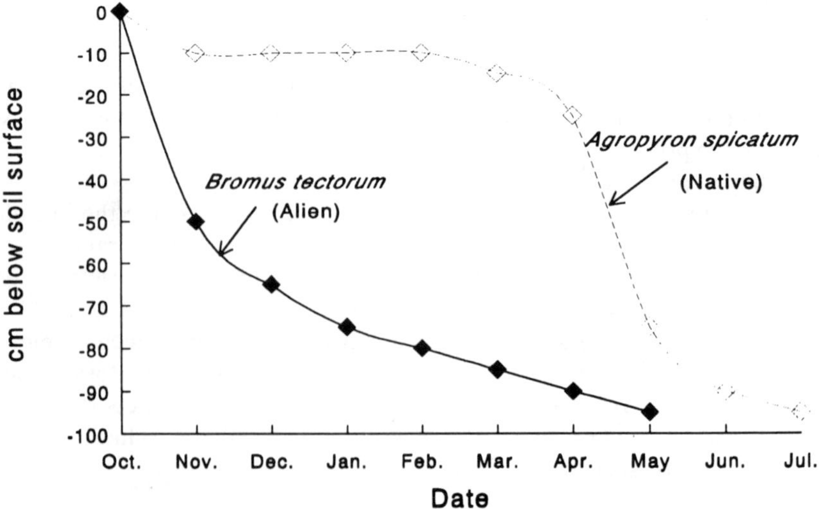

Fig. 1. Seasonal root growth of alien (*Bromus tectorum*) and native (*Agropyron spicatum*) grasses on the North American Prairies. (From: Harris, G.A. 1967. Some competitive relationships between *Agropyron spicatum* and *Bromus tectorum*. Ecol. Monogr. 37: 89-111. Copyright by the Ecological Society of America.)

ences in temperature-dependent root growth between the two species. During the winter *A. spicatum* shows no root growth, but in the spring root elongation and expansion are extremely rapid, permitting the exploitation of soil moisture before the onset of the summer droughts. The alien *Bromus tectorum* on the other hand continues root growth throughout the season and so its roots are already established in the soil profile before *A. spicatum* becomes active, enabling it to outcompete for water at all soil depths (Fig. 1).

Temperature-dependent root growth provides an explanation at a physiological level for the switch from a native to an alien grass species. Emphasis must also be placed on studies coupling a physiological approach with an ecological approach to seek explanations and general properties of invaders (Beerling 1993). This approach is exemplified by Weiss and Noble (1984). These authors studied the alien South African shrub *Chrysanthemoides monilifera* which was displacing the native *Acacia longifolia* from coastal dunes in Australia. When grown individually, it was found that the photosynthetic rate (per unit area per unit time) of *C. monilifera* was, surprisingly, lower than that of *A. longifolia*. However, when grown in competition, seedlings of *C. monilifera* outcompete those of *A. longifolia* by a more effective arrangement of leaves.

This example shows that changes in the dominance of the two species could have been predicted only by the measurement of physiological and ecological traits but even this is limited in that the question arises as to which physiological and ecological traits are the most appropriate to measure, which in turn depends upon the properties of the ecosystem itself.

The role of disturbance

Plant invasions occur in many diverse ecosystems, from tropical regions to sub-Antarctic forests (Gobbi *et al.* 1995). Susceptibility of particular ecosystems and habitats must be linked to natural disturbance phenomena, *e.g.*, fire and wind, and human impact increasing physical disturbance (Kowarik 1995b), as well as man's deliberate introduction of species. Tropical regions, for example, subject to forest destruction by logging are frequently invaded during the progression of secondary succession; indeed invading species can play a key role in this process. Riparian habitats have been referred to continually throughout this volume (Edwards *et al.* 1995; Ferreira and Moreira 1995; Bramley *et al.* 1995) as susceptible to invasion, presumably because they are subject both to naturally disturbance by flooding, which leads to the seasonal destruction of the biomass of riparian vegetation (Ellenberg 1988) and anthropogenic disturbance created in relation to flood control measures (Beerling 1991). Invasions progress quickly following disturbance because the linear river corridors and the rivers themselves aid the rapid dispersal and establishment of the propagules of many different species throughout an entire system (Beerling 1991).

These riparian processes are exemplified by the rapid expansion into the tropical wetlands of the Northern Territory of Australia by the woody weed *Mimosa pigra* L. (Lonsdale 1993) (Fig. 2). In this case *M. pigra* has spread at both local and regional spatial scales at a rate which can only be explained by the dispersal of floating seeds by water currents. Furthermore, there is a close correlation between the range extension of the plant and the previous year's rainfall confirming the importance of river

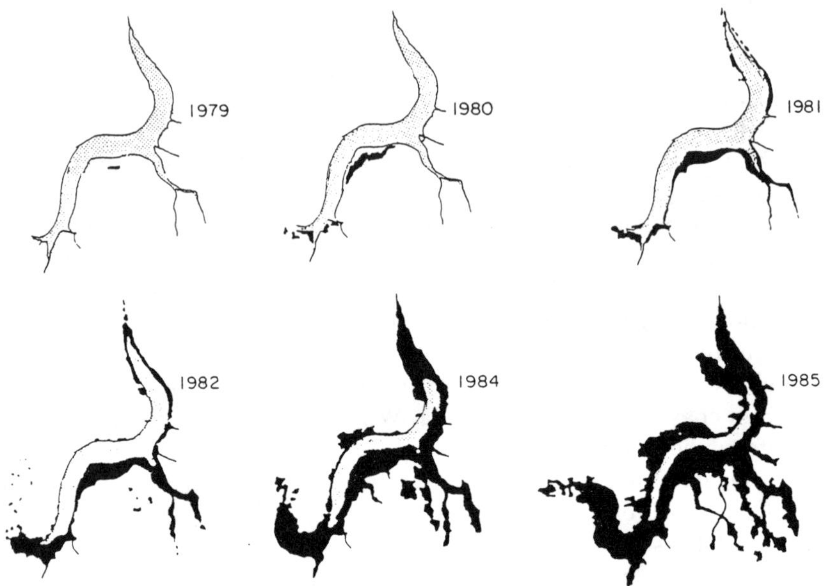

Fig. 2. The historical sequence of invasion by *Mimosa pigra* around a shallow lake on the Adelaide River flood plain in the Northern Territory of Australia. (From: Lonsdale, W.M. 1993. Rates of spread of an invading species - *Mimosa pigra* in northern Australia. J. Ecol. 81: 513-521. Published courtesy of Blackwell Science Ltd.)

flows in extending its distribution. Perhaps the key feature of this study is that predictions made at one spatial scale accurately describe events at a larger scale. Estimates of the rate of spread from an individual river system correctly predict the spread on a much larger river system 300 km away.

The wide variety of ecosystems invaded by plants and animals suggest clear differences in the rate at which different communities gain and lose species (Kowarik 1995a) which may be influenced by physical features of the environment and the individual characteristics and history of the species themselves. In the case of *M. pigra* its success results from a substantial increase in immigration overriding the fluxes between births and deaths. Models of community assembly (Case 1990; Pimm 1993) illustrate a number of features of interest here. Case (1990) showed that communities composed of strongly interacting species are resistant to invasion by alien species. Similarly the converse is also true where communities with fewer species appear easier to invade (Fig. 3a, b) especially when they have simple patterns of interspecfic interactions. As assembly proceeds, model communities become progressively harder to invade. Overall, simple communities appear to be most likely to be less persistent than complex ones. This agrees with the suggestion by Elton (1958) that more complex species-rich systems are less prone to invasion than species-poor systems, but contrasts with the conclusions of a different mathematical modelling approach (May 1974). Unequivocal evidence in support of each approach has yet to be accumulated; however interpretation of successful alien invasions on Hawaii fit with the suggestions of Elton (1958).

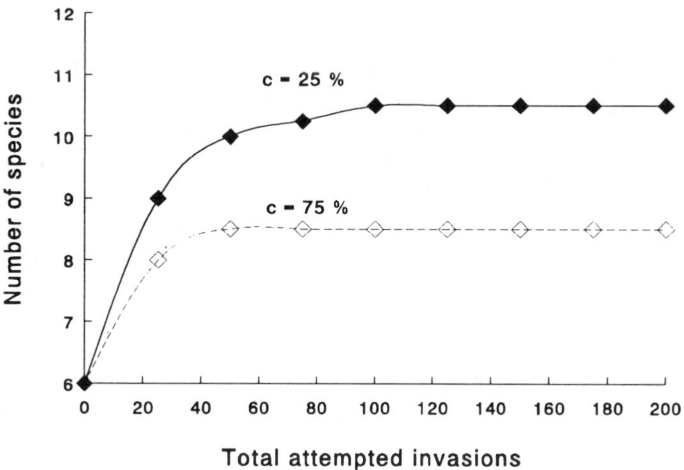

Fig. 3a.

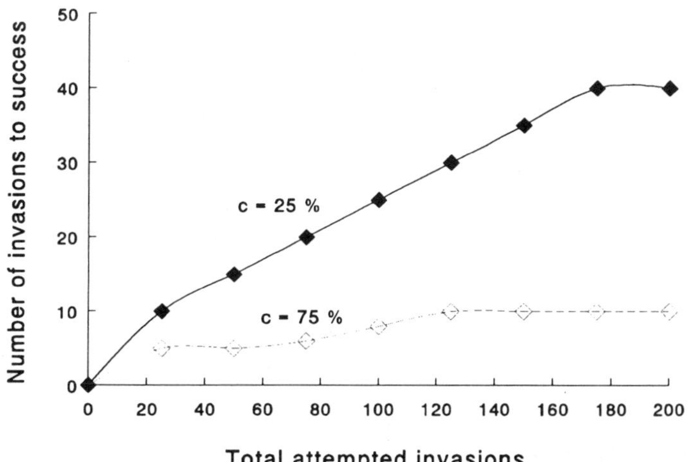

Fig. 3b.

Fig. 3. Smoothed simulations of community assembly generated from the models of Pimm (1993). *(a)* the number of species present within a community with increased number of invasions and *(b)* the number of invasions attempted before one is successful. The percentage values indicate the proportion of species interacting within the community. Note that those with a high proportion of interacting species (75%) are more difficult to invade than those with a low proportion (25%).

Relating ecosystem structure and function

It is generally considered that an increase in the disturbance of an ecosystem leads to the increased probability of invasion. When this occurs there is the clear possibility of the invader altering ecosystem structure and possibly its functioning. Fires are an extreme category of disturbance. Gobbi *et al.* (1995) have shown that in the South American Sub-Antarctic Forests region fire eliminates the natural vegetation and permits invasion of *Nothofagus antarctica* communities by alien species. In this case fires are a direct result of man's close proximity to the area. Invasions by aliens have also occurred because man's activities have increased the frequency of fires. The study of Wade *et al.* (1990) on hardwood ecosystem in the swamps of southern Florida is a good case in point highlighted recently by Woodward (1993), but worth reconsidering in the present context. By increasing the drainage and logging of the area, man's activities have also increased the frequency and intensity of fires. As a consequence the system has a reduced capacity to resist invasion, and the alien species *Melaleuca quinquenervia* has come in and drastically altered ecosystem function - because it thrives on increased fire intensity and frequency (Fig. 4). The foliage of *M. quinquenervia* is highly inflammable and the plant is capable of rapid regrowth from basal and epicormic shoots and seeds permitting it to achieve dominance and the exclusion of most of the native species.

The above examples and those discussed by Vitousek (1990) serve to illustrate how invasions can alter ecosystem structure. Altered ecosystem structure may also translate to a change in function. The important studies of Vitousek *et al.* (1987) show that the invading nitrogen fixing shrub *Myrica faya* quadruples the nitrogen fixation of the system which subsequently alters the development of vegetation within the ecosystem. One European based example where process-related studies may be made is in areas of Ireland and the UK invaded by *Rhododendron ponticum* (Cross 1981; Gritten 1995). These areas invariably suffer a drastic decline in species diversity which has been well recorded and must therefore translate into a more simple ecosystem structure. *R. ponticum* leaves are recalcitrant, form a dense mat and acidify the underlying soil, leading to expected changes in the nutrient dynamics. In Killarney oak woodlands in south-west Ireland invasion has been dramatic and rapid (Cross 1981) and these changes in litter quality may be of sufficient magnitude to be measurable (Pastor and Bockheim 1984). Opportunities for measuring these effects exist in areas currently being invaded by *R. ponticum*, such as Snowdonia National

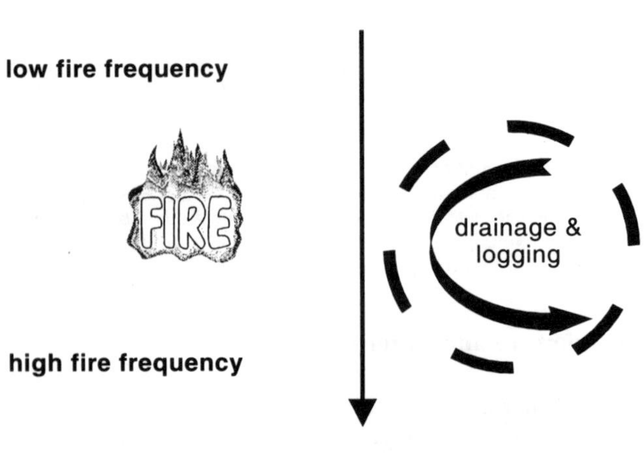

Fig. 4. The effects of man on increasing susceptibility of the hardwood ecosystems of southern Florida, USA to invasions by the alien *Melaleuca quinquenervia* (redrawn after Woodward 1993).

Park. Interestingly, by applying the conclusions of the models of Pimm (1993) and Case (1990) it can be suggested that the resulting simplified systems will be inherently unstable and not long-lived.

Global change and plant invasions

Plant invasions as discussed above are typically characterized by a decline in species diversity at all trophic levels within an ecosystem. In consequence, invasions raise many fundamental questions which parallel those concerning scientists about the effects of a loss of species diversity as a result of man's destruction of tropical forests and increased emissions of 'greenhouse' trace gases altering the global climate system (World Conservation and Monitoring Centre 1992; Peters and Lovejoy 1992). The study of plant invasions offers opportunities to address some concepts underlying our understanding of ecosystem function. With our inadequate knowledge relating ecosystem structure to function we cannot yet predict, for example, how changes in species diversity may affect biogeochemical cycles (Schulze and Mooney 1993). Where invasions occur and a single species predominates, this may have a quantifiable effect on ecosystem structure and nutrient cycling, as discussed previously. Invasions leading to a dramatic reductions in the tropic level of a system also have the potential to contribute to testing the concept of 'keystone species' - those which may drive a system but not make up a substantial fraction of the biomass or nutrients.

What are the prospects or likelihood that as the global climate system changes the rate and absolute numbers of invasions by alien species will increase? A core project of the International Geosphere-Biosphere Programme (IGBP), Global Change and Terrestrial Ecosystems (GCTE), has the objective of predicting the effects of global environmental change on terrestrial ecosystems (natural and agricultural) to determine how these effects will feedback and affect climate (Steffen *et al.* 1992). GCTE considers that species' extinctions will increase because of rising human populations, pollution, habitat degradation and destruction. This together with increased accidental and deliberate introductions of organisms will lead to "increasing the alien species component of all continents and homogenising species composition across previously distinct fauna and flora regions". These are conclusions based upon the previously stated criteria and will be exacerbated if global change increases the frequency of extreme events (Beerling and Woodward 1994) leading to greater disturbance.

For a chapter setting out possible future directions in research on plant invasions, it is of interest to consider what GCTE identifies as important for further study. Two foci from the GCTE report are of relevance to *(i)* use of knowledge about past and present distributions of organisms to predict future distributions, and hence patterns of diversity, under the impact of global change and *(ii)* evaluation of the consequences of homogenisation of biogeographic regions by the introductions of alien organisms, particularly the consequences for biodiversity of invasion alien organisms favoured by global change. The first of these two questions is currently being addressed using correlative models, and is discussed below. The second question has been touched upon in the present volume in that several papers have attempted to identify the consequences of invasions, but none has considered which groups of species are likely to be favoured by global change thereby permitting a direct focus

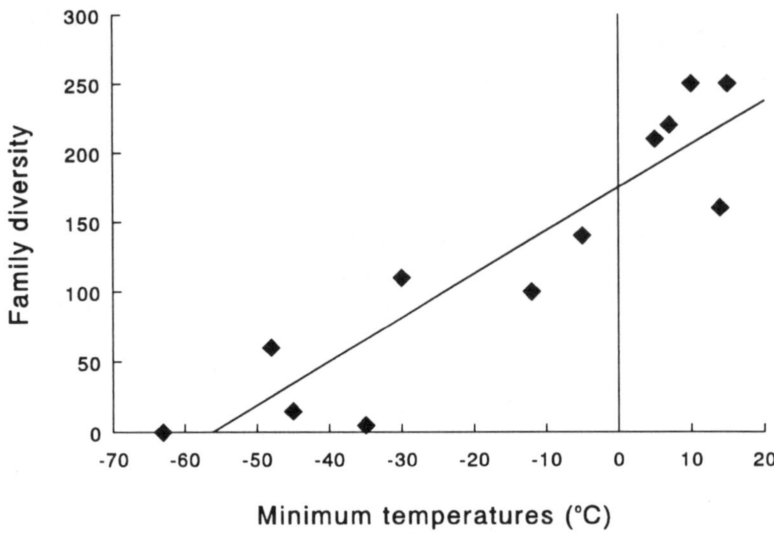

Fig. 5. Global relationship between absolute minimum temperature and plant family diversity. Both average for land surface units of 15° of latitude (after Woodward 1987).

on the potentially most threatening species. The question must be addressed if we are to "link understanding of changes in ecological complexity at regional, continental and global scales to the functioning and stability of ecosystems and the entire biosphere", an essential long-term GCTE objective (Steffen *et al.* 1992).

Based on our present knowledge, it is however possible to make some general predictions for aliens species. Earlier discussions of the models of Pimm (1993), Case (1990) and the ideas of Elton (1958) indicated that in general low species diversity leads to lower resistance to perturbations. On a global scale diversity is related to minimum temperatures (Fig. 5) (Woodward 1987). Therefore at a simplistic level in cold climates global warming may be expected to lead to an invasion of warm adapted alien species. This conclusion may be tempered by the resistance and resilience of the community in question. Analysis of the climatic range of alien species, both in their native and introduced sites, provides one means of indicating areas most likely to be invaded (Beerling 1993; Beerling *et al.*, in press; Daehler and Strong 1993). These types of analyses are of use in highlighting regions most likely to be invaded by particular species and provide an important first step in making predictions but may not be robust under different atmospheric concentrations of CO_2 and climate. More robust predictions may emerge from the use of state-of-the-art dynamic general vegetation models parametized for a few competing individual species. This achievement is presently someway off and again emphasises the need for understanding the physiological mechanisms controlling observed distribution of plants (Beerling 1993) so that general principles emerge which are widely applicable.

Conclusions

This present volume falls into several sections which broadly address the three questions originally proposed by the Scientific Committee for Problems of the Environment (SCOPE) (Holdgate 1986):

1. What properties of a species make it capable of dispersion to invade new habitats, and what features determine whether it will successfully establish itself if it does arrive in a new setting?
2. What features of a receiving habitat (physical or biological) make it prone to invasion?
3. What management strategies are appropriate to control invading species?

All three are crucial if we are to understand and control invasions. This chapter has sought to extend the consideration of alien species from species-specific cases and site-specific control measures to their use in understanding the more fundamental issues of ecology to which they may usefully contribute. The clear inter-relationships between community structure, ecosystem processes and the ability of alien species to alter both, mean that it is imperative to broaden our horizons in the use of alien invasions in ecology. In addition, by measuring processes and taking an ecophysiological approach to characterising invasive species we may stand a better chance of elucidating the underlying general mechanisms controlling invasions and therefore be in a stronger position to deal with the increased frequency of invasions anticipated to arise as a result of global environmental change. Studies of invasions by alien species will become increasingly important as species become homogenized across geographical and continental barriers as a result of global change and man's activities and it is up us to have initiated the appropriate studies to address the issues with sufficiently broad horizons.

Acknowledgments

I thank Petr Pyšek for the invitation to contribute to this volume. Comments on an earlier draft of the manuscript by Petr Pyšek, Arthur Willis and Ian Woodward were gratefully received. The author is grateful for a 1994 Royal Society University Research Fellowship.

References

Austin, M.P. 1982. Use of a relative physiological performance value in the predictions of performance in multispecies mixtures from monoculture performance. J. Ecol. 70: 559-570.

Bailey, J.P., Child, L.E. and Wade, M. 1995. Assessment of the genetic variation and spread of British populations of *Fallopia japonica* and its hybrid *Fallopia* x *bohemica*. In: P. Pyšek, K. Prach, M. Rejmánek and M. Wade (eds.), Plant Invasions: General Aspects and Special Problems, pp. 141-150. SPB Academic Publ., Amsterdam.

Beerling, D.J. 1991. The effect of riparian land use on the occurrence and abundance of Japanese knotweed *Reynoutria japonica* on selected rivers in South Wales. Biol. Conserv. 55: 329-337.

Beerling, D.J. 1993. The impact of temperature on the northern distribution limits of the introduced species *Fallopia japonica* and *Impatiens glandulifera* in north-west Europe. J. Biogeogr. 20: 45-53.

Beerling, D.J., Huntley, B. and Bailey, J.P. The use of the introduced species *Fallopia japonica* to test the predictive capacity of response surfaces. J. Veget. Sci. (in press).

Beerling, D.J. and Woodward, F.I. 1994. Climate change and the British scene. J. Ecol. 82: 391-397.

Betancourt, J.L., Long, A., Donahue, D.J., Jull, A.J.T. and Zabel, T.H. 1984. Pre-Columbian age for North American *Corispermum* L. (*Chenopodiaceae*) confirmed by accelerator radiocarbon dating. Nature 311: 653-655.

Bramley, J.L., Reeve, J.T. and Dussart, G.B.J. 1995. The distribution of *Lemna minuta* within the British Isles: identification, dispersal and niche constraints. In: P. Pyšek, K. Prach, M. Rejmánek and M. Wade (eds.), Plant Invasions: General Aspects and Special Problems, pp. 181-185. SPB Academic Publ., Amsterdam.

Brock, J.H., Child, L.E., De Waal, L.C. and Wade, M. 1995. The invasive nature of *Fallopia japonica* is enhanced by vegetative regeneration from stem tissues. In: P. Pyšek, K. Prach, M. Rejmánek and M. Wade (eds.), Plant Invasions: General Aspects and Special Problems, pp. 131-139. SPB Academic Publ., Amsterdam.

Byrne, R. and McAndrews, J.H. 1975. Pre-Columbian purslane (*Portulaca oleracea*) in the New World. Nature 253: 726-727.

Case, T.J. 1990. Invasion resistance arises in strongly interacting species-rich model competition communities. Proc. Nat. Acad. Sci. USA 87: 9610-9614.

Cross, J.R. 1981. The establishment of *Rhododendron ponticum* in the Killarney oakwoods, S.W. Ireland. J. Ecol. 69: 807-824.

Daehler, C.C. and Strong, D.R. 1993. Prediction and biological invasion. Trends Ecol. Evolut. 8: 380-381.

Di Castri, F., Hansen, A.J. and Debussche, M. 1990. Biological Invasions in Europe and the Mediterranean Basin. 463 pp. Kluwer Academic Publ., Dordrecht.

Drake, J.A., Mooney, H.A., Di Castri, F., Groves, R.H., Kruger, F.J., Rejmánek, M. and Williamson, M. 1989. Biological Invasions: A Global Perspective. 525 pp. John Wiley and Sons, Chichester.

Edwards, K.R., Adams, M.S. and Květ, J. 1995. Invasion history and ecology of *Lythrum salicaria* in North America. In: P. Pyšek, K. Prach, M. Rejmánek and M. Wade (eds.), Plant Invasions: General Aspects and Special Problems, pp. 161-180. SPB Academic Publ., Amsterdam.

Ellenberg, H. 1988. Vegetation Ecology of Central Europe. Ed. 4. 731 pp. Cambridge University Press, Cambridge.

Elton, C.S. 1958. The Ecology of Invasions by Animals and Plants. 181 pp. Methuen, London.

Ferreira, M.T. and Moreira, I.S. 1995. The invasive component of a river flora under the influence of Mediterranean agricultural systems. In: P. Pyšek, K. Prach, M. Rejmánek and M. Wade (eds.), Plant Invasions: General Aspects and Special Problems, pp. 117-127. SPB Academic Publ., Amsterdam.

Gobbi, M., Puntieri, J. and Calvelo, S. 1995. Post-fire recovery and invasion by alien plant species in a South American woodland-steppe ecotone. In: P. Pyšek, K. Prach, M. Rejmánek and M. Wade (eds.), Plant Invasions: General Aspects and Special Problems, pp. 105-115. SPB Academic Publ., Amsterdam.

Gritten, R.H. 1995. *Rhododendron ponticum* and some other invasive plants in the Snowdonia National Park. In: P. Pyšek, K. Prach, M. Rejmánek and M. Wade (eds.), Plant Invasions: General Aspects and Special Problems, pp. 213-219. SPB Academic Publ., Amsterdam.

Harris, G.A. 1967. Some competitive relationships between *Agropyron spicatum* and *Bromus tectorum*. Ecol. Monogr. 37: 89-111.

Holdgate, M.W. 1986. Summary and conclusions: characteristics and consequences of biological invasions. Phil. Trans. Roy. Soc. London B 314: 733-742.

Kornberg, H. and Williamson, M. (eds.) 1986. Quantitative Aspects of Biological Invasions. 240 pp. The Royal Society, London.

Kowarik, I. 1995a. Time lags in biological invasions with regard to the success and failure of alien species. In: P. Pyšek, K. Prach, M. Rejmánek and M. Wade (eds.), Plant Invasions: General Aspects and Special Problems, pp. 15-38. SPB Academic Publ., Amsterdam.

Kowarik, I. 1995b. On the role of alien species in urban flora and vegetation. In: P. Pyšek, K. Prach, M. Rejmánek and M. Wade (eds.), Plant Invasions: General Aspects and Special Problems, pp. 85-103. SPB Academic Publ., Amsterdam.

Lonsdale, W.M. 1993. Rates of spread of an invading species - *Mimosa pigra* in northern Australia. J. Ecol. 81: 513-521.

May, R.M. 1974. Stability and complexity in model ecosystems. Ed. 2. 265 pp. Princeton University Press, Princeton.

Noble, I.R. 1989. Attributes of invaders and the invading process: terrestrial and vascular plants. In: J.A. Drake, H.A. Mooney, F. di Castri, R.H. Groves, F.J. Kruger, M. Rejmánek and M. Williamson (eds.), Biological Invasions: A Global Perspective, pp. 301-313. John Wiley and Sons, Chichester.

Pastor, J. and Bockheim, J.G. 1984. Distribution and cycling of nutrients in an aspen-mixed hardwood podsol ecosystem. Ecology 65: 339-353.

Peters, R.L. and Lovejoy, T.E. (eds.) 1992. Global Warming and Biological Diversity. 386 pp. Yale University Press, New Haven.

Pimm, S.L. 1993. Biodiversity and the balance of nature. In: E.D. Schultze and H.A. Mooney (eds.), Biodiversity and Ecosystem Function, pp. 347-359. Springer-Verlag, Berlin.

Pyšek, P. 1995a. On the terminology used in plant invasion studies. In: P. Pyšek, K. Prach, M. Rejmánek and M. Wade (eds.), Plant Invasions: General Aspects and Special Problems, pp. 71-81. SPB Academic Publ., Amsterdam.

Pyšek, P. 1995b. Recent trends in studies on plant invasions (1974-1993). In: P. Pyšek, K. Prach, M. Rejmánek and M. Wade (eds.), Plant Invasions: General Aspects and Special Problems, pp. 223-236. SPB Academic Publ., Amsterdam.
Pyšek, P., Prach, K. and Šmilauer, P. 1995. Relating invasion success to plant traits: an analysis of the Czech alien flora. In: P. Pyšek, K. Prach, M. Rejmánek and M. Wade (eds.), Plant Invasions: General Aspects and Special Problems, pp. 39-60. SPB Academic Publ., Amsterdam.
Rejmánek, M. 1995. What makes a species invasive? In: P. Pyšek, K. Prach, M. Rejmánek and M. Wade (eds.), Plant Invasions: General Aspects and Special Problems, pp. 3-13. SPB Academic Publ., Amsterdam.
Schulze, E.D. and Mooney, H.A. (eds.) 1993. Biodiversity and Ecosystem Function. Ecological Studies Series 99. 520 pp. Springer-Verlag, Berlin.
Steffen, W.L., Walker, B.H., Ingram, J.S. and Koch, G.W. 1992. IGBP Global Change Report, 21. Global Change and Terrestrial Ecosystems. The Operational Plan. IGBP, Stockholm.
Vitousek, P.M. 1990. Biological invasions and ecosystem processes: towards an integration of population biology and ecosystem studies. Oikos 57: 7-13.
Vitousek, P.M., Walker, L.R., Whiteaker, L.D., Mueller-Dombois, D. and Matson, P.A. 1987. Biological invasion by *Myrica faya* alters ecosystem development in Hawaii. Science 238: 802-804.
Vogt Andersen, U. 1995. Comparison of dispersal strategies of alien and native species in the Danish flora. In: P. Pyšek, K. Prach, M. Rejmánek and M. Wade (eds.), Plant Invasions: General Aspects and Special Problems, pp. 61-70. SPB Academic Publ., Amsterdam.
Wade, D., Ewel, J. and Hofsetter, R. 1980. Fire in South Florida Ecosystems. US Department of Agriculture. Forest Service General Technical Report SE-17. Asheville.
Weiss, P.W. and Noble, I.R 1984. Interactions between seedlings of *Chrysanthemoides monilifera* and *Acacia longifolia*. Austral. J. Ecol. 9: 107-115.
Woodward, F.I. 1987. Climate and Plant Distribution. 174 pp. Cambridge University Press, Cambridge.
Woodward, F.I. 1993. How many species are required for a functional ecosystem? In: E.D. Schulze and H.A. Mooney (eds.), Biodiversity and Ecosystem Function, pp. 271-291. Springer-Verlag, Berlin.
World Conservation and Monitoring Centre. 1992. Global Diversity: Status of the Earth's Living Resources. 585 pp. Chapman and Hall, London.

INDEX OF AUTHORS

SUBJECT INDEX

INDEX OF TAXA